NEW RESEARCH ON GENOMIC INSTABILITY

New Research on Genomic Instability

Eleanor J. Gloscow
Editor

Nova Biomedical Books
New York

For permission to use material from this book please contact us:
Telephone 631-231-7269; Fax 631-231-8175
Web Site: http://www.novapublishers.com

LIBRARY OF CONGRESS CATALOGING-IN-PUBLICATION DATA

New research on genomic instability / Eleanor J. Gloscow, editor.
p. ; cm.
Includes bibliographical references and index.
ISBN10: 1-60021-320-0 ISBN 13: 978-1-60021-320-5
1. Mutation. 2. DNA damage. 3. Chromosome abnormalities. 4. Cancer--Genetic aspects. I. Gloscow, Eleanor J.
[DNLM: 1. Genomic Instability. 2. Neoplasms--genetics. QS 677 N532 2006]
QH462.A1N49 2006
572.8'38--dc22

2006019843

Published by Nova Science Publishers, Inc. ✢ New York

Contents

Preface

Many cancer biologists now believe that genomic instability not only initiates carcinogenesis, but also allows the tumor cell to become metastatic and evade drug toxicity. The loss of stability of the genome is becoming accepted as one of the most important aspects of carcinogenesis. One of the hallmarks of the cancer cell is the inherent instability of its genome. This book presents new research in this exciting field.

Chapter 1 - Tumor cell invasion is believed to result from an overproduction of proteolytic enzymes that cause the degradation of the basement membrane. Results from recent worldwide clinical trials with corresponding enzyme inhibitors, however, have been very disappointing, casting doubt on the validity of the proteolytic enzyme theory. Since over 90% of the tumor-related mortality results from invasion-related illness, there is an urgent need to uncover the mechanism of tumor invasion. Prompted by the fact that the absence of the ME cell layer is the most distinct sign of invasive breast lesions, the authors recent studies have attempted to identify the early signs of ME cell alterations and their impact on invasion. With multiple disciplary analyses in tissue samples from near 500 patients with pre-invasive breast tumors, these studies revealed that a subset of pre-invasive breast tumors contained focal disruptions (the absence of ME cells resulting a gap greater than the combined size of at least 3 ME cells) in their ME cell layers. Focally disrupted ME cell layers showed a significantly lower expression of tumor suppressors, whereas a significantly higher frequency of apoptosis, degeneration, and leukocyte infiltration, compared to their non-disrupted counterparts. In sharp contrast, tumor cells overlying focal ME cell layer disruptions showed a significantly higher rate of proliferation, genetic instability, and expression of invasion-related genes than adjacent cells overlying non-disrupted ME cell layers. The studies in prostate, colorectal, and cervical tumors obtained similar results. Based on these findings, we have proposed that breast tumor invasion is triggered by a focal degeneration of aged or injured ME cells and the resultant auto-immunoreactions, which result in genetic instability in overlying cells and trigger a clonal proliferation of tumor pregenitors. The hypothesis differs from the proteolytic enzyme theory in multiple aspects, including the precursor of invasive lesions, impact of immunoreactive and stromal cells, and approaches for early detection and prevention. The hypothesis currently represents the only published alternative for the enzyme theory, and the only human in vivo model of breast tumor invasion.

Chapter 2 - Numerical and structural chromosome aberrations are a common characteristic of most tumors. More than one century after scientists first described these particular lesions and suggested a link to rise of malignancy, it is still a matter of debate, whether chromosomal aberrations directly contribute to transformation of a somatic cell into a tumor cell. Chromosomal aberrations are thought to arise from various conditions that are addressed in the following chapter. Recent evidence supports the notion that numerical and structural chromosome aberrations favor generation of progressive genetic instability. Genetic instability is known to confer a selective advantage onto a particular cell, supporting outgrowth of specific cell clones. These cell populations are able to benefit from the acquired genetic alteration in their contemporary environmental context and become subject of malignant transformation. This leads to the assumption that chromosomal instability must be considered as an essential contributor in the generation of most cancers.

Chapter 3 - A characteristic feature of neoplastic cells is the increased occurrence of various genetic changes ranging from point mutations to chromosomal alterations. Based upon this general observation, it has been suggested that cancer cells are genetically unstable, and that acquisition of genomic instability may be an important prerequisite of tumorigenesis. Defects in basic cellular mechanisms maintaining genomic integrity and guarding against genomic mutations might lead to accumulation of multiple stable mutations in the genome of affected cells. Consequently, genomic instability can be regarded as any condition that predisposes a cell to the accumulation of stable genetic changes at a greater rate than normal. This relationship is well documented in the chromosomal instability syndromes, often characterized by increased incidence of different neoplasias.

Besides the spontaneous occurrence, genomic instability can also be induced or significantly increased by exposure to cytotoxic agents such as ionizing radiation. Recently, evidence has accumulated supporting a relationship between genetic background and the radiation induced genomic instability phenotype. Therefore, the monitoring of genomic instability is useful from both the oncological and radiobiological points of view. Detection of genomic instability can contribute to the effective screening of early clinical stages of neoplasia and the identification of persons with increased cancer risk.

In the present study, the authors compared stability of the genome between untreated cervical cancer patients and healthy women according to the spontaneous and γ-ray induced cytogenetic changes in their non-cancer cells (peripheral lymphocytes). Lymphocyte cultures from whole venous blood of patients with cervical neoplasia (stage T1 according to the TNM classification) and healthy female volunteers were cultivated and irradiated *in vitro* with different doses (2, 4 a 6 Gy) using ^{60}Co-gamma source. Slides were prepared using the standard air-drying procedure. The metaphases were analysed for the following structural chromosomal aberrations (CA): chromatid gaps and breaks, isochromatid gaps and breaks, and exchanges (dicentrics, rings, quadriradials, triradials and double minutes).

Generally, a dose-dependent increase of structural changes was observed in healthy women as well as in patients, while statistical analysis has shown significantly higher CA frequencies in patients as compared with healthy women only after γ-ray treatment. As the lymphocytes were cultivated in presence of 5-bromo-2'-deoxyuridine (BrdU) and further stained by the fluorescence-plus Giemsa (FPG) technique, it allowed the identification of metaphases in first, second and third mitotic divisions. Accordingly, the progression of structural chromosomal changes has been also monitored. Moreover, the simultaneous evaluation and quantification of both, CA and sister chromatid exchanges (SCE) in the same metaphases can provide information about the relationship of these two cytogenetic parameters generally accepted as the indicators of chromosomal instability (CIN).

The obtained results indicate differences in the chromosomal stability between healthy women and patients, and suggest the necessity of the mutagen treatment because of the intensification of the differences between individuals with normal and decreased genomic stability.

Chapter 4 - In addition to excess growth, the cells of most types of tumors exhibit abnormalities both of their nuclear morphology and of their specialised functions. "Malignant" tumors usually show the same changes to greater degrees, and in addition, tend to invade adjacent tissues, and spread by the blood- and lymphatic-vascular systems to distant parts of the body. However, according to tumor type, numerous exceptions to these general associations are known to occur. This chapter reviews these aspects of tumors and the theories of their possible genetic bases. The genomic effects of etiological factors and the mutations which are associated with inherited predispositions are mentioned. It is proposed that the co-mutational aspect of DNA-replicative-infidelity-type genetic instability might be involved in the morphological and behavioural abnormalities of tumor cells.

Chapter 5 - The etiology and pathogenesis of tumors of the central nervous system are still inadequately explained. This study analyses tumor suppressor gene—adenomatous polyposis coli (APC) in 28 patients with glioblastoma, the most aggressive form of glial tumors. APC protein has a structural role in adherens junctions, but also plays a signaling role as a negative regulator of the wnt pathway. The interest in the APC gene stemmed principally from the findings that the wild-type APC protein is highly expressed in the central nervous system, and upon the finding that it is critically involved in particular syndromes, among which brain tumors play a significant role. Glioblastoma samples were tested for gene instability by PCR/loss of heterozygosity using RFLP method. Two polymorphic markers were used: an Rsa I polymorphic site in exon 11, and an Msp I polymorphic site in exon 15. The results of our analysis for both markers showed allelic loss of the APC gene in 40% of our sample out of 25 heterozygous patients (informativeness 89%). Another 20% of samples demonstrated allelic imbalance of the APC allele in tumor tissue. Altogether, there were 15 samples (60%) demonstrating instability of this tumor suppressor gene. Despite increasing knowledge on glioma biology and genetics, the prognostic tools for glioblastoma still need improvement. Their findings on genomic instability of the APC gene may contribute to better understanding of glioblastoma genetic profile and could be used as a prognostic marker of disease evolution and progression.

Chapter 6 - Life on earth evolved in a low-level ionizing radiation environment comprised of terrestrial, solar, and cosmic radiations. A beneficial evolutionary consequence of the interaction of environmental low-level, sparsely ionizing radiations (e.g., X-rays, gamma rays, and beta particles) with mammalian organisms on earth is the suppression of genomic instability via radiation-hormesis-associated protective biological processes. Radiation hormesis as used here involves low-dose-induced protection from biological harm (e.g., cancer) that would otherwise occur spontaneously. High doses inhibit protection. This chapter focuses on the control of genomic instability and associated stochastic biological effects (e.g., mutant and neoplastically transformed cells and cancer) in mammals via radiation-hormesis-associated protective processes (adapted protection). Cancer risk increases as the numbers of mutant and neoplastically transformed (precancerous) cells increase. Mutant and neoplastically transformed cells increase as the number of cells with genomic instability increases. Thus, protective biological processes that suppress genomic instability also suppress mutant and neoplastically transformed cells and, thereby, prevent some spontaneous and environmental stressor-associated cancers that otherwise would have occurred.

This chapter describes how a system of cooperative protective biological processes that are activated (or stimulated) by low doses of sparsely ionizing radiation likely suppress genomic-instability-associated diseases. The protective processes that make up this adapted protection include normal high-fidelity DNA repair/apoptosis (presumed to be p53-dependent), an auxiliary apoptosis process (presumably p53-independent) that selectively removes cells with genomic instability including mutant and neoplastically transformed cells, and the immune system. The auxiliary apoptotic component relates to what is called the *p*rotective *a*poptosis-*m*ediated (PAM) process, which is considered to involve reactive oxygen and nitrogen species and specific cytokines (e.g., transforming growth factor β1). When activated by a low dose of sparsely ionizing radiation, the PAM process removes genomically unstable cells from our bodies and, along with activation of the immune system, can prevent some later occurring spontaneous cancers and other genomic-instability-associated diseases. Natural background sparsely ionizing radiation above a threshold dose activates the PAM process. The indicated hormesis-associated protective system and its activation by natural background radiation therefore guards against genomic instability and its adverse consequences to humans and other mammals. Reducing background radiation sources to the extent that radiation doses remain below the protective threshold will likely harm us rather than provide a benefit! Reduced life expectancy and significant increases in all genomic-instability-associated diseases would be the expected consequence of reducing background radiation to the degree indicated.

Chapter 7 - LET-dependent features of cell abnormality leading to death, lethal sectoring, genomic instability and cell cycle change were compared between α and X irradiated HeLa cells by means of a time-lapse photography. Cell abnormality induced by α irradiation tended to be manifested in the form of death without division (mitotic or interphase death), rather than incomplete division accompanied by anaphase bridge derived from dicentric chromosome (two-hit type damage) that was predominant for X irradiation. Frequency of incomplete division at irradiated generation was decreased remarkably (25.8% → 6.8%) by splitting X ray dose (4 Gy → 2 × 2 Gy), correlating well with the increase in cell survival (6.8% → 14.0%). However, in the case of α irradiation especially at low dose, two split doses (0.6 Gy → 2 × 0.3 Gy) significantly decreased cell survival (12.3% → 8.8%) with the accompanying increase in the frequency of incomplete division (7.2% → 13.5%) (reverse split dose effect). This suggests that an interaction between the damage by the 1st dose and the one by the 2nd dose gives rise to 2-hit type damage owing to irreparability of the damage and enhances cell killing at low dose where interphase cell death is not predominant. Lethal sectoring is the process for survival in which lethal damage remaining in irradiated cells is eliminated as lethal sector (offspring without reproductive integrity). This process occurs through the postirradiation 1st to 4th divisions, resulting in the appearance of a clonogenic progenitor harboring no lethal damage. However, nonlethal damage remaining in the clonogenic progenitors led to an elevated incidence of delayed cell death in their progeny. From an analysis of the pedigrees of cells surviving α (0.45 Gy) or X irradiation (3 Gy) (20% survival dose) which were constructed by means of a long-term (~150 h) time-lapse observation, it became clear that the mean incidence was higher for α-particle (16.3%) than X-ray survivors (8.3%), indicating the greater potentiality for genomic instability by α particles. Evidence is available to suggest the intrinsic difference in the mechanisms for the induction of genomic instability by these two radiations; the association of misrepaired clustered DNA damage (CD) with α particles and the fixation of unrepaired potentially lethal damage (PLD) with X rays. A novel phenomenon "delayed division delay (DDD)" was noticed, though occasionally (~10% per cell), with the progeny during the postirradiation 1st–3rd generations. DDD was much longer in α- (mean: ~11 h) than X-irradiated cells (~4 h). DDD might be triggered by delayed DNA double-strand breaks (DSB) at the fragile break-misrejoining sites, and the latter was supposed to be induced by a higher-order conformational change of chromatin fibers during mitosis. This demonstrates the usefulness of DDD as a new endpoint of radiation-induced genomic instability. Contrary to DDD, a significant shortening (~5 h) of cell cycle time at the postirradiation 1st generation was found with the progeny of X-ray survivors harboring no lethal nor nonlethal damage.

Chapter 8 - Chromium (Cr) compounds are regarded as a complex genotoxin. Soluble and insoluble hexavalent chromium salts have been demonstrated to induce morphological and neoplastic transformation in human and animal cells. Due to its rapid uptake by cellular anion transporters, soluble Cr(VI) has been considered as the primary carcinogenic form of chromium. Within a cell Cr(VI) reduces to lower oxidation states Cr(V/IV/III), accompanied by ROS (reactive oxygen species) production. Cr(VI) reduced forms, along with ROS, affect many cellular functions by damaging nucleic acids (as Cr-DNA adducts, DNA strand breaks, DNA-DNA and DNA-protein crosslinks), oxidizing proteins and causing lipid peroxidation. Cells have evolved many ways to either avoid or repair these injurious consequences. In cells, the toxic capacity of Cr(VI) can be decreased by: 1) activation of DNA damage checkpoints, which cause transient growth arrest to repair and prevent the transmission of damaged chromosomes; 2) antioxidant defense systems, which can neutralize the ROS generated by Cr(VI) action; 3) apoptosis as a highly organized physiological mechanism of eliminating heavily damaged cells. The majority of real life exposure to Cr(VI) is low and long-term exposure to this ecologically negative factor, which is permanently present in the atmosphere, soil and water. The human epithelial-like cells were used as a model of different possible hazardous routes of Cr(VI) exposure (e.g., ingestion, inhalation, and dermal contacts). In this chapter dose-dependent involvement of defense mechanisms in response to low level and long-term Cr(VI) treatment are analyzed in dynamic. Low level Cr(VI) doses were assigned at the boundary of cell growth disorders and apoptosis for the studied cell line. The apoptogenic and nonapoptogenic low doses initially produced ROS to a comparable level but affected differently DNA damage-induced cell cycle checkpoints and the antioxidant defense system (including glutathione cycle, catalase, superoxide dismutase). The nonapoptogenic concentration increased activity of the antioxidant system, and caused temporal cell cycle arrest elevating S phase cell number followed by the increasing of the G_2/M cell compartment that resulted in transient growth arrest. The detected imbalance of the activity of the glutathione-dependent system at the cell treatment with the toxic apoptogenic Cr(VI) concentration correlated with the permanent growth arrest at the S phase of the cell cycle, lowered DNA synthesis, irreversibly inhibited proliferation and development of apoptosis. The small compartment of viable cells that had escaped from apoptosis retained the ability to proliferate, and the resultant cell population was enriched for polyploid multinuclear cells.

Chapter 9 - Eukaryote cells maintain genomic integrity by surveillance of DNA for damage or incomplete replication. In the event of aberrant structures being detected, checkpoint mechanisms are activated that result in slowing down or arresting cell cycle progression, which allows the DNA damage to be repaired and the replication to be completed. Potentially deleterious mutations can elicit pathways leading to apoptosis. Defects in those pathways, however, seem to induce the susceptibility to DNA damage, genomic defects and resistance to apoptosis, which characterize cancer cells. In response to genotoxic, replicative stress, Chk1 kinase, activated by Atr (ataxia-telangiectasia mutated and Rad3 related), delays cell cycle progression through S to G2 phases by inhibiting Cdc25 phosphatases, which control the timing of early versus late replication origin firing during normal S phase. This pathway maintains DNA replication through structurally unfavorable stretches of DNA, such as chromosome fragile sites in mammalian cells. Common fragile sites are chromosome regions that exhibit gaps or breaks, visible in metaphase chromosomes, when cells are exposed to replicative stress conditions. FRA3B and FRA16D, at chromosome regions 3p14.2 and 16q23.3, are the two most active or most fragile sites in human genome; those are altered frequently in various tumors. In this article, recent findings of the molecular mechanism of DNA damage and replication checkpoint are discussed, linking replication stress surveillance with apoptosis induction, which play a central role in determining if DNA damage is compatible with cell survival or requires cell elimination by apoptosis.

In: New Research on Genomic Instability
Editor: Eleanor J. Glascow, pp. 1-46 ISBN: 1-60021-320-0

Chapter I

Focal Degeneration of Aged or Injured Myoepithelial Cells and Resultant Auto-Immunoreactions Trigger Genetic Instability and Breast Cancer Invasion

Yan-gao Man
Department of Gynecologic and Breast Pathology, Armed Forces Institute of Pathology and American Registry of Pathology, Washington DC 20306

Abstract

Tumor cell invasion is believed to result from an overproduction of proteolytic enzymes that cause the degradation of the basement membrane. Results from recent worldwide clinical trials with corresponding enzyme inhibitors, however, have been very disappointing, casting doubt on the validity of the proteolytic enzyme theory. Since over 90% of the tumor-related mortality results from invasion-related illness, there is an urgent need to uncover the mechanism of tumor invasion. Prompted by the fact that the absence of the ME cell layer is the most distinct sign of invasive breast lesions, our recent studies have attempted to identify the early signs of ME cell alterations and their impact on invasion. With multiple disciplary analyses in tissue samples from near 500 patients with pre-invasive breast tumors, our studies revealed that a subset of pre-invasive breast tumors contained focal disruptions (the absence of ME cells resulting a gap greater than the combined size of at least 3 ME cells) in their ME cell layers. Focally disrupted ME cell layers showed a significantly lower expression of tumor suppressors, whereas a significantly higher frequency of apoptosis, degeneration, and leukocyte infiltration, compared to their non-disrupted counterparts. In sharp contrast, tumor cells overlying focal ME cell layer disruptions showed a significantly higher rate of proliferation, genetic instability, and expression of invasion-related genes than adjacent cells overlying non-disrupted ME cell layers. Our studies in prostate, colorectal, and cervical tumors obtained similar results. Based on these findings, we have proposed that breast tumor invasion is triggered by a focal degeneration of aged or injured ME cells and the resultant auto-immunoreactions, which result in genetic instability in overlying cells and trigger a clonal proliferation of tumor pregenitors. Our hypothesis differs from the proteolytic enzyme theory in multiple aspects, including the precursor of invasive lesions, impact of immunoreactive and stromal cells, and approaches for early detection and prevention. Our hypothesis currently represents the only published alternative for the enzyme theory, and the only human in vivo model of breast tumor invasion.

INTRODUCTION

The epithelium of normal breast and pre-invasive breast tumors is physically separated from the stroma by both myoepithelial (ME) cells and the basement membrane (BM). ME cells are joined by intercellular junctions and adhesion molecules, forming a largely continuous sheet that encircles the entire epithelium (1-3) (Fig 1). The BM is composed of type IV collagens, laminins, and other molecules, constituting a continuous lining surrounding and attaching to the ME cell layer (4-6). As the epithelium is devoid of blood vessels and lymphatic ducts, its metabolism needed oxygen, nutrients, and growth factors must first pass through the BM, then the ME layer, in order to reach the epithelium. In contrast, epithelium-derived tumor cells must first pass through the ME layer, then the BM, in order to invade or metastasize.

The development of human breast cancer is believed to sequentially undergo normal, hyperplastic, *in situ*, invasive, and metastatic stages (7-9). The progression from the *in situ* to invasive stage is believed to be triggered primarily, if not solely, by the overproduction of proteolytic enzymes mainly by cancer cells, which results in the degradation of the BM (10-11). This theory alone, however, seems not to completely reflect the intrinsic mechanism of breast tumor invasion for three main reasons. First, the mechanism and process of the ME cell layer degradation are unknown. Second, results from all recent worldwide clinical trials with corresponding enzyme inhibitors have been very disappointing (12-13). Third, our previous study have detected a significantly higher level of matrix metalloproteinase-26 (MMP-26), a key enzyme for BM degradation, and its mediated pro-gelatinase B in ductal carcinoma *in situ* (DCIS) than in invasive cancers by immunohistochemistry and an integrated morphometry analysis (14). Our recent studies have further shown that the invasive front of *in situ* tumors in over 50 cases examined were completely devoid of the expression of MMP-26, but cells within the same tumor and in adjacent micro-invasive lesions were strongly positive for MMP-26 (15-16) (Fig 2).

Since *in situ* carcinomas are universally regarded as the precursor of invasive lesions, several drastic treatment methods, including local excision, local excision combined with radiation or chemotherapy, and mastectomy, are the routine treatments for all *in situ* cancers (17-21). All these methods, however, have grave drawbacks. Local excision plus radiation or chemotherapy is associated with 7-10% of recurrence (17-21). Mastectomy is overly aggressive, as over 50% of in situ cancers do not progress to invasive lesions during the patient's lifetime (22-23). As the diagnosis of *in situ* breast tumors has increased up to 10-fold in the Europe and the USA during the last 20 years (24-25), there is an urgent need to identify the specific precursor of invasive lesions. Since over 90% of cancer-related mortality results from invasion-related illness (26), there is an urgent need to uncover the mechanism of tumor invasion.

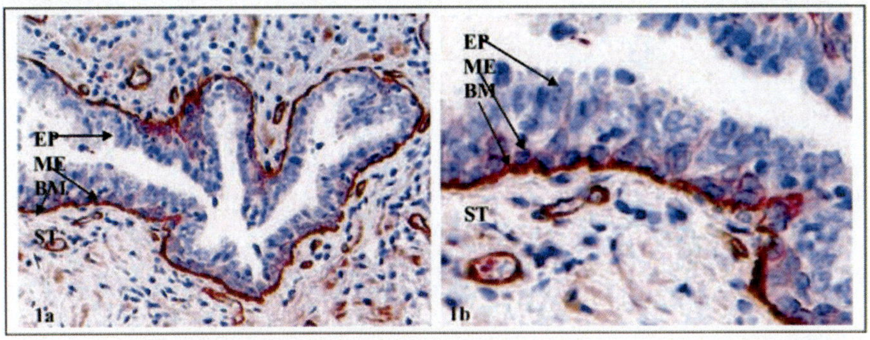

Figure 1. Structural relationship among the epithelium (EP), ME cell layer, BM, and stroma (ST). Human breast tissue section double immunostained for ME cell layer (red) and BM (brown). a: 150X; b: a higher (300X) magnification of a.

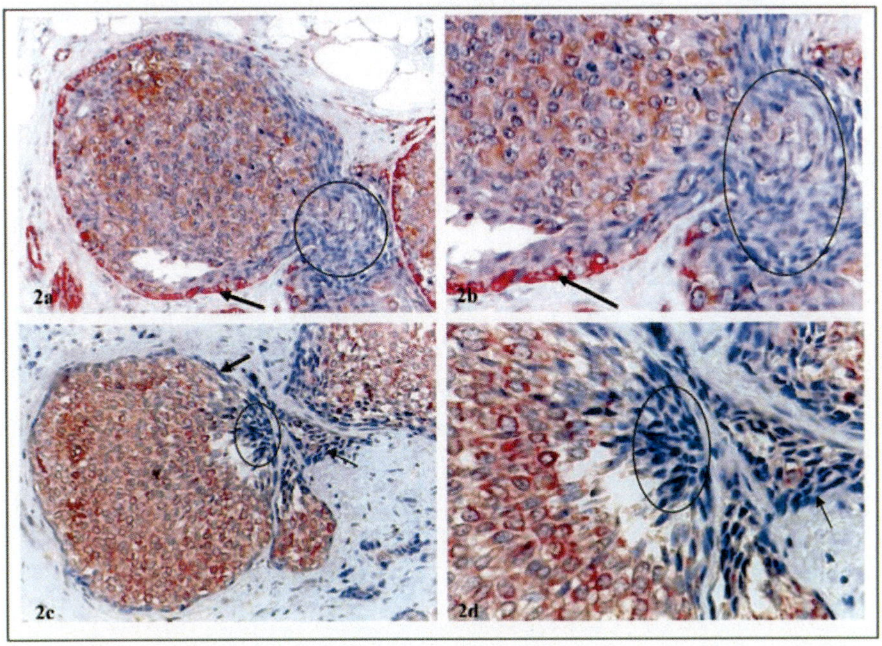

Figure 2. The lack of MMP-26 expression in the invasive front of *in situ* tumors. Doble immunostained for MMP-26 (brown) and SMA (red; a & b) or only MMP-26 (c & d). Circles identify the invasive front. Thin arrows identify micro-invasive lesions. Thick arrows identify ME cell layers. a & c: 150X. b & d : a higher magnification (400X) of a and c, respectively.

Based on three main facts: [1] epithelium-derived ductal tumor cells must first force through the ME cell layer in order to reach the stroma for invasion; [2] the absence of the ME cell layer is the most distinct sign of invasive breast lesions (1-3); [3] a number of tumor suppressors with significant inhibitory functions on tumor cell growth and invasion are exclusively present in ME cells (see below for detail), the author strongly believes that the structural and functional status of ME cells may be closely correlated with the genetic and functional stability of associated tumor cells. Thus, this article attempts to elucidate early signs of ME cell alterations and their potential impact on tumor invasion and progression. In addition, this article presents several unique findings from our recent studies correlating the structural integrity of ME cell layers with the genetic and immunohistochemical profiles in adjacent tumor cells. These unique findings have led to the introduction of a novel hypothesis for breast tumor invasion. The rationale and postulated steps for our hypothesis are summarized and discussed.

MORPHOLOGIC FEATURES OF NORMAL ME CELLS

The normal breast consists of two major compartments: the epithelium and the stroma. The epithelium contains two cell types, luminal and ME cells. Luminal cells form both secretory lobules and branching ducts. ME cells are positioned between the luminal cells and basement membrane, attaching to luminal cells by desmosomes and the basement membrane by hemidesmosomes (1-3). In clear-cut normal human breast tissues, ME and epithelial cells are approximately equal in number. ME cells are of a cuboid or spindle shape, with pale cytoplasm and nuclei (Fig 3). The ME cell layer is continuous in the normal duct system and a vast majority of the lobular units. In the terminal ductal-lobular unit (TDLU), the ME cell layer is occasionally discontinuous, with about 20% of luminal cells directly lie on the BM (27).

The number and morphology of ME cells, however, may vary substantially at different stages of tumor development. In hyperplastic and *in situ* lesions, ME cell layers are often substantially attenuated, with little or no distinct cytoplasm, morphologically resembling the smooth muscle cells of small blood vessels or fibroblasts (Fig 4).

In invasive and metastatic lesions and a rare benign lesion, microglandular adenosis, ME cells are either completely absent (Fig 5), or are occasionally appreciable in immunohistochemically stained sections in the residual normal tissues trapped in these lesions.

BIOCHEMICAL FEATURES OF NORMAL ME CELLS

Under electronic microscope, normal breast ME cells showed a substantially fewer or no distinct rough reticulum endoplasm, Golgi complex, and secretory granules, along with a very low level of proliferating activities, compared to the associated epithelial cells (3). Thus, ME cells have been traditionally believed to belong to an idle and passive physical barrier with no significant active functions (1-3). Recent studies, however, have shown that a number of proteins are exclusively or preferentially present in ME cells. Based on their biochemical compositions, sub-cellular localizations, and potential functions, these molecules can be roughly classified into two groups.

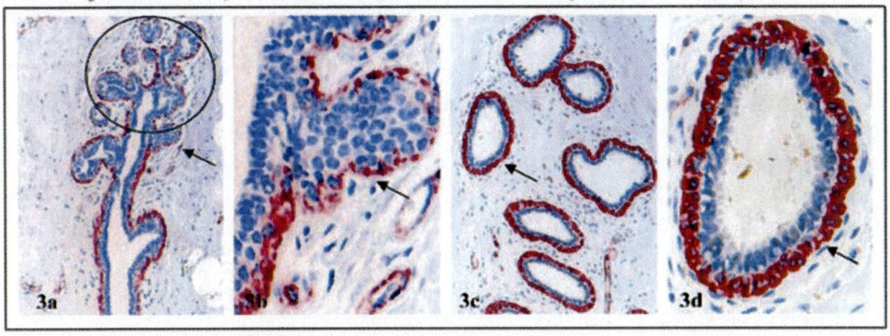

Figure 3. The distribution of normal ME cells in acini and ducts. Immunohistochemically stained for smooth muscle actin (red). a & b: a longitudinal section of a small duct and associated TDLU (circles). c & d: the cross sections of several small ducts. Arrows identify ME cell layers. a & c: 150X. b & d: a higher magnification (400X) of a & c, respectively.

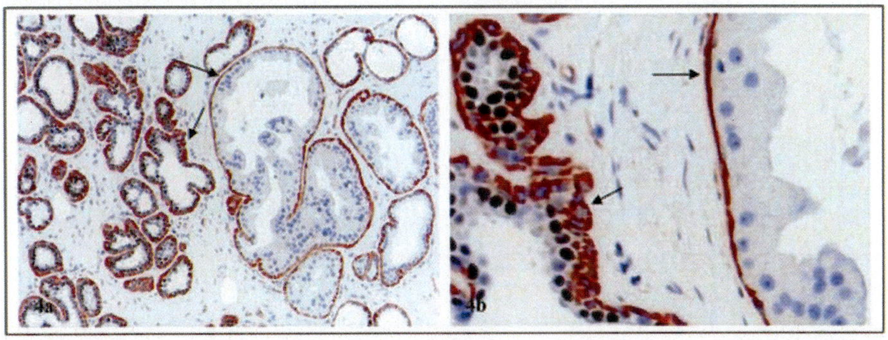

Figure 4. The attenuation of ME cell layers in hyperplastic and *in situ* breast lesions. Immunohistochemically stained for SMA (red). Note that although the ME cell layers are morphologically distinct and continuous, they are substantially attenuated in hyperplastic and *in situ* breast lesions (arrows). a: 150X. b: a higher magnification (300X) of a.

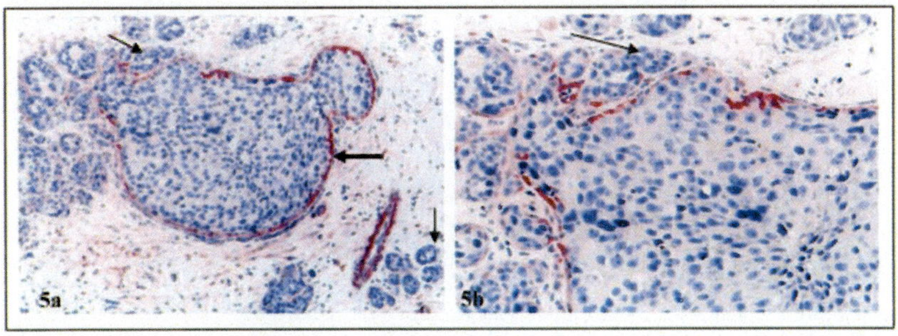

Figure 5. The absence of ME cells in invasive breast lesions. Immunohistochemically stained for SMA (red). Note that the ME cell layer is present in the *in situ* lesion (thick arrows), whereas it is absent in invasive lesions (thin arrows). a: 100X. b: a higher magnification (300X) of a.

A. Structure-specific proteins (related to microfilaments and intermediate filaments)

1. Smooth muscle actin (SMA). SMA is a cytoplasmic protein and is detectable in all or nearly all of the ME cells in normal ducts and lobules, as well as in non-invasive breast lesions (28-29). Invasive lesions and a very rare benign lesion, microglandular adenosis, are devoid of SMA immunoreactivity due to the lack of ME cells surrounding these lesions. SMA expression, however, is also present in other cell types with myogenous features, including smooth muscle cells, myofibroblasts, and vascular endothelial cells.

2. Smooth muscle myosin heavy chain (SMMHC). SMMHC, a cytoplasmic structural protein, is a major component of the contractile apparatus of smooth muscle cells, and is specific for the early development of smooth muscles. It is also a myoepithelium-associated protein, presenting heavily in the ME cells of normal and benign breast tissues, and has been reported to be a very useful marker in differentiation between noninvasive and invasive breast lesions (30-33).

3. Calponin. Calponin is a calmodulin, which binds tropomyosin and F-actin, and is thought to be involved in the regulation of smooth muscle contraction. The expression of calponin is mainly restricted to smooth muscle cells, but has also been demonstrated in human breast ME cells. It is a very useful marker in the differentiation between ME cells and spindle cells of the stroma, and also between non-invasive and invasive lesions (33-34). However, calponin staining in ME cells may occasionally be discontinuous or absent in *in-situ* ductal lesions [33-34].

4. H-caldesmon (HCD). HCD is a cytoskeleton-associated actin-binding protein. In human breast tissues, HCD is predominantly expressed in ME cells, but a high level of HCD expression is also often seen in the smooth muscle cells of small blood vessels (33-34).

5. P-cadherin. P-cadherin, like E-cadherin, is a Ca^{2+} –dependent cell adhesion molecule with a fundamental role in the maintenance of the integrity of multicellular structures and the phenotype of epithelial cells (35-37). In breast tissues, p-cadherin is preferentially expressed in the ME cells of normal ducts, lobules, and sclerotic lesions (35-37). However, p-cadherin is also occasionally seen in hyperplastic tissues and in some tumor cells of *in situ* carcinomas (37).

6. Cytokeratins (CK) 5, 7, 14, and 17. Cytokeratins 5, 7, 14, and 17 are a group of cellular structural proteins found in epithelial cells and cells with epithelioid features (38-41). In human breast tissues, these molecules are predominantly present in the myoepithelial cell layer (38-41). CK-17 is also related to early skin development and wound repair (41).

B. Non-structural specific molecules (not related to microfilaments and intermediate filaments)

1. Maspin. Maspin, or mammary-specific serpin, is a tumor suppressor protein belonging to the serine proteinase inhibitor family (42-44). It is present in both breast and prostate glands. In the breast, maspin is predominantly a soluble cytoplasmic protein that is consistently detectable in normal human ME cells and ME-derived tumors (42-44). The expression of maspin decreases with increasing malignancy of breast tumors (42-44). Both *in vitro* and *in vivo* studies have shown that maspin exhibits significant inhibition of tumor growth and invasion (42, 43).

2. Wilms' tumor 1 (WT-1). WT-1 is as a transcription factor regulating gene expression in a fashion similar to that of p53 (45). The expression of WT-1 in the breast appears to correlate with the behavior and progression of breast tumors (46). Our previous study showed that WT-1 expression was consistently detectable in ME cells, and that the level of WT-1 expression was inversely correlated with breast tumor progression (47-49). The immunohistochemical detection of this protein, however, requires an optimal pre-antigen retrieve, and only the antibody carried by Cell Marqu (Hot Springs, AR) showed consistent results.

3. p63 and p73. p63 is a nuclear protein that shares significant amino acid identity with the p53 protein in the transactivation domain, the DNA-binding domain, and the oligomerization domain (50-53). The expression of p63 in the breast is exclusive to the nuclei of ME cells, and neither stromal fibroblasts nor vascular smooth muscle cells show p63 immunostaining (52). p63 is believed to be critical for maintenance of the progenitor cell populations that are necessary to sustain epithelial development and morphogenesis (50-53). p73 is also a homologue of p53, and is believed to have a function and distribution similar to that of p63 (54).

4. 14-3-3Sigma. This recently introduced ME cell marker is a candidate tumor suppressor gene transactivated by p53 in response to DNA damage (55). It is consistently present in benign and pre-invasive breast lesions, and is preferentially expressed by myoepithelial cells. It might also serve as a novel prognostic factor for breast cancer patients (55).

5. Neuropilin-1 (NRP-1). NRP-1 is a recently identified specific receptor for the vascular endothelial growth factor (56-57). ME cells in both hyperplastic and neoplastic lesions displayed a higher level of NRP-1 expression than those in normal breast tissues. NRP-1 expression is also detectable in vascular smooth muscle and endothelial cells (56-57).

6. CD 10. CD10 is a 100 kDa cell surface metalloendopeptidase, also called neprilysin, which inactivates a variety of biologically active peptides. This protein was initially classified as a common acute lymphoblastic leukemia antigen. Subsequent studies have shown that CD10 was consistently positive in the ME cells of all normal breast tissues, whereas the number of ME-positive cells and the intensity of the immunostaining were substantially reduced in distended ducts and ductal adenomas [58-59].

7. S100. The S100 gene family comprises more than 20 members whose protein sequences encompass at least one EF-hand Ca^{2+} –binding motif. The expression of individual family members appears to be tissue-specific (60). S100 proteins are consistently present in human breast ME cells, but are often expressed in luminal epithelial cells, thus, they are not specific markers for myoepithelial cells (61-63). Molecular analysis of breast tumors has revealed that several S100s, including S100A2, S100A4, and S100A7, exhibit altered expression levels during breast tumorigenesis and progression (60). S100A4 is considered a tumor promoter gene (64). Elevated levels of S100A4 are associated with poor survival rates in breast cancer patients and induce metastasis in rodent models, as well as increased cell motility and invasion *in vitro*(66). S100A9 expression in human invasive breast adenocarcinoma is closely associated with poor differentiation (65-66). Further studies are needed to better understand the important roles of S100s in human breast cancer.

8. CXCL14. A recent study, which used cell type-specific surface markers and magnetic beads to separate the ME, tumor, and stromal cells for genetic comparison using SAGE, has detected an elevated expression of CXCL14 and CXCL12 in ME and myofibroblasts, respectively (67).

9. Other important molecules. A number of growth factors, growth factor receptors, and other biologically important molecules have also been reported to be present in ME cells, but their clinical significance has yet to be determined (68-70).

Postulated Functions of the Myoepithelial Cell Layer

In contrast to the traditional belief that ME cells belong to an idle and passive physical barrier with no significant functions (1-3), recent studies have suggested that ME cells may possess active functions.

1. Production of tumor suppressor proteins. A number of tumor suppressor proteins, including maspin, Wilms' tumor 1, p63, p73, and 14-3-3 Sigma have been detected exclusively or preferentially in ME cells (42-55). *In vitro* and *in vivo* studies have demonstrated that these proteins have significant inhibitory functions against tumor growth and invasion (42-55, 71).

2. Paracrine down-regulation of matrix metalloproteinase (MMP) expression. It has been suggested that ME cells may modulate breast tumor invasion by controlling the expression of MMPs in both tumor and surrounding fibroblasts. A recent study quantitatively compared the MMP expression levels in breast cancer cell lines and fibroblasts co-cultured with and without ME cells purified from normal breast tissues (72). Results revealed that both cancer cells and fibroblasts co-cultured with ME cells had significantly lower expression of MMP-2, MMP-9 and MT1-MMP when compared to those cultured without ME cells, and this reduction of MMP expression was accompanied by reduced invasion (72).

3. Turnover of extracellular matrix (ECM). Although most ECM-degrading enzymes found in breast tissues are manufactured by cancer or stromal cells, several such enzymes, including a recently described angiogenesis-related matrix metalloproteinase, MMP-19, are reported to be produced by normal ME cells (73), and are expected to participate in the turnover of ECM (73).

4. Regulation of normal development. During normal development, a number of growth factors and their corresponding receptors, including the epidermal growth factor, basic fibroblast growth factor and its receptor, Tcf transcription factors, and transforming growth factors, are differentially expressed in the luminal and ME cells (69-70,74). A perturbation in the expression of these and other related molecules in ME cells led to abnormal breast morphogenesis and development (69-70,74).

5. Establishment and stabilization of the polarity of the bilayer epithelial structure. In tissue cultures, luminal cells alone could form acinus-like structures with a reversed polarity

(75). The addition of ME cells from normal breast tissues could establish and stabilize a correct polarity, whereas ME cells from breast tumors failed to correct the reversed polarity, suggesting that normal ME cells play important roles in the establishment and stabilization of bi-layer epithelial structures (75).

6. Steroid hormone metabolism. A recent study assessed the mRNA expression and enzymatic activity of steroid sulfatase (STS) in normal human breast ME and human adenocarcinoma (MCF-7) cells, and the effects of 17-beta estradiol on the activity of STS (76). The study revealed that sulfatase activity was about 120 times greater in the ME cells than in the MCF-7 cells, and that exposure to 17-beta estradiol was associated with a 70% reduction of sulfatase activity in the MCF-7 cells, but a 9% increase in ME cells, suggesting a potential role for ME cells in the conversion of hormone precursors into active steroid hormones (76).

Previously Defined Pathologic Alterations in ME Cells

Compared to epithelial lesions, ME cell lesions are very rare, which frustrates a detailed biological and clinical profiling of this entity. Morphologically, ME cell lesions could be roughly classified into three primary categories (77-78).

1. Myoepitheliosis. This lesion is characterized by a mild proliferation of ME cells within and around ducts. The proliferating cells are of spindle or cuboid shape, often distending or occluding the involved ducts. These ME cells can be easily distinguished from the adjacent ductal epithelial cells by immunohistochemical staining because antibodies against microfilament- and intermediate filament-related molecules will recognize ME cells, while antibodies against epithelial-specific antigen (ESA) will uniformly identify the epithelial cells.

2. Adenomyoepithelioma. This lesion is characterized by extensive proliferation of ME cells. Morphologically, these proliferating cells can be divided into spindle cells, tubular cells, and lobular variants. The common feature of this abnormality is the formation of solid masses that distend or occlude the ductal structures. Similar to myoepitheliosis, these hyperplastic lesions can be easily differentiated from epithelial cells by immunohistochemical staining with antibodies against microfilament- and intermediate filament-related molecules or epithelial-specific antigen (ESA).

3. Myoepithelial carcinoma. These lesions are rare, accounting for less than 1% of breast malignancies. Based on published reports, the most commonly seen type of this lesion is the spindle cell type. This type of lesion is not typically clinically aggressive, although examples of metastases from pure ME carcinomas have been reported (79-81). Breast carcinomas composed in part or entirely of myoepithelial cells have been diagnosed in well-defined categories: benign tumors, tumors with low malignant potential, and malignant tumors, including adenomyoepitheiloma, low-grade adenosquamous carcinoma, adenoid cystic carcinoma, and malignant myoepitheilioma (77-78, 82). Pure myoepithelial carcinoma of the breast is rare, and little is known about its natural history or long-term outcome following treatment. A recent report showed that it adopts an aggressive clinical course, with an outcome comparable to poorly differentiated adenocarcinoma of the breast (83). Comparative genomic hybridization analysis of myoepithelial carcinomas of the human breast compared to ductal carcinomas revealed that the mean number of genetic alterations was 2.1 for myoepithelioma and 8.6 for ductal carcinoma (84). The relatively few genetic alterations found in otherwise aggressive neoplasms indicate that myoepithelial tumors may be useful models for the delineation of genes important in breast tumorigenesis.

Recently Revealed ME Cell Alterations and the Possible Mechanism

Promoted by the fact that the loss of ME cell layer is the most distinct sign of invasive and metastatic breast lesions (1-3), our recent studies have attempted to identify the early signs of ME cell alterations. Our studies have revealed two unique ME cell alterations that have not been previously reported.

1. A high frequency of focal ME cell layer disruptions in a subset of patients with pre-invasive tumors. Using a double immunostaining method, our initial study examined the structural integrity of ME cell layers in 220 patients with estrogen receptor (ER) positive tumors. Of 5,698 ducts examined, 405 (7.1%) contained focal ME layer disruptions (the absence of ME cells resulting in a gap greater than a combined size of at least 3 ME cells). Of the 405 focal disruptions, 367 (90.4%) occurred in DCIS, atypical ductal hyperplasia, and lobular neoplasia, and 38 (9.4%) occurred in normal or hyperplastic ducts (85) (Fig 6).

The frequency of focal ME cell layer disruptions among cases varied significantly, form none in 126 (57.3%) to over 30% in 33 (15%) cases (85) (Table 1).

Patients carrying tumors with a high frequency (>30%) of focal ME cell layer disruptions often contained large clusters of normal or hyperplastic ducts or acini, in which all were with focally disrupted ME cell layers. As shown in Figure 7, none of over 15 ducts in case 1 (a & b) shows focal disruptions in their ME cell layers. In sharp contrast, each of over 15 morphologically comparable ducts in case 2 (c & d) was with focally disrupted ME cell layers.

The frequency of focal ME cell layer disruptions appeared to be independent of the size of the tumors and duct lumen, and also be independent of the histological type and grades (86-88) (Fig 8).

2. The complete lack of all ME cell phenotypic markers in morphologically distinct ME cells. We have found that some patients contained clusters of immunohistochemically altered ME cells. These cells were morphologically distinct in H & E staining, and formed a clearly appreciable layer surrounding tumor clusters (Fig 9). These ME cells in the entire layers, however, completely lacked the expression of all ME cell markers. The adjacent normal ME cells in these cases also lacked the expression of all the tumor suppressors, although they were positive for structure related proteins (89-90) (Fig 9).

3. Possible mechanism of ME cell alterations. Since ME cells belong to a self-renewing population that has to consistently undergo a normal cellular replenishment process to replace aged or injured cells (91-93), we hypothesized that focal ME cell layer disruptions result from elevated ME cell degeneration. To test our hypothesis, ME cell proliferation and differentiation related status was assessed with a panel of markers. Our studies showed that, compared to their non-disrupted counterparts, focally disrupted ME cell layers had several unique features:

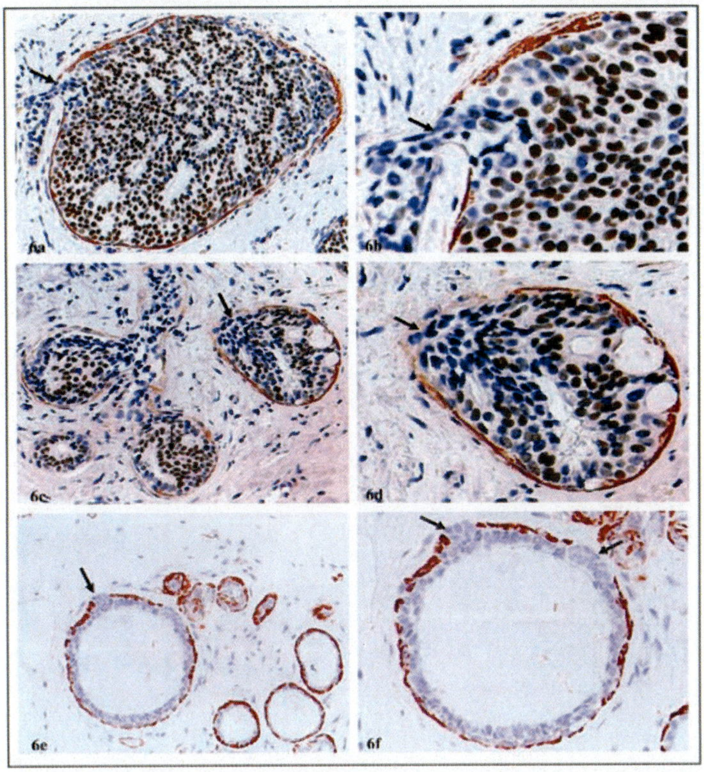

Figure 6. Focal ME cell layer disruptions in DCIS, hyperplastic and normal ducts. Double immunostained for estrogen receptor (a-d, brown) and SMA (red) or only for SMA. Arrows identify focal ME cell layer disruptions in DCIS (a-b), atypical ductal hyperplasia (c-d) and normal duct (e-f). a, c, & e: 150X. b, d, & f: a higher magnification (400X) of a,c, & e, respectively.

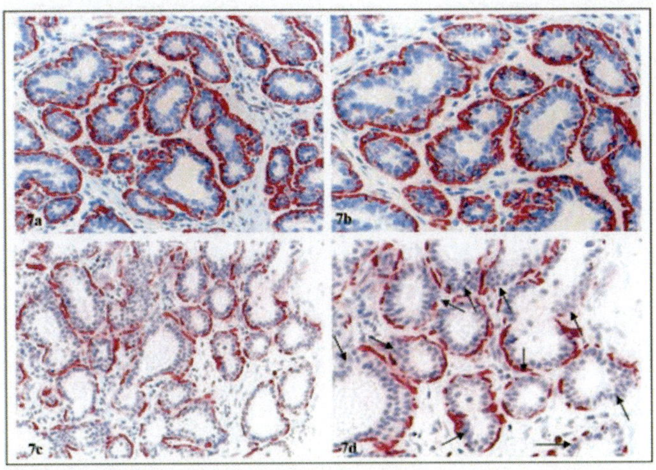

Figure 7. A high frequency of focal ME cell layer disruptions in a subset of patients. Immunostained for SMA (red). Arrows identify focal ME cell layer disruptions. a & c: 150X. b & d: a higher magnification (300X) of a & c, respectively.

Table 1. The frequency of focal ME cell layer disruptions among different cases

Total case number	No disruption	< 30% disruptions	> 30% disruptions	p
220	126 (57.3%)	61 (27.7%)	33 (15%)	< 0.05

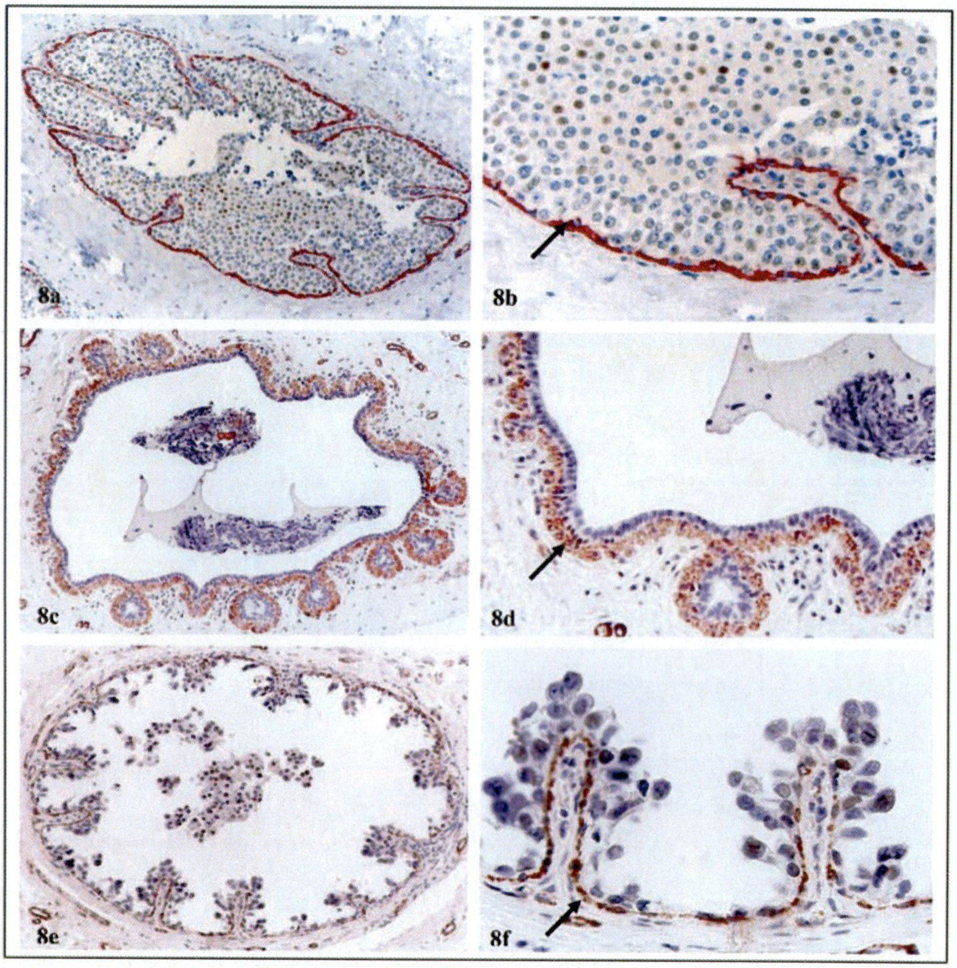

Figure 8. Focal ME layer disruptions are independent of tumor size, duct lumen, or histological grades. Immunostained for ER (brown) and SMA (a & b; red), or SMA (c-f). Arrows identify ME cell layers in a low grade DCIS (a & b), a duct with large lumen (c & d), and a grade DCIS. a, c & e: 40X. b, d & f: a higher magnification (300X) of a, c & e, respectively.

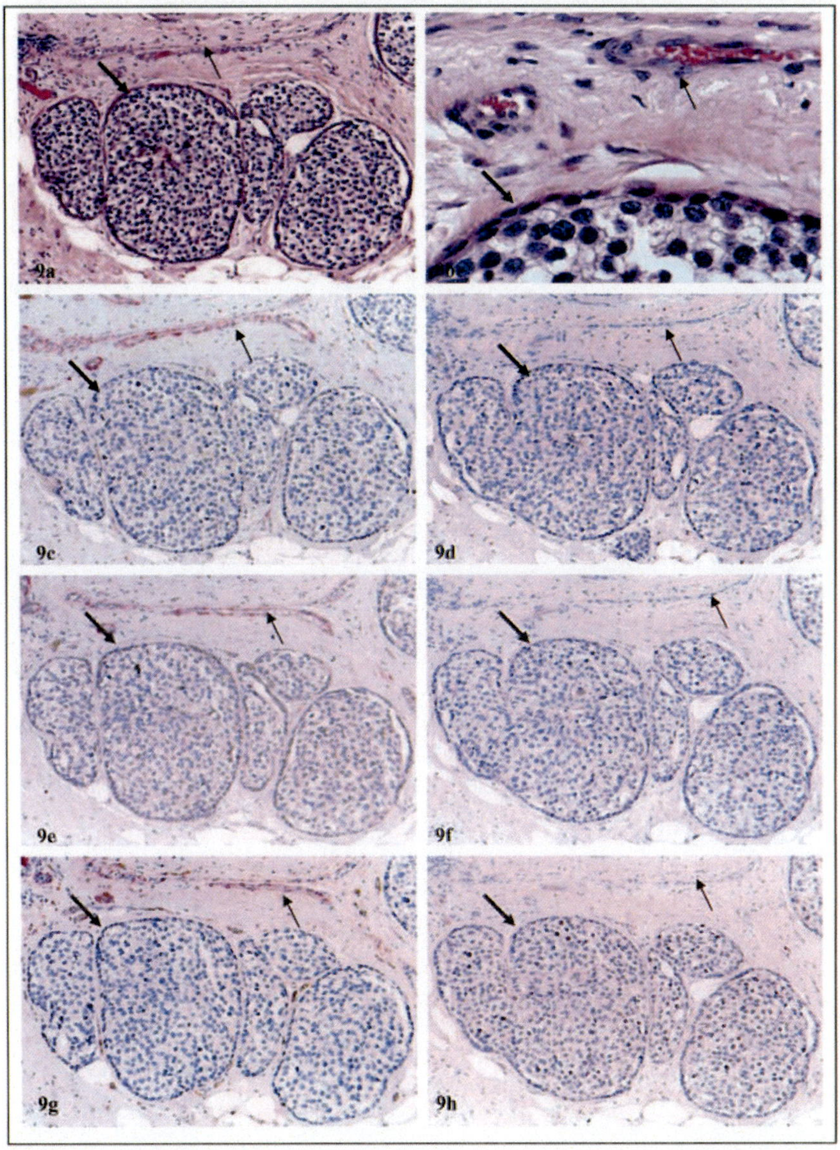

Fig 9. The complete lack of all ME cell phenotypic markers. A set of consecutive sections stained with H & E (a & b) or with different ME cell markers. Thick arrows identify altered ME cell layers. Thin arrows identify adjacent normal ME cells.

a, c-h: 200X. b: a higher magnification (500X) of a.

1. A significantly lower expression of tumor suppressor p63.

In sections double immunostained for SMA and p63, a vast majority of the normal ME cells in cases with no or low rate of focal ME cell layer disruptions were positive for both markers, whereas most normal ME cells in cases with a high rate of ME cell layer disruptions were only positive for SMA (Fig 10).

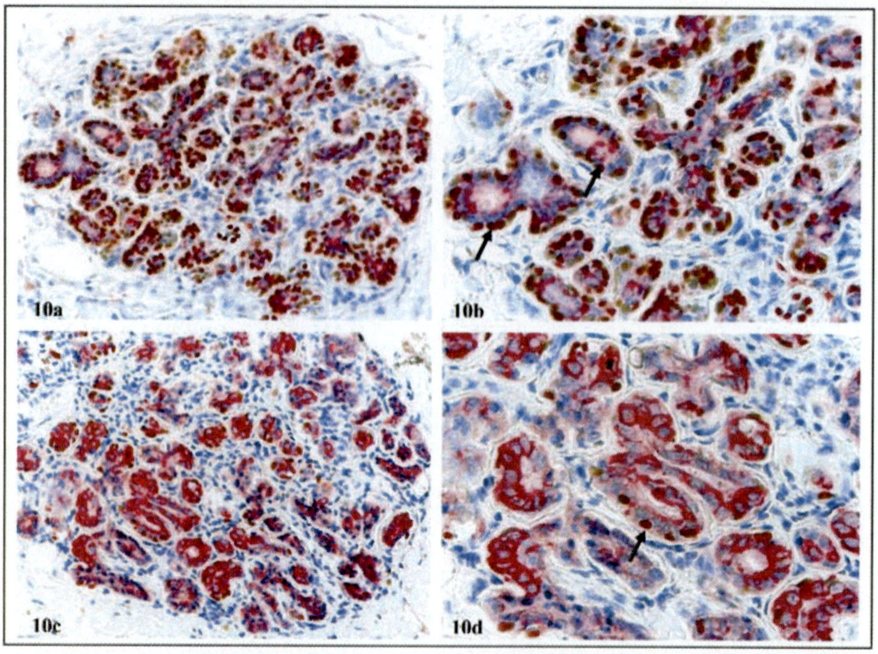

Figure 10. A significantly lower p63 expression in focally disrupted ME cell layers. Double immunostained for SMA (red) and p63 (brown). Arrows identify examples of ducts with both p63 and SMA expression. Note that all the ducts in case 1 (a & b) are with p63, whereas only a very few ducts in cases 2 (c & d) were with p53. a & c: 100X. c & d: a higher (200X) of a & c, respectively.

2. A significantly higher apoptosis Focally disrupted ME cells layers had a significantly higher apoptosis and most apoptotic cells were located at or near focal ME cell layer disruptions (Fig 11).

3. A significantly higher frequency of leukocyte infiltration. In sections double immunostained for SMA and leukocyte common antigen (LCA), ducts or acini with focally disrupted ME cell layers showed a significantly higher frequency of LCA infiltration than their comparable counterparts with non-disrupted ME cell layers (94) (Table 2). Our recent studies have further revealed that the most predominant cells associated with focal ME cell layer disruptions are cytotoxic T-lymphocytes and mast cells, which are consistently detectable at or near the site of focal ME cell layer disruptions (95-101) (Fig 12).

4. A significantly higher frequency of ME cell degeneration. In sections double immunostained for ME cell phenotypic markers and LCA, ME cells surrounded by or adjacent to leukocytes often showed a variety of signs of degenerations, including shrinkage, swelling, condensation of the nuclei, and the loss of normal shape and association (Fig 13).

Table 2. Frequency of leukocyte infiltration in ducts with and without ME cell layer disruptions

	Section number	With WBC	Without WBC	P
Disrupted	191	186 (97.4%)	5 (2.6%)	
Non-disrupted	207	46 (22.2%)	161 (77.8%)	< 0.01

Together, these findings suggest that focal ME cell disruptions may result from a localized degeneration of aged or injured ME cells and resultant auto-immunoreactions. We have hypothesized that the degradation products of degenerated ME cells or diffusible molecules from adjacent tumor cells may function as self-epitopes, which attract the migration and infiltration of cytotoxic T-lymphocytes and mast cells to the affected sites. The infiltrated immunoreactive cells discharge their digestive enzymes upon direct physical contact with altered ME cells, causing a focal structural destruction or disruption in the ME cell layer and BM. Our hypothesis is consistent with previous reports that cytotoxic T-lymphocytes and mast cells are the two majority players in auto-immunoreactions and that these cells could directly result in the destruction of both altered cells and BM (102-108).

The significance of focal ME cell layer disruption is unknown. However, given the fact that several tumor suppressors are exclusively present in ME cells and that the absence of ME cell layer is the most distinct sign of invasive and metastatic lesions (1-3; 42-55), a focal disruption in the ME cell layer could have several potential consequences: [1] a focal loss of tumor suppressors and paracrine inhibitory function; [2] a focal increase of permeability for oxygen, nutrients, or growth factors; [3] a focal increase of leukocyte infiltration; [4] a direct tumor-stromal cell contact. These alterations could promote or facilitate breast cancer invasion through different mechanisms. The loss of tumor suppressors confers tumor cell growth advantage to escape programmed cell death (109-111). The change of oxygen to the physiologic levels selectively favors proliferation of progenitor or stem cells (112-114). The infiltrated leukocytes directly export growth factors to tumor cells (115-119). Direct tumor-stromal cell contact augments the expression of stromal matrix metalloproteinases and facilitates epithelial-mesenchymal transition (120-122).

Therefore, if a disruption happens to occur near a tumor progenitor or stem cell, a clonal proliferation is expected, which forms a tumor cell cluster that may differ genetically and immunohistochemically from adjacent cells within the same duct, but distant from the disruption. In contrast, decreased proliferation and increased apoptosis in ME cells disfavor the replenishment process to repair of disruptions. Therefore, the size of focal disruptions is likely to increase with time. Eventually, the entire ME cell layer might disappear or become immunohistochemically altered, as those seen in Fig 11.

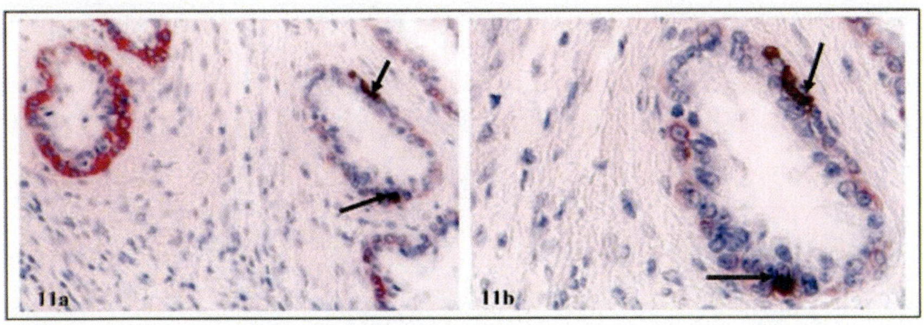

Figure 11. A significantly higher apoptosis in focally disrupted ME cell layers. Assessed for apoptosis and immunostained for SMA. (red). Arrows identify apoptotic ME cells. a: 150X. b: is a higher magnification (300X) of a.

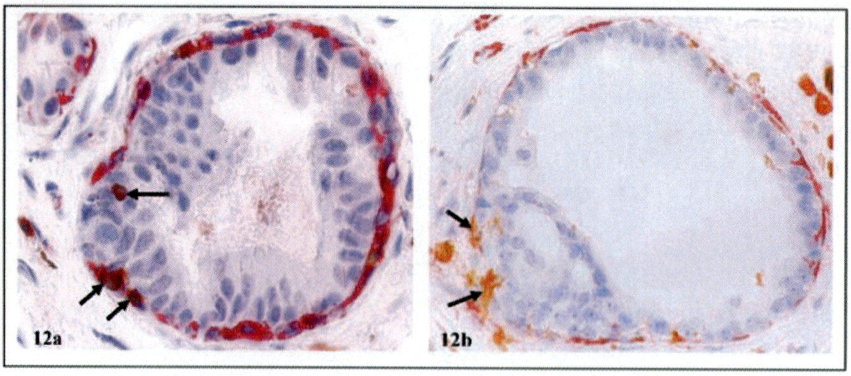

Figure 12. A significantly higher rate of leukocyte infiltration in focally disrupted ME cell layers. Double immunostained for ME cells (red) and LCA (brown). Arrows identify leukocytes associated with focal ME cell layer disruptions. 400X.

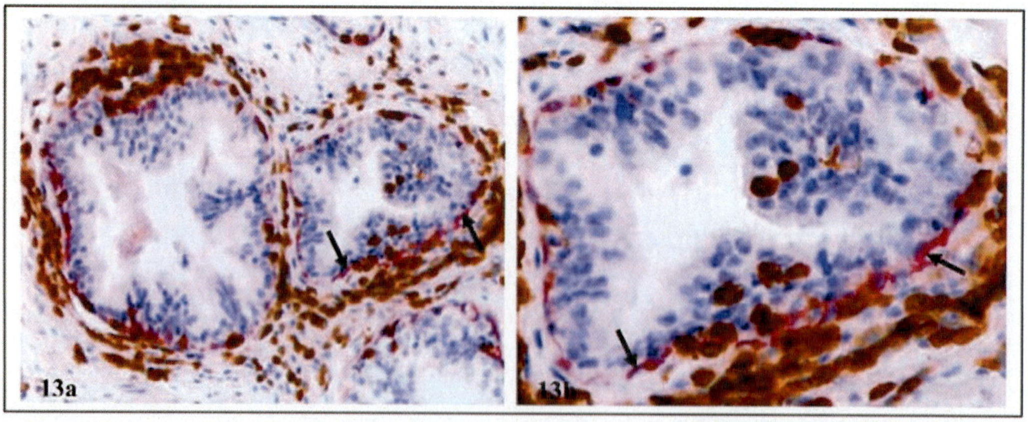

Figure 13. Degeneration of ME cells surrounded by or adjacent to leukocytes. Double immunostained for LCA (brown) and SMA (red). Arrows identify degenerated ME cells. a: 300X. b: a higher magnification (600X) of a.

IMMUNOHISTOCHEMICAL CHANGES ASSOCIATED WITH FOCAL ME CELL LAYER DISRUPTIONS

Consistent with our hypothesis, epithelial cells overlying focally disrupted ME cell layers displayed a number of unique morphological and immunohistochemical alterations, compared to their adjacent counterparts within the same duct, but distant from the disruption.

1. A total loss or substantial reduction of ER expression. Of the 405 focal ME cell layer disruptions detected in our initial study, a vast majority of them were overlaid by cells that completely lacked the expression of ER, in sharp contrast to adjacent cells within the same duct, which were strongly positive for ER (85) (Table 3).

Table 3. Association of focal ME cell layer disruptions with ER negative cells

Number of disruptions	With ER (-) cells	With ER (+) cells	p
405	350 (86.4%)	55 (13.6%)	< 0.001

Cells overlying focal ME cell layer disruptions were generally arranged as clusters. Cells within a given cluster or among clusters, in DCIS or normal ducts were morphologically and immunohistochemically similar (Fig 14). These ER negative cell clusters (ER NCC), however, differed substantially from their adjacent counterparts within the same duct in cell size, shape, density, and polarity (Fig 14).

2. The loss of topoisomerase II alpha expression. A majority of ER NCC lacked the expression of topoisomerase II alpha (TOPOIIa), whereas adjacent cells within the same duct were positive (Fig 15). The loss of TOPOIIa) expression is believed to be a main mechanism for drug resistance (123).

3. A significantly higher tumor cell proliferation index. Tumor cells in ducts with focally disrupted ME cell layers had a significantly higher proliferation index than those in morphologically similar ducts with non-disrupted ME cell layers, about 15% versus 4%, respectively, in twp separate studies (124-125). In sections double immunostained for proliferation markers (cyclin D1 or Ki-67), a vast majority of the proliferating cells were located at or near the site of focal ME cell layer disruptions (94) (Fig 16).

In tumor clusters with immunohistochemically altered ME cell layers, the proliferation index in the tumor cells could be up to over 50% (Fig 17). In sharp contrast, all the immunohistochemically altered ME cell layers were completely devoid of any proliferating cells (Fig 17).

4. A significantly higher expression of c-erb-B2. In contrast to adjacent cells within the same duct, most ER NCC exclusively showed atypical expression of c-erbB2 in their cytoplasm (15-16) (Fig 18).

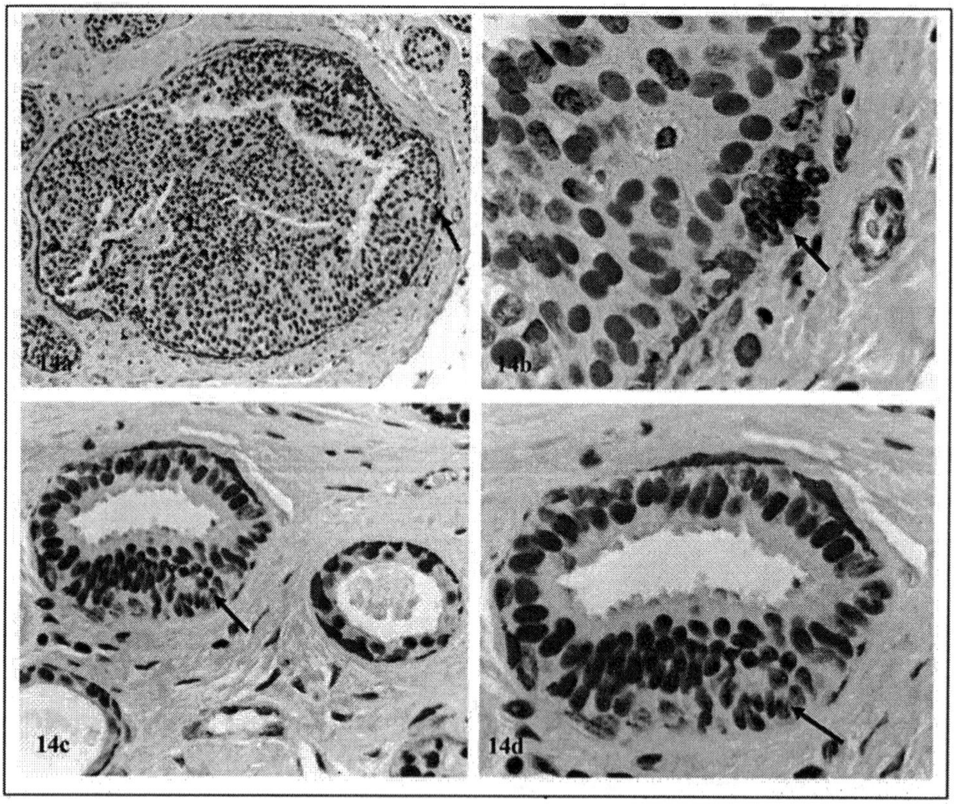

Figure 14. Loss of ER expression in cell clusters overlying focally disrupted ME cell layers. Double immunostained for SMA (red) and ER (brown). Arrows identify ER NCC in DCIS (a & b) and a normal duct (c & D). a & c: 150X. b & d: a higher magnification (400X) of a and b, respectively.

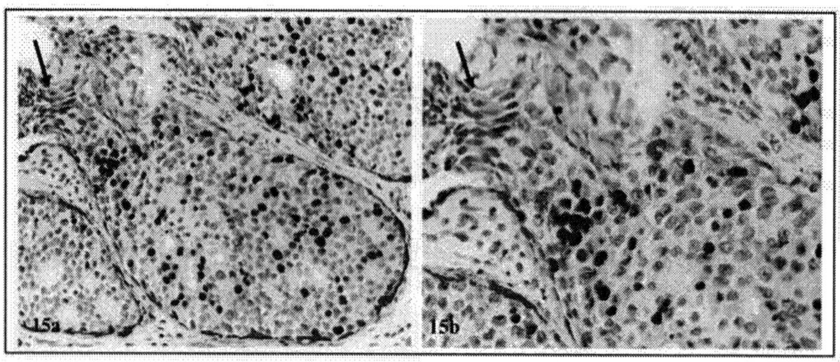

Figure 15. The loss of TOPOIIa expression in ER NCC. Double immunostained for SMA (red) and TOPOIIa (brown). Arrows identify ER NCC. a: 150X. b: a higher magnification (300X) of a.

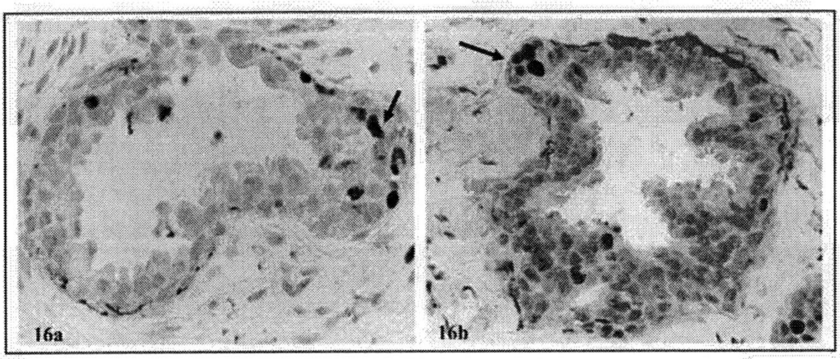

Figure 16. A significantly higher tumor cell proliferation in ducts with focally disrupted ME cell layers. Double immunostained for Ki-67 (brown) and SMA (red). Arrows identify focal ME cell layer disruptions and associated proliferating cells. 400X.

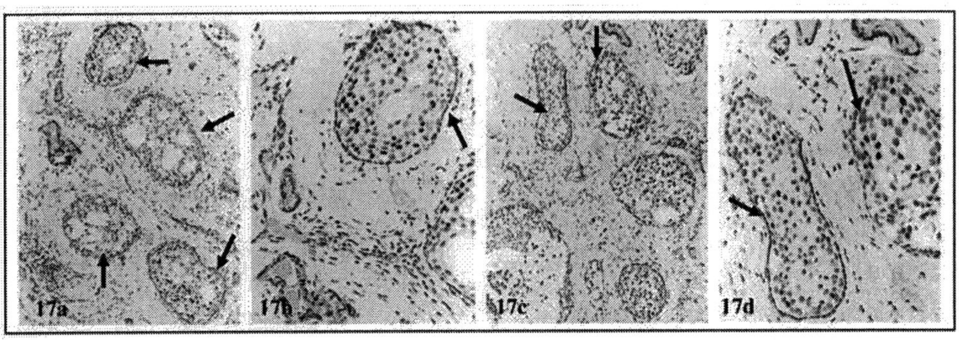

Figure 17. Elevated cell proliferation in tumors with immunohistochemically altered ME cell layers. Double immunostained for SMA (red) and Ki-67 (brown). Note that about 50% of the tumor cells are positive for Ki-67, whereas no proliferating cells are seen in altered ME cell layers (arrows). a & c: 100X. b & d: a higher magnification (200X) of a & c, respectively.

5. p53 expression in normal appearing ducts. Distinct p53 expression was exclusively detected in a subset of normal appearing ducts in patients with a high frequency of focal ME cell layer disruptions. These ducts were distributed as clusters, morphologically indistinguishable from adjacent ducts. In immunostained sections with ME markers, all these ducts were with focally disrupted ME cell layers (126-127) (Fig 19).

6. Expression of epithelial-mesenchymal transition (EMT) and tumor progression related molecules. Our recent immunohistochemical studies have shown that several EMT and tumor progression related molecules, including vinculin, P-Tyr-100, Tyr-416, Tyr-527, and beta-catenin (128-130) were aberrantly expressed in cells overlying focally disrupted ME cell layers. These molecules appeared to be exclusively localized in some cells of the invasive front (Fig 20).

7. Elevated expression of tenascin expression. Our immunohistochemical studies also detected high levels of tenascin, a high molecular weight glycoprotein containing domains homologous to epidermal growth factor (131), exclusively or preferentially in the stroma near focally disrupted ME cell layers (Fig 21).

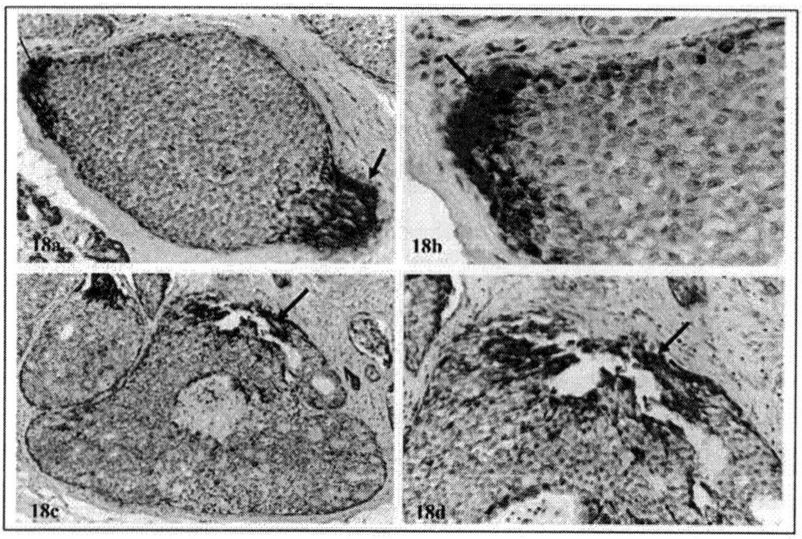

Figure 18. Atypical c-erbB2 expression in ER NCC. Double immunostained for SMA (red) and c-erbB2 (black). Arrows identify ER NCC. a & c: 150X. b & d: a higher magnification (300X) of a & c, respectively.

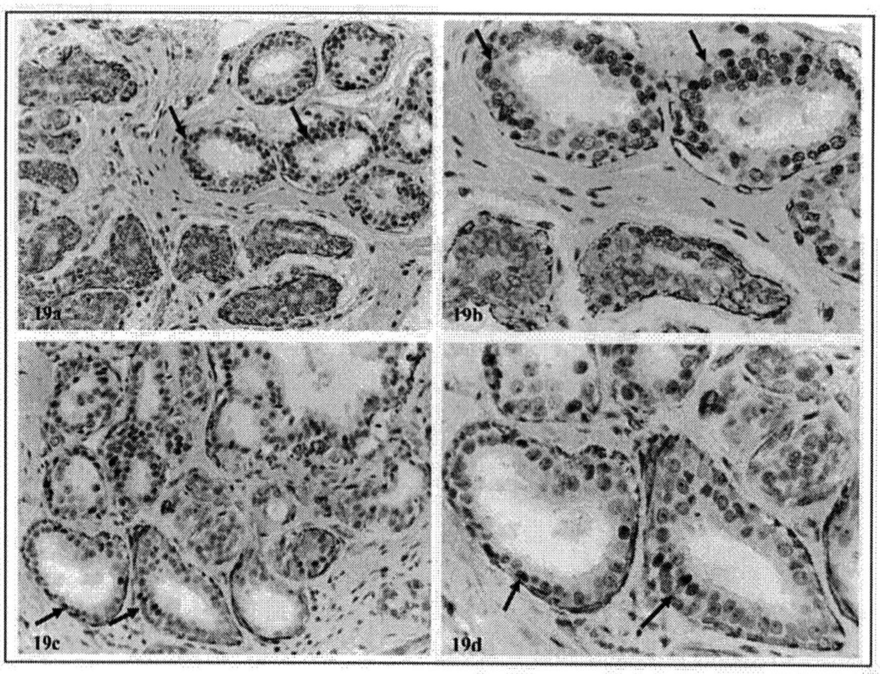

Figure 19. Atypical p53 expression in cell clusters with focally disrupted ME cell layers. Double immunostained for SMA (red) and p53 (black or brown). Arrows identify focal ME cell layers and associated p53 positive cells. a & c: 100X. b & d: a higher magnification (300X) of a & c, respectively.

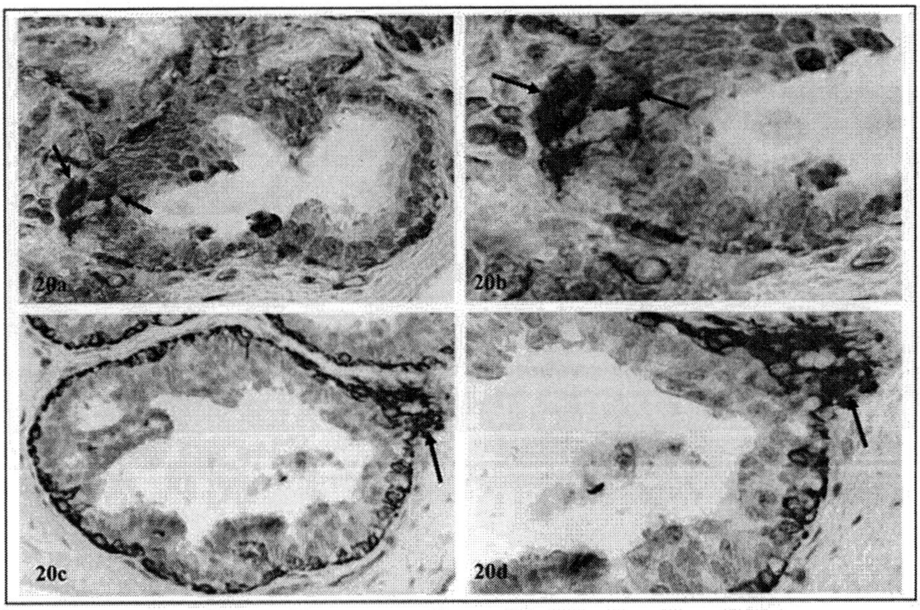

Figure 20. Elevated expression of EMT and tumor progression related molecules in cells overlying focally disrupted ME cell layers. Double immunostained for SMA (brown) plus P-Tyr-100 (red; a & b) or Tyr416 (red; c & d). Arrows identify cells with elevated expression of these molecules overlying focal ME cell layer disruptions. a & c: 200X. b & d: a higher magnification (400X) of a & c, respectively.

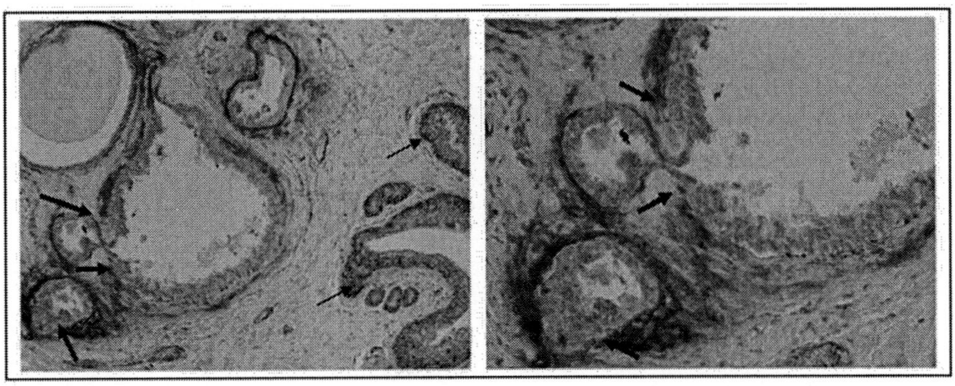

Figure 21. Elevated expression of tenascin in the stroma near focally disrupted ME cell layers. Double immunostained for SMA (red) and tenascin (black). Arrows identify focal ME cell layer disruptions. Note that no distinct tenascin expression is seen in the stroma near non-disrupted ME cell layers (thin arrows). a: 100X. b: a higher magnification (300X) of a.

Although a vast majority of the ER NCC showed a significantly higher proliferation index and expression of tumor progression related molecules than their adjacent counterparts within the same duct, about 10% of ER NCC completely lacked the expression of all tumor cell proliferation and tumor progression related markers (Fig. 22). These ER NCC, however, showed distinct sign of stromal and vascular invasion (15-16).

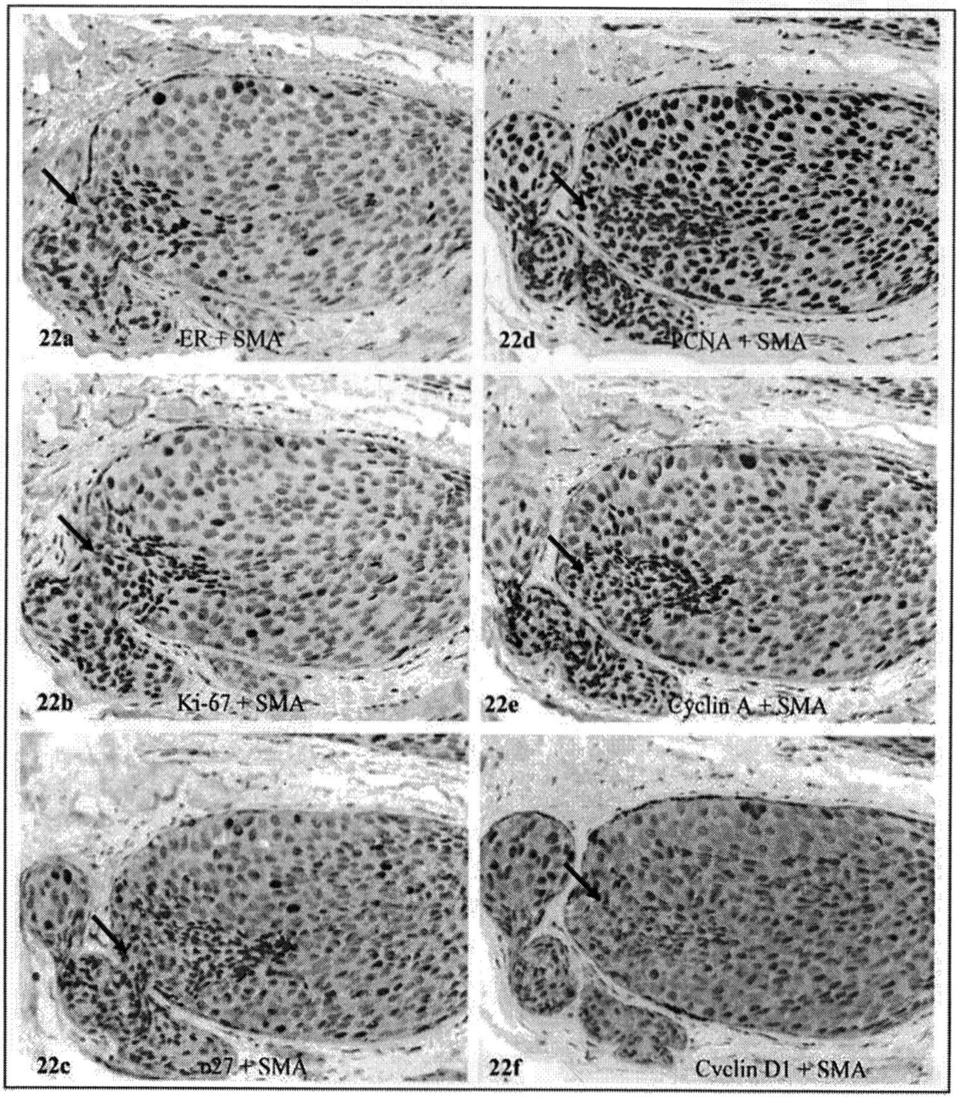

Figure 22. The lack of all proliferation related markers in some ER NCC. Double immunostained for SMA (red) and ER (brown; a) or different proliferation related markers (brown or black). Note that most tumor cells are positive for these markers, while cells in the ER NCC (arrows) are devoid of the expression of these markers. 300X.

GENETIC ALTERATIONS ASSOCIATED WITH FOCAL ME CELL LAYER DISRUPTIONS

Since ER NCC and their adjacent counterparts within the same duct displayed a substantially different immunohistochemical profile, we speculated that they might also have a different genetic profile. Using a microdissection technique, we physically separated ER NCC and their adjacent counterparts within the same duct and compared their genetic profiles relating to tumor progression and invasion. Our studies revealed that ER NCC showed the following unique features, compared to their adjacent ER positive counterparts:

1. A significantly higher frequency of loss of heterozygosity (LOH). Our studies compared the frequency and pattern of LOH in these microdissected cells using 15 DNA satellite markers from 5 different chromosomes (85,132-133). Results showed that ER NCC had a markedly higher or different LOH in 10 of the 15 markers, including those at chromosomes 3p, 1e1p, 13q, and 16q, which harbor different tumor suppressors (85, 132-133) (Fig 23).

2. A significantly higher expression of cell cycle control and tumor invasion-related genes. Using a focused cDNA microarray containing 96 tumor progression and invasion related genes, our subsequent studies further showed that ER NCC had a significantly higher expression of multiple cell cycle control and tumor-invasion related genes(134-135) (Fig 24).

EVIDENCE OF THE TRANSFORMATION OF ER NCC TO INVASIVE BREAST LESIONS

Together, the above findings suggest that focal ME cell layer disruptions could substantially impact the biological presentation of the overlying epithelial cells, represent a trigger factor or early sign of breast tumor invasion. These findings also suggest that ER NCC overlying focally disrupted ME cell layers are likely to represent the direct precursor of invasive breast lesions. Our findings are in total agreement with previous findings that breast tumor progression is paralleled by a progressive hormonal independence, and that ER negative tumors have a substantially more aggressive clinical presentations and worse prognosis than ER positive tumors (136-137). Our findings are also consistent with previous reports that human breast tumor progression from one stage to another is driven by sequential expression of stage-specific molecules and the selection of biologically more aggressive cell clones (7-9). Our findings, however, are hard to reconcile with the fact that over 80% of invasive breast cancers are reported to be ER positive (138-139).

Our recent studies have attempted to identify the potential factors accounting for this discrepancy. Based on three reasons: [1] most malignant tumors arise from a single cell (140-142), [2] epithelial cell "budding" is a common event shared by normal development and tumor invasion, which all occurred exclusively at the site of focally degraded BM (143-144), and [3] our previous studies have demonstrated that a vast majority of the proliferating cells are located at or near the site focal ME cell layer disruptions (94), we hypothesized that ER NCC and their "budding" derivatives were derived from a monoclonal proliferation of the same progenitor, whereas they are at different differentiation stages with a different ER expression status. In addition, as our previous studies in normal and surgically operated adult submandibular glands showed that newly formed cell clusters by stem cells completely lacked the expression of several proteins that were heavily expressed in their adjacent adult counterparts, whereas they gradually gained the expression of these molecules with time (145-146), we further hypothesized that ER NCC may represent tumor progenitors that are not mature enough to manufacture ER, or are with a "resting" ER gene. The derivatives of ER NCC, however, could gradually gain ER expression after "budding" from focally disrupted ME layers and invading deeper into the stroma, due to a programmed sequence or direct interactions with stromal or immunoreactive cells.

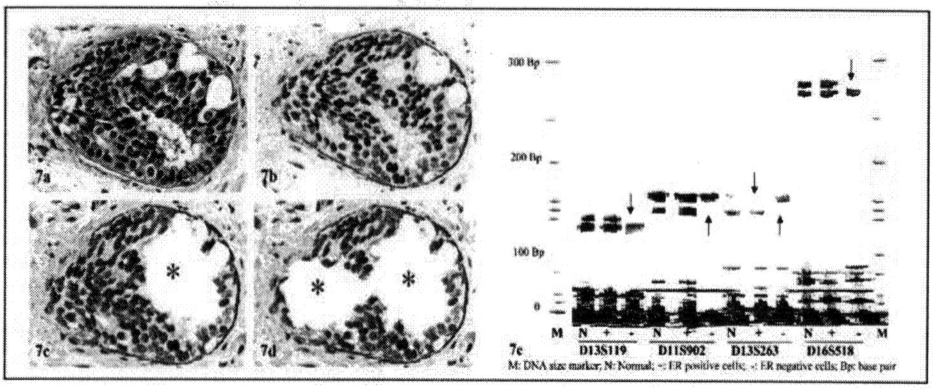

Fig 23. Higher rate of genetic instabilities in ER NCC. Double immunostained for ER (brown) and ME cells (red). ER(+) and ER(-) cells within the same duct were microdissected for genetic analysis. Arrows identify LOH.

(Adopted from Man et al. Breast Cancer Res 5: R231-141, 2003)

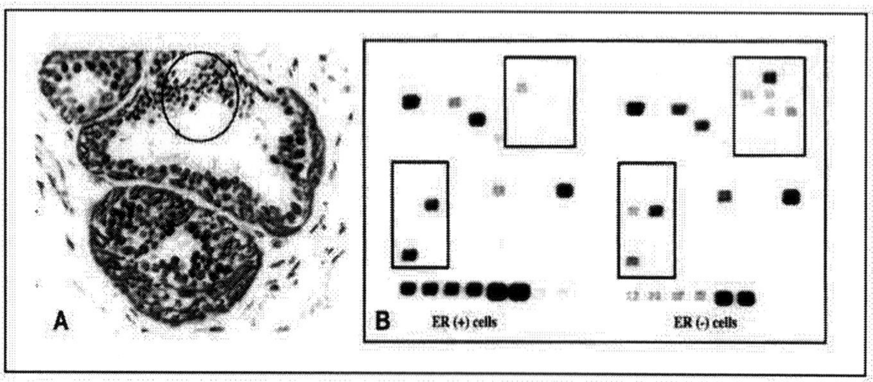

Figure 24. Higher expression of cell cycle control and tumor invasion-related genes in ER negative cells. Double immunostained for ER (brown) and SMA (red). A circle identifies dissected ER negative cells. Squares identify differentially expressed genes.

Using a double immunohistochemical technique to compare the ER expression profiles in ER NCC and their derivatives in tumors with co-existing *in situ* and invasive components, our studies have obtained several unique findings.

1. ER NCC arranged as tongue-like projections. Of 100 selected cases, 17 were found to contain large ER NCC with "budding" derivatives that were arranged as tongue-like projections, "puncturing" into the stroma (147-148) (Fig 25).

Table 4. Comparison of the frequency and level of mRNA expression in ER (+) and (-) cells

Gene group	Gene name	Higher in ER (-)	Higher in ER (+)	Equal in both	Case number
Adhesion	CD44	13	4	1	18
	CDH1	13	4	3	20
	MUC18L	5	5	2	12
Angiogenesis	EGFR	5	6	1	12
	IFNA1	10	5	2	17
	TNF	5	9	6	20
Apoptosis	BAX	9	4	0	13
	CASP9	9	5	5	19
	CFLAR	10	5	1	16
Cell cycle	CDC25A	9	1	7	17
	CDKN2 A	10	4	2	16
	RAD53	9	1	2	12
Inv & metastasis	NME4	10	4	6	20
	TIMP1	6	6	3	15
Signal transduct	NFKB1	11	6	3	20
Total		134 (54.3%)	69 (27.9%)	44 (17.8%)	247
P	< 0.01				

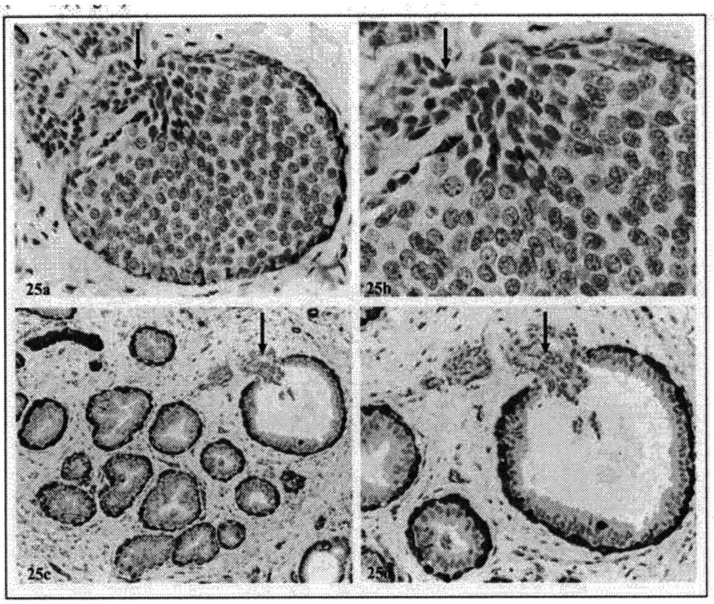

Figure 25. Tongue-like cell clusters at focally disrupted ME cell layers. Immunostained for SMA (red). Arrows identify tongue-like cell clusters in DCIS (a & b) and a normal duct (c & d). a & c:200X. b & d: a higher magnification (400x) of a & c, respectively.

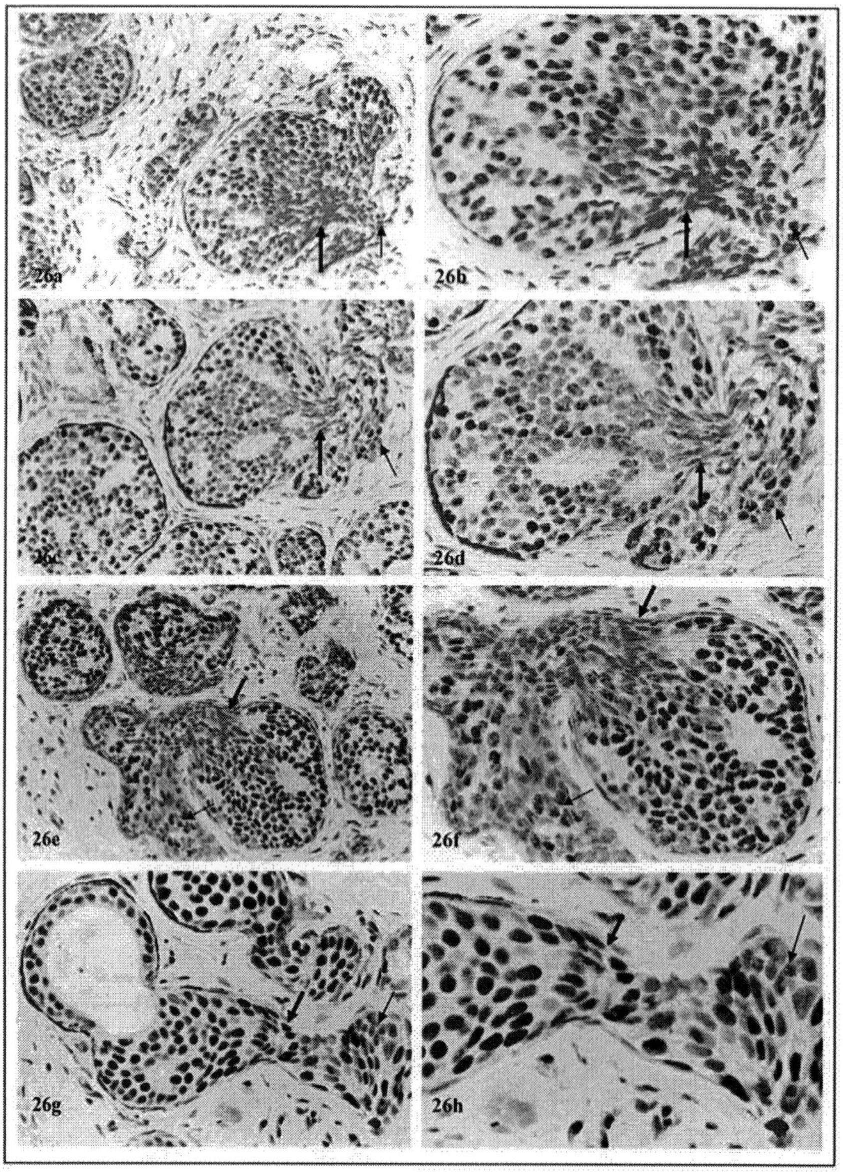

Figure 26. ER expression in ER NCC and their derivatives. Double immunostained for SMA (red) and ER (Brown). Thick arrows identify ER NCC, thin arrows identify their derivatives. a & c: 150X. b & d: a higher magnification (400X) of a & c, respectively.

2. Increasing ER expression in direct derivatives of ER NCC. All or nearly all cells of each ER NCC completely lacked ER expression, in contrast to their adjacent counterparts within the same duct, which were strongly positive for ER. The number of ER positive cells and intensity of ER immunostaining in "budding" derivatives, however, increased linearly as cells invaded deeper into the stroma (147-148) (Fig 26).

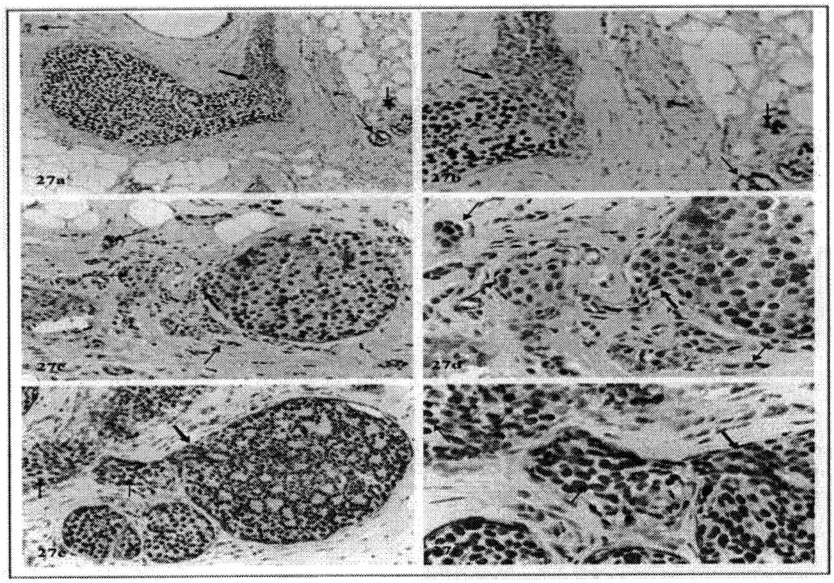

Figure 27. ER expression in ER NCC and adjacent micro-invasive lesions. Double immunostained for SMA (red) and ER (brown). Thick arrows identify ER NCC. Thin arrows identify micro-invasive lesions. a & c: 150X. b & d: a higher magnification (400X) of a & c, respectively.

3. Increasing ER expression in micro-invasive lesions. Of the 100 selected cases, there were also three cases containing micro-invasion that were immediately adjacent to, or in direct continuity with ER NCC. The patter of ER expression in these micro-invasive was very similar to those seen in ER NCC (Fig 27).

4. Vascular invasion of ER NCC. Some ER NCC appeared to be in direct continuity with blood vessel-like structures (15-16, 149-150). Some of these vessel-like structures contained a large number of red blood cells (Fig 28). Immunohistochemical studies with specific markers to blood vessel and lymphatic ducts showed that some of these are indeed are blood vessels and lymphatic ducts (Fig 29).

Similar Focal Disruptions and Associated Alterations in Other Tumors

Our subsequent studies in human prostate, cervical, and colorectal tumors have revealed a similar high frequency of focal disruptions in their surrounding basal cell layers or the BM in a subset of patients with pre-invasive and invasive lesions (151-153) (Fig 30). Similar to focally disrupted ME cell layers, focally disrupted basal cell layers showed a significantly lower frequency of p63 expression (Fig 31), whereas they showed a significantly higher frequency of apoptosis (Fig 32) and leukocyte infiltration (154-160) (Fig 33). Tumors with focally disrupted basal cell layers showed a significantly higher cell proliferation index than their morphologically similar counterparts (the same histological type and grade, and similar size and location) with non-disrupted basal cells. In sections double immunostained for basal cell specific and proliferation markers, a vast majority of the proliferating cells were located at or near the site of focal basal cell layer disruptions (154-160) (Fig 34).

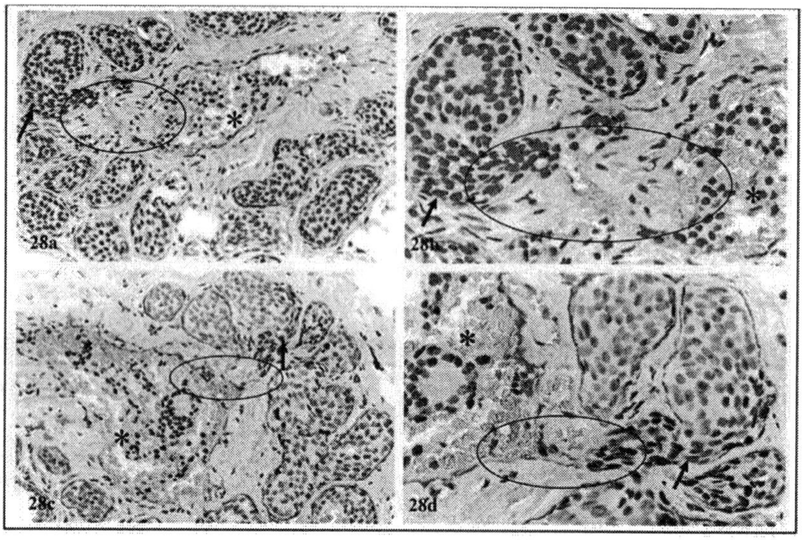

Figure 28. Invasion of ER NCC into vessel-like structures. Double immunostained for SMA (red) and ER (brown). Arrows identify ER NCC. Circles identify the continuity of ER NCC with vessel-like structures. Asterisks identify vessel-like structures with red blood cells. a & c: 150X. b & d: a higher magnification (400X) of a & c, respectively.

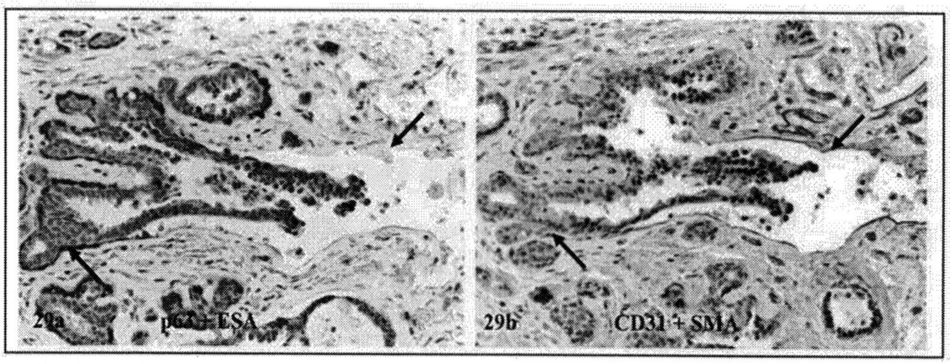

Figure 29. Expression of CD31 in vessel-like structures. Double immunostained for p63 (brown) and ESA (red; a), or CD31 (brown) and AE1/AE3 (red; b). The wall of vessel-like structures are negative for ME cell marker p63, but are positive for vessel marker CD31 (arrows). 200X.

Our Hypothesis of Breast Tumor Invasion and Genetic Instability

Based on the above and other findings, we have proposed that breast tumor invasion is triggered by a localized degeneration of altered ME cells and the resultant immunoreactions, which facilitates a clonal proliferation of the overlying tumor progenitor or stem cells. Based on our hypothesis, the major steps leading to the initiation and progression of tumor invasion are as follows:

1. A subset of patients carries the predisposition of cell replenishment or cycle control related genetic defects in the ME cell population, which impair the normal process to replace aged or injured ME cells, resulting in structural and functionally instable ME cell layers with an elevated apoptosis and degeneration in ME cells.

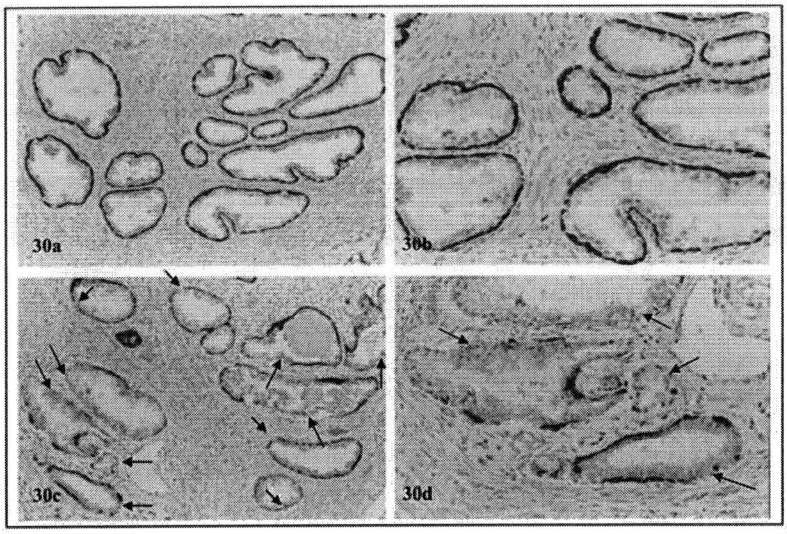

Figure 30. A high frequency of basal cell layer disruptions in a subset of patients. Immunostained for cytokeratin (CK) 34βE12 (red). Note that none of the over 10 acini in case 1 (a & b) shows focal disruptions in the basal cell layers, but each of the 10 acini in case 2 (c & d) are with focally disrupted basal cell layers. a & c: 100X. b & d: a higher magnification (200X) of a and c, respectively.

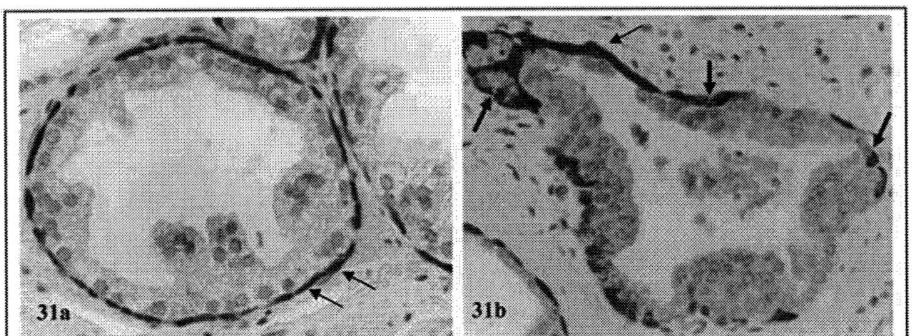

Figure 31. A significantly lower p63 expression in focally disrupted basal cell layers. Double immunostained for CK 34βE12 (red) and p63 (black). Note that nearly all the basal cells in the non-disrupted layer are positive for both CK 34βE12 and p63 (thin arrows), whereas many basal cells in the disrupted layer are positive for only CK 34βE12 (thick arrows). 300X.

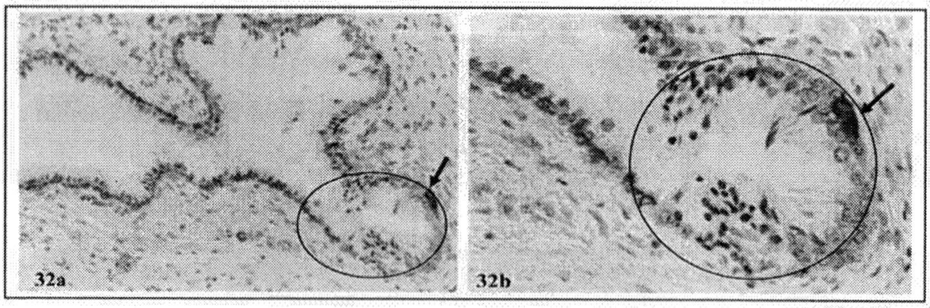

Figure 32. A significantly higher apoptosis in focally disrupted basal cell layers. Assessed for apoptosis and then immunostained for CK 34βE12 (red). Arrows identify apoptotic basal cells. Note that tumor cells near focal basal cell layer disruptions show elevated proliferation (circles). a: 100X. b: a higher magnification (400X) of a.

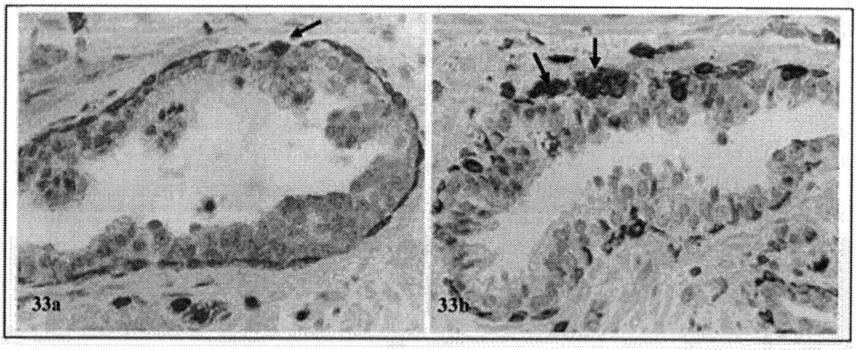

Figure 33. A significantly higher leukocyte infiltration in focally disrupted basal cell layers. Double immunostained for leukocyte common antigen (LCA, brown) and CK 34βE12 (red). Note that most leukocytes (arrows) are located at or near focal basal cell layer disruptions. 300X.

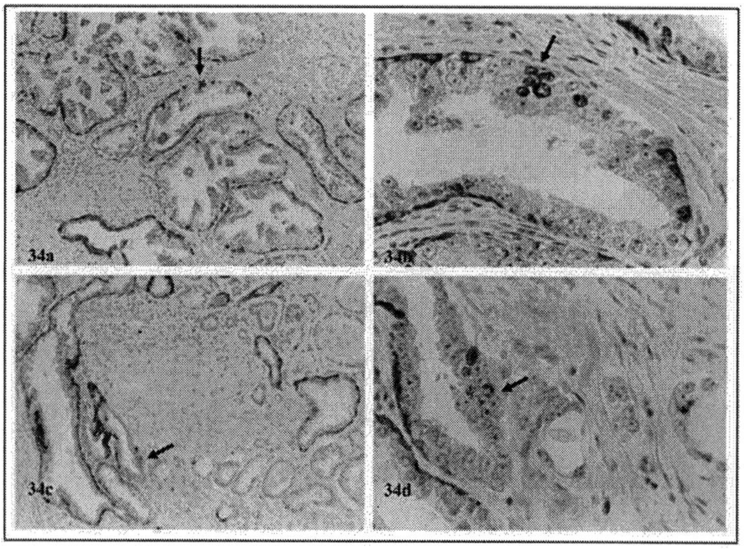

Figure 34. A significantly higher tumor cell proliferation at or near focal basal cell layer disruptions. Double immunostained for CK 34βE12 (red) and Ki-67 (brown). Note that a vast majority of the proliferating tumor cells are located at or near focal basal cell disruptions (arrows). a & c: 100x. b & d: a higher magnification of a & c, respectively.

2. Elevated apoptosis and degeneration in the ME cell population result in the reduction or loss of the production of tumor suppressors, which confer growth advantage upon associated epithelial cells.

3. The degradation products of damaged or dying ME cells or diffusible molecules from associated epithelial cells function as self-epitopes that attract the migration of cytotoxic T-lymphocytes and mast cells to the affected sites.

4. A focal external insult, such as a localized trauma, inflammation, or excessive stimulation, aggravates or accelerates the above alterations, or directly causes structural damages, which attract infiltration of immunoreactive cells into the affected ME cell layers and associated epithelial cells.

5. The direct physical contact of altered ME or epithelial cells with immunoreactive cells stimulates the discharge of digestive enzymes from immunoreactive cells, which cause physical destruction of both the BM and degenerated ME cells, resulting a focal disruption.

6. A focal disruption in the BM and ME cell layer could have at least four potential consequences:

a: a focal loss of tumor suppressors and paracrine inhibitory function,

b: a focal increase of permeability for oxygen, nutrients, and growth factors,

c: a focal increase of immunoreactive cell infiltration, and

d: a direct tumor-stromal cell contact, which could promote or facilitate breast cancer invasion through different mechanisms (see the above for detail).

7. The above consequences could individually or collectively impact the biological presentation of the overlying epithelial cells. The pattern and extent and the subsequent consequences in these epithelial cells, however, depend on the genetic and biochemical composition of these cells.

a: if the overlying cells are fully differentiated, no substantial alteration may be detectable.

b: if these cells are partially differentiated, a limited increase in cell proliferation might be seen, which may lead to a localized stromal invasion.

c: if they are tumor progenitor or stem cells, the combined impact of these potential consequences could further lead to aberrant expression of c-erbB2 and p53, which consequently activates different cell proliferation pathways and facilitate epithelial-mesenchymal transition, resulting in the formation of ER NCC overlying focally disrupted ME cell layers.

8. ER NCC overlying focally disrupted ME cell layers are not mature enough to expression ER, or are with a “resting” ER gene. However, the derivatives of ER NCC could gain ER expression after invading deep into the stroma, due a pre-programmed genetic sequence or interactions with stromal or immunoreactive cells.

9. The derivatives of ER NCC undergo differentiation, releasing invasion-associated molecules, which triggers expression of tenascin, angiogenesis, tissue remodeling, and increased production of growth factors in the stroma, providing a favorable environment for tumor cell growth.

10. The interactions between tumor and stromal cells lead to further destruction of ME layers and degradation of the basement membrane, accompanied by deeper and wider migration and invasion of the newly formed cell clusters.

Our hypothesis could not only reasonably interpret our experimental findings, but is also consistent with previous reports that patients carrying BRCA1 and BRCA2 gene mutations are at a significant higher risk for development of breast and ovarian tumors (161-162), and that mutations of the E-cadherin gene not only impact the expression of cellular adhesion molecules and cell morphology, but also markedly influence the beta-catenin localization and susceptibility to cytoskeletal alterations or the accumulation of apoptosis related proteins (163-165). In addition, our hypothesis is in total agreement with previous reports that a mutated stem cell is the common cellular origin of teratocarcinomas and cancers, and that a functional mammary can be formed by a single cell, and that tumor stem cells and other aggressive tumor cells are capable of generating their own vascular structures (140-142). Our hypothesis and supporting date has been published recently in several peer-reviewed journals (15, 85, 94, 134, 147,151).

MAIN DIFFERENCES BETWEEN OUR HYPOTHESIS AND THE PROTEOLYTIC ENZYME THEORY

In addition to the central difference in the direct cause for triggering tumor invasion, our hypothesis also differs from the proteolytic enzyme theory at the following aspects:

1. The stage of tumor invasion. According to the enzyme theory (10-11), the level of proteolytic enzymes linearly increases with tumor progression and reaches the highest level at the *in situ* cancer stage in which tumor invasion occurs. Based on our hypothesis, tumor invasion could potentially take place at any stage of tumor evolution, including the normal or hyperplastic stage, if the focal disruption happens to occur at or near tumor progenitor or stem cells. In other words, breast tumor invasion might not always follow the hypothesized sequences.

2. The cellular origin of invasive lesions. According to the proteolytic enzyme theory, all the *in situ* cancer cells could equally contribute to invasive lesions. Based on our hypothesis, invasive lesions are solely or predominantly derived from stem or progenitor cells overlying focal ME cell layer disruptions. It is also possible that the normal and hyperplastic duct clusters in which all their surrounding ME cell layers are focally disrupted may represent the direct precursor of invasive lesions, as they might be derived from a common tumor progenitor and have accumulated substantial genetic alterations.

3. The significance of immunoreactive cells. Based on the proteolytic enzyme theory, the degradation of the BM is caused by elevated proteolytic enzymes produced by mainly by cancer cells. Based on our hypothesis, the degradation of the BM and ME cell layers are caused by interactions between degenerated ME cells and immunoreactive cells. Although the enzymes of immunoreactive cells also belong to the proteolytic enzyme family, these enzymes are discharged only upon a physical contact. Therefore, their impact is focal and independent of the tumor stage and histological grades.

4. The significance of stromal cells. According to the proteolytic enzyme theory, the main function of stromal cells is the production of proteolytic enzymes that involved in the degradation of the BM. Based on our hypothesis, stromal cells may also play other important roles, including the production of tenascin, facilitation of EMT and angiogenesis. More importantly, stromal cells may significantly influence the fate and gene expression of epithelial cells. Our previous studies in human breast, lung, and cervical tumors have consistently showed that the tumor cells and their surrounding stromal cells shared a high degree of common genetic and immunohistochemical alterations (166-172).

5. The putative consequences of focal basal cell layer and basement membrane disruptions. According to the proteolytic enzyme theory, as a destructive level of proteolytic enzymes could be reached only at the *in situ* cancer stage, the potential consequence of the basement membrane degradation is likely to be one directional-tumor invasion. Based on our hypothesis, the consequence of a focal basal cell layer and basement membrane disruption is primarily depended on the genetic composition of the overlying cells.

6. The potential approaches for treatment and prevention. According to the proteolytic enzyme theory, using corresponding enzyme inhibitors is the primary, if not the sole, regime. Based on our hypothesis, however, there are at least five approaches: [1] to neutralize the molecules that attract immunoreactive cells, [2] to reduce the number of immunoreactive cells associated with ME cell layers, [3] to prevent the formation of genetically altered clusters overlying focal ME layer disruptions, and [4] to control the extent of stromal reactions, [5] to stimulate ME cell growth.

IMPLICATIONS OF OUR HYPOTHESIS

Our hypothesis that localized ME cell death and subsequent immunoreactions are the direct cause for ME cell layer degradations and subsequent tumor invasion is predominantly based on the following reasons. First, both breast epithelial and ME cells belong to self-renewing populations, with a life-span of about 300 days (91-93)). Therefore, the ME cell layers in each individual normally contain aged or degenerated ME cells. Second, all proteolytic enzyme inhibitor-based human clinical trials have been disappointing, indicating that blocking basement membrane-degrading proteases alone is not sufficient to inhibit tumor invasion and metastasis, or additional factors are involved in the processes (12-13). Third, our previous studies had shown that a subset of morphologically distinct ME cells completely lacked the expression of all ME cell specific phenotypic markers, suggesting that ME cells may also be the target of pathological alterations (89-90). Fourth, the frequency of focal ME cell layer disruption appears to be independent of the size, length, architecture, and overall ER negativity of the ducts, and also of the histological grade of the lesions (85-88). Fifth, ducts with focal ME cell layer disruptions have a significantly higher leukocyte infiltration, and most leukocytes are located at or near the disruptions (94-101). Similar focal disruptions, leukocyte infiltration, and associated tumor cell proliferation have been observed in human prostate, cervical, and colorectal tumors (151-160). It appears that the leukocyte infiltration is in response to ME cell death, with inflammatory responses following ME cell death rather than preceding it.

It is not yet known, however, whether or to what extent, our hypothesis reflects the intrinsic mechanisms of breast tumor invasion for two main reasons: [1] the underlying mechanisms and detailed processes for each of our hypothesized steps have not been elucidated, and [2] the conclusive correlation between our findings and clinical outcomes has not been established. Furthermore, it is possible that ME cell death might be triggered by the primary lesion during cancer development, inflammatory changes of the stroma, or other factors. This question would be resolved through the study of developing disruptions as opposed to existing disruptions, but as yet, the formation of these disruptions cannot be predicted. Therefore, this article concentrates on the elucidation of the significance of ME cell layer disruptions in tumor invasion and progression rather than the causes of ME cell death.

On the other hand, given the fact that [1] results from recent worldwide clinical trials with corresponding proteolytic enzyme inhibitors have been very disappointing, casting doubt on the validity of the traditional theory for tumor invasion; and [2] the absence of ME cell layers is the most distinct sign of invasive and metastatic breast lesions, our hypothesis appears to have important scientific and clinical values. In the scientific research field, our hypothesis currently represents the only published alternative for the proteolytic enzyme theory of breast tumor invasion. As the proteolytic enzyme theory appears to be increasingly losing popularity, our hypothesis may lead to a new direction for further exploration of the mechanism of breast tumor invasion. Furthermore, as our hypothesis is predominantly based on findings from human clinical samples, it represents the first feasible *in vivo* model of human breast tumor invasion. Because none of the currently available tissue culture systems or animal models could completely mimic the structural and functional features of ME cell layer, our proposed model is likely to be more useful and more representative. In addition, as the prostate and salivary glands contain cells that are morphologically and functionally similar to ME cells, and all the epithelia contain the BM, our hypothesis might also be applicable to all epithelium-derived tumors.

In the clinical field, our hypothesis might bring several immediate benefits for the early detection and prevention of tumor invasion. First, as ME cell degeneration products and associated immunoreactive cells (cytotoxic T-lymphocytes) are likely to enter the blood stream, a quantitative assessment of these products or cell types may lead to the identification of the specific individual at a greater risk for invasive breast lesions. Second, as genetic alterations not only determine the scope and extent of, but also precede, both biochemical and morphologic changes, microdissection of these ER NCC and their adjacent ER-positive counterparts for genetic and biochemical comparisons could potentially lead to the identification of specific molecules associated with the early events of ME cell layer disruptions and tumor invasion. The development of antibodies or chemical reagents to target ER NCC may provide a more effective and less toxic means to block tumor invasion at the earliest stages. Third, microdissection of these ER NCC in frozen tissues for tissue culture may lead to the establishment of useful cell lines for the benefit of cancer and stem cell researchers. More importantly, as ME cells are capable of accommodating the massive increase of acini and ducts during pregnancy to maintain the normal structural relationship, it is very likely that the functional status of ME cells could be clinically manipulated. Therefore, the development of therapeutic approaches to stabilize the basal cell layer integrity or to stimulate ME cell growth might have significant values in cancer prevention and treatment.

Acknowledgment

This study was supported in part by grants DAMD17-01-1-0129, DAMD17-01-1-0130, PC051308 from Congressionally Directed Medical Research Programs, and 05AA from AFIP/ARP initiative fund to Yan-gao Man, MD., PhD.

References

1. Tsubura A, Shikata N, Inui T, et al. Immunohistochemical localization of myoepithelial cells and basement membrane in normal, benign and malignant human breast lesions. *Virchows Arch* 413: 133-139, 1988.
2. Jolicoeur F, Seemayer TA, Gabbiani G, et al. Multifocal, nascent, and invasive myoepithelial carcinoma(malignant myoepithelioma) of the breast: an immunohistochemical and ultrastructural study. *Int. J. Surg. Pathol* 10: 281-291, 2002.
3. Murad TM, von Haam E, Ultrastructure of myoepithelial cells in human mammary gland tumor. *Cancer* 21: 1137-1149, 1986.
4. Slade MJ, Coope RC, Gomm JJ, et al. The human mammary gland basement membrane is integral to the polarity of luminal epithelial cells. *Exp Cell Res* 247: 267-278, 1999.
5. Nerlich A. Morphology of basement membrane and associated matrix proteins in normal and pathological tissues. *Veroff Pathol* 145: 1-139, 1995.
6. Miosge N. The ultrastructural composition of basement membrane *in vivo. Histol Histopathol* 16: 1239-1248, 2001.
7. Beckmann MW, Niederacher D, Schnurch HG, et al. Multistep carcinogenesis of breast cancer and tumour heterogeneity. *J Mol Med* 75:429-439, 1997.
8. Schmitt FC. Multistep progression from an oestrogen-dependent growth towards an autonomous growth in breast carcinogenesis. Eur J Cancer 31A:2049-2052, 1995.
9. Clarke R, Brunner N, Katzenellenbogen BS. Progression of human breast cancer cells from hormone- dependent to hormone-independent growth both in vitro and in vivo. *Proc Natl Acad Sci* USA 86:3649- 3653, 1989.
10. Goldfarb RH, Liotta LA. Proteolytic enzymes in cancer invasion and metastasis. *Semin Thromb Hemost* 12: 294-307, 1986.
11. Duffy MJ, Maguire TM, Hill A, et al. Metalloproteinases: role in breast carcinogenesis, invasion and metastasis. Breast Cancer Res 2: 252-257, 2000.
12. Coussens LM, Fingleton B, Matrisian LM. Matrix metalloproteinase inhibitors and cancer: trials and tribulations. *Science* 295: 2387-2392, 2002.
13. Matrisian LM, Sledge GW Jr., Mohla S. Extracellular proteolysis and cancer: meeting summary and future directions. *Cancer Res* 63: 6105-6109, 2003.
14. Zhao YG, Xiao AZ, Park HI, Newcomer RG, Yan M, Man YG, Heffelfinger SC, Sang QX. Endometase in human breast carcinoma, selective activation of progelatinase B and inhibition by tissue inhibitors of metalloproteinases-2 and –4. *Cancer Res* 64: 590-598, 2004

15. Man YG, Shen T, Weisz J, Berg PE, Schwartz AM, Mulshine JL, Sang QXA, Nieburgs HE. A subset of *in situ* breast tumor cell clusters lacks expression of proliferation and progression related markers but shows signs of stromal and vascular invasion. *Cancer Detect Prev* 29:323-331, 2005.
16. Man YG, Berg PE, Sang QXA. Differential expression of tumor invasion related proteins in cells overlying focally disrupted myoepithelial cell layers and adjacent cells within the same duct. *Proceedings of Department of Defense Breast Cancer Research Program Meeting* P10-5: 75, 2005
17. Rosen PP. *Rosen's Breast Pathology*. Lippncott Williams & Wilkins, Philadelphia, PA, p257-323, 2001.
18. Fisher B, Land S, Mamounas E, et al. Prevention of invasive breast cancer in women with ductal carcinoma in situ: an update of the national surgical adjuvant breast and bowel project experience. *Semin Oncol* 28: 400-418, 2001.
19. Simmons RM, Osborne MP. The evaluation of high risk and pre-invasive breast lesions and the decision process for follow up and surgical intervention. *Surg Oncol* 8: 55-65, 1999.
20. Sakorafas GH, Farley DR. Optimal management of ductal carcinoma in situ of the breast. *Surg Oncol* 12: 221-240, 2003.
21. Baxter NN, Virning BA, Durham SB, Tuttle TM. Trends in the treatment of ductal carcinoma in situ of the breast. *J Natl Cancer Ins* 96: 443-448, 2004.
22. Page DL, Dupont WD, Rogers LW, Jensen RA, Schuyler PA. Continues local recurrence of carcinoma 15-25 years after a diagnosis of low grade ductal carcinoma in situ of the breast treated only by biopsy. *Cancer* 76:1197-2000, 1995.
23. Eusebi V, Feudale E, Foschini MP, et al. Long-term follow-up of in situ carcinoma of the breast. *Semin Diagn Pathol* 11:223-235, 1994.
24. Leonard GD, Swain SM. Ductal carcinoma in situ, complexities and challenges. *J Nat Cancer Inst* 96: 906-920., 2004.
25. Zaugg K, Bodis S. Is there role for molecular prognostic factors in the clinical management of ductal carcinoma in situ (DCIS) of the breast? *Radiother Oncol* 55: 95-99, 2000.
26. Parker SL, Tong T, Bolders S, Wingo PA. Cancer statistics. *Cancer J Clin* 47: 5-27, 1997
27. Stirling JW, Chandler JA. The fine structure of the normal, resting terminal ductal-lobular unit of the female breast. *Virchows Arch A Pathol Anat Histo* 372: 205-226, 1976.
28. Bose S, Derosa CM, Ozzello L. Immunostaining of type IV collagen and smooth muscle actin as a aid in the diagnosis of breast lesions. *Breast J* 5:194-201, 1999.
29. Joshi MG, Lee AK, Pedersen CA, et al. The role of immunohistochemical markers in the differential diagnosis of proliferative and neoplastic lesions of the breast. *Mod Pathol* 9: 57-62, 1996.
30. Yoshida, Immunohistochemical studies on expression of human vascular smooth muscle myosin heavy chain isoforms in normal mammary glands, benign mammary disorders and mammary carcinomas. *Pathol Int* 48: 433-439, 1998.

31. Ohyabu I, akasaki T, Akiba S, et al. Immunohistochemical studies on expression of human vascular smooth muscle myosin heavy chain isoforms in normal mammary glands, benign mammary disorders and mammary carcinomas. *Pathol Int* 48: 433-439, 1998.
32. Damiani S, Ludvikova M, Tomasic G, et al. Myoepithelial cells and basal lamina in poorly differentiated *in situ* duct carcinoma of the breast. An immunocytochemical study. *Virchows Arch* 434: 227-234, 1999.
33. Dabbs DJ, Gown AM. Distribution of calponin and smooth muscle myosin heavy chain in fine-needle aspiration biopsies of the breast. *Diagn Cytopathol* 20: 203-207, 1999.
34. Miettinen MM, Sarlomo-Rikala M, Kovatich AJ, et al. Calponin and h-caldesmon in soft tissue tumors: consistent with h-caldesmon immunoreactivity in gastrointestinal stromal tumors indicates traits of smooth muscle differentiation. *Mod Pathol* 12:756-762, 1999.
35. Shimoyama Y, Yoshida T, Terada M. Molecular cloning of a human Ca –dependent cell-cell adhesion molecule homologous to mouse placental cadherin: its low expression in human placental tissues. *J Cell Biol* 109:1787-1794, 1989.
36. Jarrard DF, Paul R, van Bokhoven A, et al. P-cadherin is a basal cell-specific epithelial marker that is not expressed in prostate cancer. *Clin Cancer Res* 3: 2121-2128, 1997.
37. Soler AP, Knudsen KA, Salaxar H, et al. P-cadherin expression in breast carcinoma indicates poor Survival. *Cancer* 86: 263-272, 1999.
38. Moll R, Franke WW, Schiller DL, et al. The catalog of human cytokeratins: Patterns of expression in normal epithelia, tumors and cultured cells. *Cell* 31: 11-24, 1982.
39. Wetzels RH, Holland R, Van Haelst UJ, et al. Detection of basement membrane components and basal cell keratin 14 in noninvasive and invasive carcinomas of the breast. *Am J Pathol* 34: 571-579, 1989.
40. Wetzels RH, Kuijpers HJ, Lane EB, et al. Basal cell-specific and hyperproliferation-related keratins in human breast cancer. *Am J Pathol* 138: 751-763, 1991.
41. McGowan KM, Coulumbe PA. Onset of keratin 17 expression coincides with the definition of major epithelial lineages during skin development. *J Cell. Biol* 143: 469-486, 1998.
42. Maass N, Hojo T, Zhang M, et al. Maspin-a novel protease inhibitor with tumor suppressing activity in breast cancer. *Acta oncol* 39: 931-934, 2000.
43. Zhang M, Volpert O, Shi YH, et al. Maspin is an angiogenesis inhibitor. *Nature Med.* 6: 196-199, 2000.
44. Reis-Filho JS, Milanezi F, Silva P, et al. Maspin expression in myoepithelial tumors of the breast. *Pathol Res Pract* 197: 817-821, 2001.
45. Scharnhorst V, van der Eb AJ, Jochemsen AG. WT1 proteins: functions in growth and differentiation. *Gene* 273: 141-161, 2001.
46. Fabre A, McCabb AH, O'Shea D, et al. Loss of heterozygosity of the Wilms' tumor suppressor gene (WT-1) in *in-situ* and invasive breast carcinoma. *Hum Pathol* 30: 661-665, 1999.

47. Zhang R, Man YG, Vang RS, et al. A subset of morphologically distinct mammary myoepithelial cells lacks corresponding immunophenotypic markers. *Breast Cancer Res* 5: R151-156, 2003.
48. Man YG, Barner R, Vang R, et al. Non-smooth muscle restricted proteins exclusively or preferentially expressed in mammary myoepithelial cells: a programmed or induced phenomenon? *Mod Pathol* 17 (supplement 1): 40a, 2004
49. Man YG, Vang RS, Saenger JS, et al. Co-expression of Maspin and Wilms' tumor 1 proteins in mammary myoepithelial cells---implication for tumor progression and invasion. *Proceedings of Department of Defense Breast Cancer Research Program Meeting* 1: P9,16, 2002.
50. Barbareschi M, Pecciarini L, Cangi MG, et al. p63, a p53 homologue, is a selective nuclear marker of myoepithelial cells of the human breast. *Am J Surg Pathol* 25:1954-1060, 2001.
51. Yang A, Schweitzer R, Sun D, et al. p63 is essential from regenerative proliferation in limb, craniofacial and epithelial development, *Nature* 398: 714-718,1999.
52. Milis AA, Zheng B, Wang XL, et al. p63 is a p53 homologue required for limb and epidermal Morphogenesis. *Nature* 398:708-713, 1999.
53. Batistatou A, Stefanou D, Arkoumani E, Agnantis NJ. The usefulness of p63 as a marker of breast myoepithelial cells. *In Vivo* 17: 573-576, 2003.
54. Yamamoto T, Oda K, Miyazaki K, et al. p73 is highly expressed in myoepithelial cells and in carcinomas with metaplasia. *Int J. Oncol* 19: 271-276, 2001.
55. Simpson PT, Gale T, Reis-Filho JS, et al. Distribution and significance of 14-3-3sigma, a novel myoepithelial marker, in normal, benign, and malignant breast tissue. *J Pathol* 202: 274-285, 2004.
56. Stepheson JM, Banerjee S, Saxena NK, et al. Neuropilin-1 is differentially expressed in myoepithelial cells and vascular smooth muscle cells in preneoplastic and neoplastic human breast: a possible marker for the progression of breast cancer. *Int J Cancer* 101: 409-414, 2002.
57. Fakhari M, Pullirsch D, Abraham D, et al. Selective upregulation of vascular endothelial growth factor receptors neuropilin-1and -2 in human neuroblastoma. *Cancer* 94: 258-263, 2002.
58. Moritani S, Kushima R, Sugihara H, et al. Availability of CD10 immunohistochemistry as a marker of breast myoepithelial cells on paraffin sections. *Mod Pathol* 15: 397-405, 2002.
59. Iwaya K, Ogawa H, Izumi M, et al. Stromal expression of CD10 in invasive breast carcinoma: a new predictor of clinical outcome. *Virchows Arch* 440: 589-593, 2002.
60. Emberley ED, Murphy LC, Watson PH. S100A7 and the progression of breast cancer. *Breast Cancer Res* 6: 153-159, 2004.
61. Gillett CE, Bobrow LG, Millis RR. S100 protein in human mammary tissue—immunoreactivity in breast carcinoma, including Paget's disease of the nipple, and value as a marker of myoepithelial cells. *J Pathol* 160:19-24, 1990.
62. Dwarakanath S, Lee AK, Delellis RA, et al. S-100 protein positivity in breast carcinomas: a potential pitfall in diagnostic immunohistochemistry. *Hum Pathol* 18:1144-1148, 1987.

63. Lunde S, Nesland JM, Holm R, Johannessen JV. Breast carcinomas with protein S-100 immunoreactivity. An immunocytochemical and ultrastructural study. *Pathol Res Pract* 182: 627-631, 1987.

64. Grey SR, Dlay SS, Leone BE, et al. Prediction of nodal spread of breast cancer by using artificial neural network-based analyses of S100A4, nm23 and steroid receptor expression. *Clin Exp Metastasis* 20: 507- 514, 2003.

65. Jenkinson SR, Barraclough R, West CR, Rudland PS. S100A4 regulates cell motility and invasion in an *in vitro* model for breast cancer metastasis. *Br J Cancer* 90: 253-262, 2004.

66. Arai K, Teratani T, Kuruto-Niwa R, et al. S100A9 expression in invasive ductal carcinoma of the breast: S100A9 expression in adenocarcinoma is closely associated with poor tumor differentiation. *Eur J Cancer* 40:1179-1187, 2004.

67. Allinen M, Beroukhim R, Gai L, et al. Molecular characterization of the tumor microenviroment in breast cancer. *Cancer Cell* 6:17-31, 2004.

68. Deugnier MA, Farado MM, Rousselle P, et al. Cell-extracellular matrix interactions and EGF are important regulators of the basal mammary epithelial cell phenotype. *J Cell Sci* 112:1035-1044, 1999.

69. Gomm JJ, Browne PJ, Coope RC, et al. A paracrine role for myoepithelial cell-derived FGF2 in the normal human breast. *Exp Cell Res* 234:165-173, 1997.

70. Roose J, Huls G, van Beest M, et al. Synergy between tumor suppressor APC and the beta-catenin-Tcf4 target Tcf1. *Science* 285:1923-1926, 1999.

71. Barsky SH, Karlin NJ. Mechanism of disease: breast tumor pathogenesis and the role of the myoepithelial cell. *Nat Clin Pract Oncol* 3:138-151, 2006.

72. Jones JL, Shaw JA, Pringle JH, R.A. Walker, Primary breast myoepithelial cells exert an invasion- suppressor effect on breast cancer cells via paracrine down-regulation of MMP expression in fibroblasts and tumor cells. *J Pathol* 201: 562-572, 2003.

73. Djonov V, Hogger K, Sedlacek R, et al. MMP-19: cellular localization of a novel metalloproteinase within normal breast tissue and mammary gland tumors. *J Pathol* 195:147-155, 2001.

74. Moller P, Mechtersheimer G, Kaufmann M, et al. Expression of epidermal growth factor receptor in benign and malignant primary tumours of the breast. *Virchows Arch A Pathol Anat Histopathol* 414: 157-164, 1989.

75. Gudjonson T, Ronnov-Jessen L, Villadsen R, et al. Normal and tumor-derived myoepithelial cells differ in their ability to interact with luminal breast epithelial cells for polarity and basement membrane deposition. *J Cell Sci* 115:39-50, 2002.

76. Tobacman JK, Hinkhouse M, Khalkhali-Ellis Z. Steroid sulfatase activity and expression in mammary myoepithelial cells. *J Steroid Biochem Mol Biol* 81: 65-68, 2002.

77. Foschini MP, Eusebi V. Carcinomas of the breast showing myoepithelial cell differentiation. A review of the literature. *Virchows Arch* 432:303-310, 1998.

78. Coyne JD, Dervan PA, Barr L. High-grade carcinomas of the breast showing patterns of mixed ductal and myoepithelial differentiation (including myoepithelial cell-rich carcinoma of the breast). *Histopathology* 44:580-584, 2004.

79. Tavassoli FA. Myoepithelial lesions of the breast: myoepitheliosis, adenomyoepithelioma, and myoepithelial carcinoma. *Am J Surg Pathol* 15:554-568, 1991.
80. Schurch W, Potvin C, Seemayer TA. Malignant myoepithelium (myoepithelial carcinoma) of the breast: An ultrastructural and immunohistochemical study. *Ultrastruct Pathol* 8:1-11,1985.
81. Thorner PS, Kahn HJ, Baumal R, et al. Malignant myoepithelioma of the breast: An immunohistochemical study by light and electron microscope. *Cancer* 57:745-750, 1986.
82. Tamai M. Intraductal growth of malignant mammary myoepithelioma. *Am J Surg Pathol* 16: 1116-1125, 1992.
83. Behranwala KA, Nasiri N, A'Hern R, Gui GP. Clinical presentation and long-term outcome of pure myoepithelial carcinoma of the breast. *Eur J Surg Oncol* 30:357-361, 2004.
84. Jones C, Foschini MP, Chaggar R, et al. Comparative genomic hybridization analysis of myoepithelial carcinoma of the breast. *Lab Invest* 80: 831-836, 2000.
85. Man YG, Tai L, Barner R, Vang R, et al. Cell clusters overlying focally disrupted mammary myoepithelial cell layers and adjacent cells within the same duct display different immunohistochemical and genetic features: implications for tumor progression and invasion. *Breast Cancer Res* 5: R231-241, 2003.
86. Man YG, Tai L, Barner R, et al. Focal losses of ER expression in epithelial cells and disruptions of subjacent myoepithelial cell layers are correlated events in ER (+) ductal intraepithelial neoplasia. *Proceedings of Department of Defense Breast Cancer Research Program Meeting*. 1: P9,17, 2002,
87. Man YG, Shekitka KM, Bratthauer GL, Tavassoli FA. Immunohistochemical and genetic alterations in mammary epithelial cells overlying focally disrupted myoepithelial cell layers. *Breast Cancer Res Treat* 76 (sup 1): S143, 569, 2002
88. Man YG, Saenger JS, Vang R, et al. Identification of invasive precursor cells in normal and hyperplastic breast tissue. *Proceedings of the American Association for Cancer Research. Cancer Res* 44: 68, 357, 2003
89. Zhang R, Man YG, Vang RS, et al. A subset of morphologically distinct mammary myoepithelial cells lacks corresponding immunophenotypic markers. *Breast Cancer Res* 5: R151-156, 2003
90. Zhang R, Man YG, Strauss BL, et al. A subset of morphologically identifiable mammary myoepithelial cells lacks expression of corresponding phenotypic markers. *Mod Pathol* 16(1): 52A, 2003
91. Woodward WA, Chen MS, Behbod F, Rosen JM. On mammary cells. *J Cell Sci* 118:3585, 2005.
92. Ponti D, Costa A, Zaffaroni N,et al. Isolation and in vitro propagation of tumorigenic breast cancer cells with stem/progenitor cell properties. *Cancer Res* 65:5506-5511, 2005.
93. Dontu G, Jackson KW, McNicholas E, et al. Role of Notch signaling in cell-fate determination of human mammary stem/progenitor cells. *Breast Cancer Res* 6:R605-615, 2004.

94. Yousefi M, Mattu R, Gao C, Man YG. Mammary ducts with and without focal myoepithelial cell layer disruptions show a different frequency of white blood cell infiltration and growth pattern: Implications for tumor progression and invasion. *Appl Immunohistochem Mol Morphology*13:30-37, 2005

95. Man YG, Vinh T, Zhao C, Walker A, Barner R. Potential roles of T-lymphocytes and natural killer cells in human myoepithelial cell layer disruptions and tumor invasion. *Mod Pathol* 18: 42A, 179, 2005

96. Wang LP, Mannion C, Man YG. Reduction of tumor suppressors and elevation of cytotoxic cells in myoepithelial cell layers of inflammation breast carcinoma: Implication for tumor aggressiveness. *Mod Pathol* 18: 54A, 238, 2005

97. Zhao C, Mannion C, Man YG. Potential roles of cytotoxic T-lymphocytes and nature killer cells in prostate basal cell layer disruptions and tumor invasion. *Mod Pathol* 18: 175A, 809, 2005

98. Man YG. CD8 and mast cell tryptase positive cells are preferentially associated with focal myoepithelial cell layer disruptions: implications for breast tumor invasion. *Proceedings of Department of Defense Breast Cancer Research Program Meeting* P38-13,263, 2005

99. Man YG. Focal degenerations in surrounding structures and infiltration of immunoreactive cells are a potential trigger for invasion of breast and other epithelium-derived tumors. *Proceedings of Department of Defense Breast Cancer Research Program Meeting* P10-7: 75-76, 2005

100. Man YG. Genetically different primary bilateral breast tumors show similar signs of potential progression and invasion. *Proceedings of Department of Defense Breast Cancer Research Program Meeting* P10-8: 76, 2005

101. Man YG. CD8 and mast cell tryptase positive cells are differentially distributed in benign and malignant breast tissues with and without myoepithelial cell layers. *Proceedings of Department of Defense Breast Cancer Research Program Meeting* P10-9: 76, 2005

102. Dighiero G, Rose NR. Critical self-epitopes are key to the understanding of self-tolerance and autoimmunity. *Immunology Today* 1999;20:423-8.

103. Nevo U, Kipnis J, Golding I, Shaked I, Neumann A, Akselrod S, Schwartz M. Autoimmunity as a special case of immunity: removing threats from within. *Trends in Mol. Med* 2003; 9:88-93.

104. Karlsen AE, Dyrberg T. Molecular mimicry between non-self, modified and self in Autoimmunity. *Semin Immunol* 1998;10:25-34.

105. Barker RN, Elson CJ. Multiple self epitopes on the Rhesus polypeptides stimulate immunologically ignorant human T cells in vitro. *Eur J Immunol* 24:1578-82, 1994.

106. Zhang XD, Schiller GD, Gill PG, Coventry BJ. Lymphoid cell infiltration during cancer growth: a syngeneic rat model. *Immunol Cell Biol* 76: 550-555, 1998.

107. Steadman R, Irwin MH, John PL Sr., et al. Laminin cleavage by activated human neutrophils yields proteolytic fragments with selective migratory properties. *J Leukoc Biol* 53: 354-365, 1993.

108. Heck LW, Blackburn WD, Irwin MH, et al. Degradation of basement membrane laminin by human neutrophil elastases and cathepsin G. *Am J Pathol* 136:1267-1274, 1990.

109. Brummer T, Schramek D, Hayes VM, et al. Increased proliferation and altered growth factor dependence of human mammary epithelial cells overexpressing the Gab2 docking protein. *J Biol Chem* 281:626-631, 2006.

110. Oliveira AM, Ross JS, Fletcher JA. Tumor suppressor genes in breast cancer: the gatekeepers and the caretakers. *Am J Clin Pathol* 124(Suppl):S16-28, 2005.

111. Yanochko GM, Eckhart W. Type 1 insulin-like growth factor overexpression induce proliferation and anti-apoptotic signaling in a three-dimensional culture model of breast epithelial cells. *Breast Cancer Res* 8(2): R18 [Epub ahead of print], 2006.

112. Chakravarthy MV, Spangenhurg EE, Booth FW. Culture in low levels of oxygen enhances in vitro proliferation potential of satellite cells from old skeletal muscles. *Cell Mol Life Sci* 58:1150-1158, 2001.

113. Csete M, Walikonis J, Slawny N, et al. Oxygen-mediated regulation of skeletal muscle satellite cell proliferation and adipogenesis in culture. *J Cell Physical* 189:189-196, 2001.

114. Studer L, Csete M, Lee SH. Enhanced proliferation, survival, and dopaminergic differentiation of CNS precursors in lowed oxygen. *J Neurosci* 20: 7377-7383, 2000.

115. Asano-kato N, Fukagawa K, Okada N. Tryptase increase proliferative activity of human conjunctive fibroblasts through protease-activated receptor. *Invest Ophthalmol Vis Sci* 46:4622-4626, 2005.

116. Freeman MR, Schneck FX, Gognon ML, et al. Peripheral blood T-lymphocytes and lymphocytes infiltrating human cancers express vascular endothelial growth factor: a potential role for T cells in angiogenesis. *Cancer Res* 55: 4140-4145, 1995.

117. Nienartowicz A, Sobaniee-Lotowska ME, Jarocka-Cyrta E, Lemancewicz D. Mast cells in neoangiogenesis. *Med Sci Monit* 12(3):RA53-56, 2006.

118. Qu Z. Immunohistological detection of growth factors and cytokines in tissue mast cells. *Methods Mol Biol* 315:257-272, 2006.

119. Takahashi Y, Bucana CD, Liu W, et al. Platelet-derived endothelial cell growth factor in human colon cancer angiogenesis: role of infiltrating cells. *J Natl Cancer Inst* 88:1146-1151, 1996.

120. Kang Y, Massague J. Epithelial-mesenchymal transition: twist in development and metastasis. *Cell* 118:277-279, 2004.

121. Sato T, Sakai T, Noguchi Y, Takita M, Hirakawa S, Ito A. Tumor-stromal cell contact promotes invasion of human uterine cervical carcinoma cells by augmenting the expression and activation of stromal matrix metalloproteinases. *Gynecol Oncol* 92: 47-56, 2004.

122. Strizzi L, Bianco C, Normanno N. Epithelial mesenchymal transition is a characteristic of hyperplasias and tumors in mammary gland from MMTV-Criptol-1 transgenic mice. *J Cell Physiol* 201:266-276, 2004.

123. Houlbrook S, Addison CM, Davies SL, et al. Relationship between expression of topoisomerase II isoforms and intrinsic sensitivity to topoisomerase inhibitors in breast cancer lines. *Br J Cancer* 72: 1454-1461, 1995.

124. Man YG, Saenger JS, Strauss B, Vang RS, et al. Focal alterations of p27 expression and subjacent myoepithelial cell layer disruptions are correlated events in ER (-) ductal intraepithelial neoplasia. *Proceedings of Department of Defense Breast Cancer Research Program Meeting.* 1: P9,14, 2002.

125. Man YG, Shekitka KM, Bratthauer GL, Tavassoli FA. Immunohistochemical and genetic alterations in mammary epithelial cells overlying focally disrupted myoepithelial cell layers. *Breast Cancer Res Treat* 76 (sup 1): S143, 569, 2002

126. Man YG, Nieburgs HE. A subset of morphologically normal and hyperplastic breast tissues contains cell clusters with malignant features: Implification for tumor progression and invasion. *Cancer Detect Prev*, in press.

127. Zhao C, Barner R, Vinh T, Walker A, Man YG. A subset of normal and hyperplastic appearing human breast cell clusters exhibited similar immunohistochemical and cytological alterations to cancer cells. *Mod Pathol* 18: 57A, 250, 2005.

128. Desouki MM, Rowan BG. SRC kinase and mitogen-activated protein kinases in the progression from normal to malignant endometrium. *Clin Cancer Res* 10: 546-555, 2004.

129. Strzzi L, Bianco C, Normanno N, et al. Epithelial mesenchymal transition is a characteristic of hyperplasias and tumors in mammary gland from MMTV-Criptol-1 transgenic mice. *J Cell Physiol* 201: 266-276, 2004.

130. Saito T, Oda Y, Tanaka K, et al. +β-catenin nuclear expression correlates with cyclin D1overexpression in sporadic desmoids tumours. *J Pathology* 195:222-228, 2001.

131. Mackie EJ, Chiquet-Ehrismann R, Pearson CA, et al. Tenascin is a stromal marker for epithelial malignancy in the mammary gland. *Proc Natl Acad Sci* 84: 4621-4625, 1987.

132. Man YG, Strauss B, Saenger JS, et al. Genetic alterations in ER (-) mammary epithelial cells overlying focally disrupted myoepithelial cell layers. *Proceedings of Department of Defense Breast Cancer Research Program Meeting.* 1: P9,15, 2002

133. Man YG, Saenger JS, Vang RS, et al. Identification of invasive precursor cells in normal and hyperplastic breast tissue. Proceedings of the American Association for Cancer Research. *Cancer Res* 44: 68, 357, 2003.

134. Man YG, Zhang Y, Shen T, et al. cDNA expression profiling identifies elevated expressions of tumor progression and invasion related genes in cell clusters of in situ breast tumors. *Breast Cancer Res Treat* 89:199-208, 2005.

135. Man YG, Zeng X, Shen T, et al. Cell clusters overlying focally disrupted myoepithelial cell layers and their adjacent counterparts within the same duct display a different pattern of mRNA expression. *Mod Pathol 17* (supplement1): 40-41a, 2004.

136. Rochefort H, Glondu M, Sahla ME, et al. How to target estrogen receptor-negative breast cancer. *Endocr Relat Cancer* 10: 261-266, 2003.

137. Sheikh MS, Garcia M, Pujol P, Fontana JA, Rochefort H. Why are estrogen receptor negative breast cancers more aggressive than the estrogen receptor positive breast cancers? *Invasion Metastasis* 14: 329- 336, 1994-1995.

138. Swain SM, Wilson JW, Mamounas EP, et al. Estrogen receptor status of primary breast cancer is predictive of estrogen receptor status of contralateral breast cancer. *J Natl Cancer Inst* 96: 497-499, 2004.

139. Luna-More S, Casquero S, Perey-Mellado A, et al. Importance of estrogen receptors for the behavior of invasive micropapillary carcinoma of the breast. Review of 68 cases with follow-up of 54. *Pathol Res Pract* 196: 35-39, 2000.

140. Nowell PC. The clonal evolution of tumor cell populations. *Science* 194(4260): 23-28, 1976.

141. Kordon EC, Smith GH. An entire functional mammary gland may comprise the progeny from a single cell. *Development* 125: 1921-1930, 1998.

142. Middleton LP, Vlastos G, Mirza NQ, Eva S, Sahin AA. 2000. Multicentric mammary carcinoma: evidence of monoclonal proliferation. *Cancer* 94:1910-1916.

143. Jourquin J, Yang N, Kamy, et al. Dispersal of epithelial cancer cell colonies by lysophosphatidic acid (LPA). *J Cell Physiol* 206: 337-346, 2006.

144. Lu J, Izyolsky KI, Qian J, et al. Identification of FGF10 targets in the embryonic lung epithelium during bud morphogenesis. *J Biol Chem* 280: 4834-4841, 2005.

145. Man YG, Ball WD, Culp AJ, Hand AR, Moreira JE. Persistence of a perinatal cellular phenotype in the ducts of adult glands. *J Histochem Cytochem* 43(12):1203-1215, 1995

146. Man YG, Ball WD, Marchetti L, Hand AR. Contributions of intercalated duct cells to normal parenchyma of submandibular glands of adult rats. *Anat Rec* 263:202-214, 2001

147. Man YG, Sang QXA. The significance of focal myoepitehlial cell layer disruptions in breast tumor invasion: a paradigm shift from the “protease-centered” hypothesis. *Exp Cell Res* 301:103-118, 2004.

148. Man YG. Cell clusters overlying focally disrupted myoepithelial cell layers change status of estrogen receptor expression during tumor invasion. *Proceedings of Department of Defense Breast Cancer Research Program Meeting* P60-12, 423, 2005

149. Man YG, Zhang R, Mattu R, Shen T, Sang QXA. A subset of mammary epithelial cells overlying focally disrupted myoepithelial cell layers shows an unusual immunostaining pattern for proliferation -related proteins. *Breast Cancer Res Treat* 82 (supplement 1): S163-164, 2003

150. Man YG. Stromal and vascular invasion of normal and hyperplastic appearing human breast ductal Cells. *Proceedings of Department of Defense Breast Cancer Research Program Meeting* P10-4: 74- 75, 2005.

151. Man YG, Shen T, Zhao YG, Sang QX. Focal prostate basal cell layer disruptions and leukocyte infiltration are correlated events: A potential mechanism for basal cell layer disruptions and tumor invasion. *Cancer Detect Prev* 29: 161-169, 2005.

152. Nelson A, Man YG. Mast cells in high grade lesions of patients co-infected with human papilloma virus and human immunodeficiency virus. Accepted for poster presentation at the 2006 USCAP Annual Meeting. *Lab Investigation* 86:257A; 1190, 2006.

153. Wang HL, Man YG. Potential roles of focal basement membrane disruptions and lymphocyte infiltration in colorectal cancer invasion. *Am J Clin Pathol* 124(4):647, 2005.

154. Man YG, Sang QXA, Zhao CQ, Mannion C, Gardner WA. Focal basal cell degeneration induced lymphocyte infiltration is a potential trigger factor for prostate tumor invasion: implications for treatment and prevention. *Cancer Detect Prev*, in press.

155. Man YG, Zhao CQ, Wang J, XL Chen. A subset of morphologically distinct prostate basal cells lacks corresponding immunophenotypic markers: Implications for clinical diagnosis. *Pathology-Research & Practice*, Conditionally accepted.

156. Man YG, Shen T, Zhao YG, Sang QX. Focal prostate basal cell layer disruptions and leukocyte infiltration are correlated events: Implications for basal call layer degradation and tumor invasion. *Cancer Detection & Prevention, 2004 Symposium Issue* S-51: 15, 2004.

157. Man YG, Shen T, Zhao YG, Sang QX. Morphologically comparable prostate acini and ducts with and without a focal basal cell layer disruption have a different cell proliferation rate: Implications for tumor invasion. *FASEB* 18(5): A1193, 2004.

158. Zhao C, Mannion C, Man YG. Potential roles of cytotoxic T-lymphocytes and nature killer cells in prostate basal cell layer disruptions and tumor invasion. *Mod Pathol* 18: 175A, 809, 2005.

159. Sang QXA, Zhao YG, Man YG. Mechanism of human prostate cancer invasion: Basement membrane degradation and basal cell layer disruption. *Proceedings of American Association for Cancer Research* 46: 1106, 4689, 2005.

160. Man YG, Chen XL, Garcia FU, Gardner WA. Reduced p63 expression and elevated apoptosis in focally disrupted prostate basal cell layers. *Lab Investigation* 86:292A; 1358, 2006.

161. Larson PS, Schlechter BL, de las Morenas A, et al. Allele imbalance, or loss or heterozygosity, in normal breast epithelium of sporadic breast cancer cases and BRCA1 gene mutation carries is increased compared with reduction mammoplasty tissues. *J Clin Oncol* 23: 8613-8619, 2005.

162. Bramley M, Clarke KB, Howell A, et al. Effects of oestrogens and anti-oestrogens on normal breast tissue from women bearing BRCA1 and BRCA2 mutations. *Br J Cancer* 94: 1021-1028, 2006.

163. Handschuh G, Candidus S, Luber B, et al. Tumor-associated E-cadherin mutations alter cellular morphology, decrease cellular adhesion and increase cellular motility. *Oncogene* 8: 4301-4312, 1999.

164. Luber B, Candidus S, Handschuh G, et al. Tumor-derived mutated E-cadherin influences beta-catenin localization and increases susceptibility to actin cytoskeletal changes induced by pervanadate. *Cell Adhes Commun* 7: 391-408, 2000.

165. Fricke E, Keller G, Becker I, et al. Relationship between E-cagherin gene mutation and p53 gene mutation, p53accumulation, Bcl-2 expression and Ki-67 staining in diffuse-type gastric carcinoma. *Int J Cancer* 104: 60-65, 2003.

166. Moinfar F, Man YG, Arnould L, Bratthauer GL, Ret al. Concurrent and independent genetic alterations in the stromal and epithelial cells of mammary carcinoma: Implications for tumorigenesis. *Cancer Res* 60: 2562-2566, 2000

167. Man YG, Martinez A, Avis IM, et al. Phenotypically different respiratory epithelial cells with hnRNP A2/B1 over-expression display similar genetic alterations. *Am J Respir Cell Mol Biol* 23:636-645, 2000

168. Man YG, Mannion C, Albores-Saaverdra J, et al. Allelic losses at 3p and 11p are detected in both epithelial and stromal components of cervical small cell neuroendocrine carcinoma. *Appl Immunohistochem Mol Morphology* 9:340-345, 2001

169. Man YG, Magrane GG, Lininger RA, et al. Morphologically similar epithelial and stromal cells in primary bilateral breast tumors display different genetic profiles: implications for treatment. *Appl Immunohistochem Mol Morphology* 12:305-314, 2004

170. Moinfar F, Kremser KL, Man YG, et al. Allelic imbalances in endometrial stromal neoplasms: frequent genetic alterations in the normal-appearing endometrial and myometrial tissues. *Gynecol Oncol* 95:662- 671, 2004

171. Halbwedi I, Ullmann R, Kremser ML, Man YG, et al. Chromosomal alterations in low-grade endometrial stromal sarcoma and undifferentiated endometrial sarcoma as detected by comparative genemic hybridization. *Gynecol Oncol* 97: 582—587, 2005.

172. Moinfar F, Gogg-Kamerer M, Sommersacher A, Regitnig P, Man YG, et al. Endometrial stromal sarcomas frequently express epidermal growth factor receptor (EGFR, HER-1): potential basis for a new therapeutic approach. *Am J Surg Pathol* 29: 485-489, 2005.

In: New Research on Genomic Instability
Editor: Eleanor J. Glascow, pp. 47-76

ISBN: 1-60021-320-0

Chapter II

Chromosomal Instability and Cancer

Dominik Schnerch and Ralph Wäsch*
Department of Hematology and Oncology,
Albert-Ludwigs University Medical Center, Freiburg, Germany

Abstract

Numerical and structural chromosome aberrations are a common characteristic of most tumors. More than one century after scientists first described these particular lesions and suggested a link to rise of malignancy, it is still a matter of debate, whether chromosomal aberrations directly contribute to transformation of a somatic cell into a tumor cell. Chromosomal aberrations are thought to arise from various conditions that are addressed in the following chapter. Recent evidence supports the notion that numerical and structural chromosome aberrations favor generation of progressive genetic instability. Genetic instability is known to confer a selective advantage onto a particular cell, supporting outgrowth of specific cell clones. These cell populations are able to benefit from the acquired genetic alteration in their contemporary environmental context and become subject of malignant transformation. This leads to the assumption that chromosomal instability must be considered as an essential contributor in the generation of most cancers.

* Correspondence: Ralph Wäsch, M.D. Dept. Hematology and Oncology, Albert-Ludwigs University Medical Center, Hugstetter Str. 55, D-79106 Freiburg, Germany. Tel: +49-761-2707289. Fax: +49-761-2703318; E-mail: ralph.waesch@uniklinik-freiburg.de

Introduction

Numerical (aneuploidies) and structural (translocations) chromosome aberrations are found in most cancers. However, the exact role of chromosomal instability in tumorigenesis is still not clear. Several mechanisms are thought to be responsible for the generation of such genetic alterations. Defects in cell division control such as compromised mitotic signaling, supernumerary centrosomes or a defective tetraploidy checkpoint are proposed to contribute to gain and loss of whole chromosomes. Moreover, telomere shortening and defective DNA-damage signaling appeared to be powerful driving forces of genomic instability. The loss of telomere sequences at the end of each chromosome and DNA double-strand breakage accompanied by compromised damage signaling favor fusion of chromosomes and generation of translocations and aneuploidy. Gene dosage changes that result from chromosomal instability may explain several aspects of rise and progression of tumor cells.

Aneuploidy

Aneuploidy is defined by a set of chromosomes of more or less than 46 within a single cell. Hyperdiploidy describes a chromosome number of more than 46, hypodiploidy is characterized by less than 46 chromosomes. As hypodiploidy is frequently associated with cell death, most aneuploid cells present with more than 46 chromosomes. Different mechanisms may lead to aneuploidy.

Defects in Mitotic Control

The cell division cycle consists of a replication step in order to double the genetic material. This process is followed by equal distribution of chromosomes to the two developing daughter cells in mitosis. During a cell division cycle not only DNA but also centrosomes undergo duplication. Each of the two centrosomes form a pole of the bipolar spindle during the following mitosis. Generation of more than two centrosomes can cause multipolar mitotic spindles and may lead to missegregation of chromosomes. Moreover, missegregation can occur if a cell is facing defects during mitosis in the presence of impaired monitoring. The cell may thereby loose its ability to recognize improper attachment and spindle tension. This can ultimately result in the onset of anaphase even if chromosomes are not properly attached to the mitotic spindle.

Mitotic Control by the Spindle Assembly Checkpoint

A defective spindle assembly checkpoint (SAC) is considered to be one of the main reasons of chromosome missegregation during a bipolar anaphase. SAC function is provided by a highly complex network of different proteins that is localized at the centromere region of the chromosome [1]. This network controls important aspects of mitosis such as surveillance of chromosome attachment to the mitotic spindle [2] and generation of tension across the kinetochores [3]. Proper monitoring guarantees faithful segregation and delays mitotic progression in the presence of chromosomal misalignment (Figure 1). Due to its complexicity the SAC is highly vulnerable towards malfunction. The causes of SAC malfunction are due to damage or loss of participating proteins and in consequence perturbed interaction of the various components. Aneuploidy is a state that may potentially originate from mitotic chromosome misalignment in the presence of SAC malfunction.

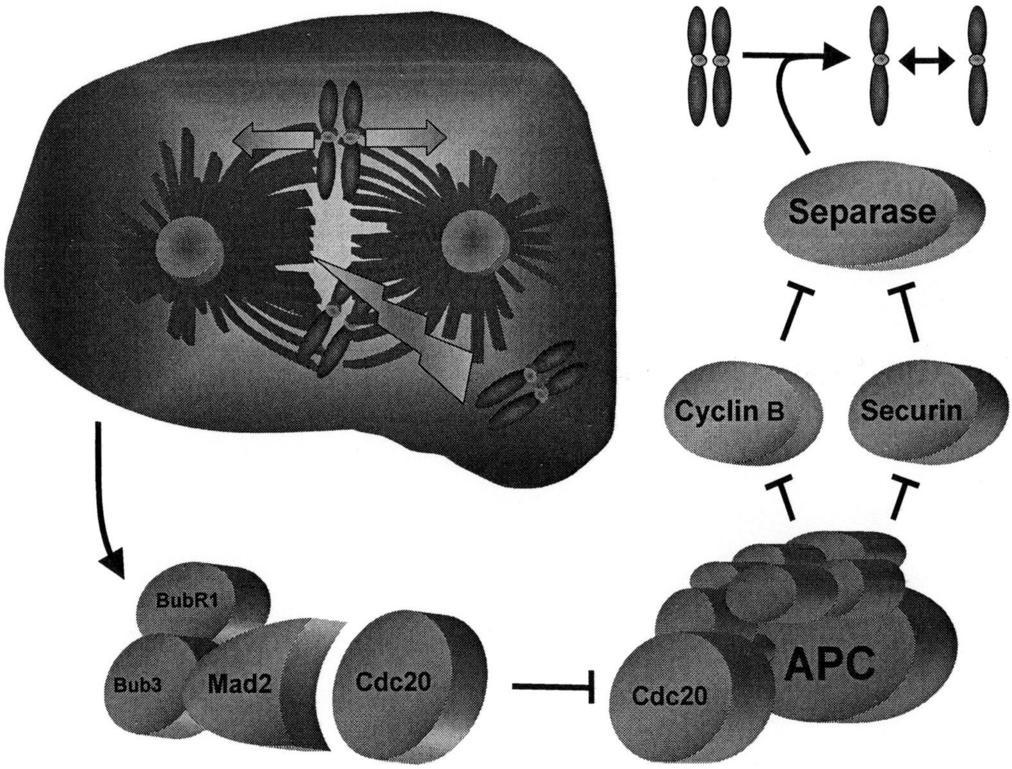

Figure 1. The SAC is responsible for proper segregation of the genetic material. The SAC is a surveillance mechanism required to delay anaphase onset in order to provide time for recovery from mitotic failure. Unattached chromosomes or lack of tension across the kinetochores is sensed by a highly sophisticated network that includes a multitude of different proteins. In case of inappropriate chromosome alignment, specific inhibitory complexes (mitotic checkpoint complexes = MCCs) are generated at the centromere region of the concerned chromosomes. The MCCs, that consist of BubR1, Mad2 and Bub3, are known to bind and thus inactivate the APC subunit Cdc20 leading to stabilization of the APC target proteins cyclin B1 and securin, the latter one being an inhibitor of the protease separase. Cleavage of the cohesion subunits that keep sister chromatids together is a requirement for anaphase onset. Tight control on separase activity by APC-mediated proteolysis of securin and cyclin B1 thus guarantees precise timing for proper chromosome separation.

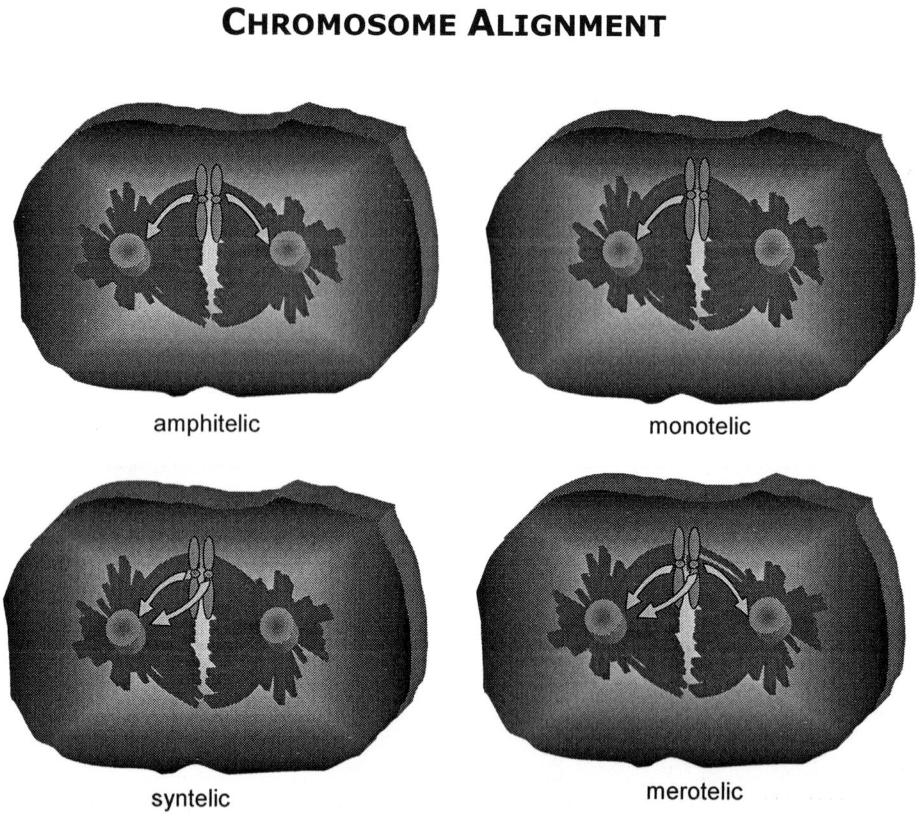

Figure 2. Chromosomal missegregation is preceded by attachment failure. Proper alignment of the chromosomes at the metaphase plate of the mitotic spindle and amphitelic kinetochore attachment is a prerequisite for accurate chromosome segregation. Amphitelic attachment is the only condition to generate proper tension across the kinetochores by the opposite pulling forces to the spindle poles. There are various possibilities of chromosome attachment which have to be transformed into an amphitelic attachment. Deficient monitoring might allow anaphase onset to occur even in the presence of improper chromosome attachment which ultimately leads to chromosomal missegregation and aneuploidy.

The SAC is able to sense the interactions between the microtubule structures of the mitotic spindle and the chromosome. The functional SAC remains activated until all chromosomes have established a amphitelic connection to the microtubule of the mitotic spindle (Figure 2). SAC signaling delays anaphase onset until every single unattached chromosome has recovered from failure in order to guarantee correct chromosomal segregation. Proper attachment and tension across the kinetochore are prerequisites for reliable segregation. A highly complex signaling network covering a large variety of proteins evolved to sense correct interaction [4]. The mitotic kinase Aurora B is a main regulator of kinetochore-microtubule attachment. Several studies identified Aurora B as an important protein in recognition and recovery from inappropriate kinetochore-microtubule connection [5-8]. Aurora B was found to be associated with a number of proteins all of which turned out to behave as chromosome passenger proteins (INCENP, survivin, borealin) that typically localize to the inner centromere during metaphase [9]. Depletion of the component borealin resulted in significantly enhanced attachment failure [10]. Loss of function of Aurora B kinase was reported to abrogate mitotic checkpoint signaling in the presence of chromosome misalignment and lack of tension across the kinetochore [11, 12]. It is believed that Aurora B dissolves kinetochores in case of misalignment to promote appropriate amphitelic attachment. Interestingly, Aurora B (Ipl1) was also shown to play an important role in so called NoCut monitoring in yeast. This pathway induces delay in cytokinesis in presence of spindle-midzone defects. Chromosomes have to be completely removed from the cleavage plane to trigger onset of cell separation. Impairment of NoCut surveillance was shown to give rise to DNA damage [13].

Not surprisingly, recent data gained from mice suggest a correlation between overexpression of Aurora B and aggressiveness of the concerned tumor. Consistently, frequent development of pulmonary and liver metastases were detected in mice that were injected with transformed cells characterized by Aurora B overexpression [14]. However, the exact mechanism of Aurora B functioning and its possible impact on tumorigenesis is still uncertain.

Chromosomal Separation

The inhibitory effector response of the SAC was identified as the mitotic checkpoint complex (MCC) (Figure 1). This protein complex is formed by association of the three proteins Mad2 (mitotic arrest deficient), Bub3 (budding uninhibited by benzimidazole) and BubR1 [15]. The participating components are recruited from the cytoplasm and are brought into close proximity at the kinetochores of chromosomes that are not properly attached to the mitotic spindle. The MCC is able to bind and thus inactivate Cdc20 leading to inhibition of proteolysis of mitotic regulators [1, 15, 16]. In dependence on SAC signaling, MCC formation is triggered until all chromosomes are attached to the metaphase plate of the mitotic spindle. This leads to stabilization of the anaphase inhibitor securin. Securin is known to prevent cleavage of the cohesin bonds that keep sister chromatids together (Figure 1) [17]. As cleavage of cohesin is a requirement for anaphase onset, inhibition of the cleaving enzyme separase is essential for metaphase delay [18]. Loss of Pds1p or Esp1p function, which are securin and separase homologues in budding yeast, were found to result in chromosomal aberrations [19, 20]. Lack of securin also leads to extensive chromosomal instability in human cells [21]. Overexpression of human securin was shown for numerous carcinoma cell lines. Further examination revealed a direct correlation between the expression level of securin and the invasiveness of pituitary tumors and finally led to in vitro transformation and in vivo tumor formation [22].

Targeted Degradation of Cell Cycle Regulators

The SAC interferes with mitotic progression by prevention of ubiquitin-mediated proteolysis of cell cycle regulators [1]. This specific biochemical pathway is known to play a critical role not only in regulation of mitosis but within the entire cell cycle. Degradation of Cyclin B1 and Securin is one of the key requirements of anaphase onset (Figure 1) [23]. As the regulation of mitosis has to be performed in a highly sensitive manner, a specific protein complex is needed for proper execution. This specific protein complex is known as the ubiquitin ligase anaphase-promoting complex (APC). The APC is a multi subunit E3 ubiquitin ligase which is needed to attach polyubiquitin chains to specific target proteins [24, 25]. This allows recognition by the proteasome and leads to proteolytic degradation. Besides the SAC network, additional components are involved in mitotic control. Emi1 and Rassf1A are proteins that also inhibit APC-dependent proteolysis in order to delay mitotic progression [26, 27].

Cell Cycle Regulation by Proteolysis

Activation of the SAC inhibits the activating APC subunit Cdc20, a specificity factor for selective degradation of particular proteins. Targeted inhibition of Cdc20 prevents ordered proteolysis of cyclin B1 and securin (Figure 1), which is necessary for metaphase delay [24]. Ordered proteolysis of cell cycle regulators is one of the main aspects to generate a characteristic activity pattern of protein kinases. Kinase activity is fine-tuned by subsequent steps of positive and negative regulation to facilitate proper control of cell cycle transitions. Negative regulation of cyclin-dependent kinase 1 (CDK1), mainly by targeted proteolytic degradation of cyclin B1 during mitosis is essential for mitotic exit [23, 28]. Proteolysis guarantees cell cycle directionality since the degraded regulators have to be newly synthesized during the following cell division cycle. Degradation of cell cycle regulators prevents anti-clockwise cell cycle regression [28]. The precise alternation of expression of different cyclins leads to sequential association with CDKs. This specific association is a necessity for kinase activation. To prepare mitotic entry, Cyclin B1 is expressed during G2 and forms a complex with CDK1. However, onset of CDK1 activity at the G2-M transition is retained by inhibitory phosphorylation until cells are completely prepared for entry into mitosis. Another mechanism to generate sudden rise in CDK-activity is precisely timed degradation of small inhibitory proteins (CKIs) that keep CDKs in an inactive state [29, 30]. Accordingly, regulatory phosphorylation and tight control of CKIs provide fine-tuning of CDK activity at several transitions within the entire cell cycle. Importantly, those proteins involved in fine-tuning are subject of extensive regulation themselves by a variety of converging checkpoint pathways. As discussed below, recent data also implicate important roles for localization of regulatory proteins (e.g. to centrosomes, which might provide interaction platforms for cell cycle control). The two major aspects of cell cycle regulation, kinase activation by defined subunits (cyclins) and fine-tuning by targeted kinase inhibition (CKIs and phosphorylation/dephosphorylation), are not only found during increase but are also common mechanisms during decline of kinase activity. Mitotic exit is mainly regulated by ubiquitin dependent proteolysis of the mitotic cyclin B1 [23, 28]. Persisting CDK activity in G1 might further be diminished by CKIs such as p21 and/or p27 [23, 24 ,31] to establish a stable G1 phase.

Generation of Aneuploidy as a Consequence of SAC Malfunction

Aneuploidy may arise, if the MCC cannot associate or is unable to bind and inactivate the APC-activating subunit Cdc20. Even malfunction of single components of this highly complex signaling network seem to compromise the integrity and the stability of the karyotype. Recently a germline mutation in the SAC protein BubR1 was shown to be associated with the mosaic variegated aneuploidy (MVA) syndrome [32]. This disorder is characterized by an abnormal karyotype that is accompanied by a predisposition for the development of malignant tumors. Studies in order to identify mutations in SAC components were conducted by analysing cells from patients that suffered from a malignant tumor. Moreover, tumor cell lines generated from various malignancies were screened for mutations. Analyses focused on tumors that were characterized by an aneuploid set of chromosomes, suggesting SAC deficiency as the reason. Mutations of different SAC components were found among different malignancies. Mad2 – breast cancer [33], bladder cancer [34]; Mad1 – lung cancer [35], lymphoma [36], prostatic gland cancer [36], breast cancer [36]; Bub1 – lung cancer [37], ATLL [38, 39], pancreatic cancer [40], colorectal cancer [41]; BubR1 – colorectal cancer [41], ATLL [38]. However, only few mutations were found among these tumor cells. Further analyses confirmed a large discrepancy between the number of cases exhibiting SAC malfunction and the number of cells which showed a mutation in a SAC gene. These examinations led to the conclusion that mutations are very rare among SAC deficient cells [42].

As particular mutations are known to act in a dominant negative way, the resulting ablation of SAC signaling might lead to severe destabilization of the karyotype. Recent experiments suggest, that complete loss of central SAC players leads to a catastrophe-prone phenotype [43]. It seems very unlikely that such extensive genome instability represents a frequent factor to favor clonal expansion. This might provide an explanation for the low frequency of mutations among SAC-deficient cells. Different mechanisms such as transcriptional and posttranscriptional modulation of SAC components might rather account for interference with mitotic signaling in a more gentle way. Accordingly, transcriptional downreglation of murine BubR1 in diploid cells resulted in an aneuploid phenotype [44]. Reduced expression of the SAC component Mad2 was described as a characteristic in breast, nasopharygeal and ovarian cancer [45-47]. Transformation of a diploid into an aneuploid cell was also described to occur as a result of Mad2 overexpression possibly due to inactivation of the retinoblastoma protein (see below) [48].

Despite frequent demonstration that SAC disruption is able to induce aneuploidy, a growing body of knowledge suggest impaired SAC function to collaborate with different defects as part of concerted oncogenic aberrations. Either Bub1 or BubR1 were found to be mutated in thymic lymphomas that arose from BRCA-2 deficient mice [49]. In contrast, additional data underscore the importance of proper BRCA-2 function for cell separation. Inactivation of BRCA-2 led to abnormal cytokinesis which might account for the frequently observed aneuploidy among BRCA-2 deficient tumors [50]. Certain oncogenes and tumor suppressor genes appeared to directly interfere with regulation of spindle checkpoint genes. The DNA damage recognition gene BRCA-1 was shown to transcriptionally decrease Mad2-

levels [51]. Mad2 expression was found to be increased in Rb-negative tumours as a consequence of E2F-dependent transcriptional enhancement [48]. Mechanisms such as generation of tetraploidy following Mad2 overexpression with subsequent aneuploidization during the upcoming mitoses are discussed [52]. Various findings further suggest involvement of p53 in the regulation of Mad1 [53, 54]. An additional tumor associated gene, BCSG-1 (breast cancer specific gene 1), was suggested to be connected with SAC regulation. This molecule was shown to bind and degrade BubR1 in advanced stage breast carcinoma [55]. As a viral oncogene, the tax oncoprotein from HTLV-1 was found to bind to the SAC component Mad1 thus preventing Mad2 activation [56]. In analogy, the simian virus 40 large T antigen appeared to bind and compromise Bub1 function [57].

In vitro experiments confirm that weakening of the mitotic checkpoint itself leads to an increased percentage of cells that exhibit an aneuploid set of chromosomes. However, in vivo tumorigenesis might be driven by concerted action of different oncogenic pathways. Apart from oncogenic modulation of intrinsic proliferative pathways, induction of SAC deficiency by particular oncogenes or by loss of tumor suppressors might be a powerful enhancer for cellular transformation [57]. This suggests a rather supportive role for SAC deficiency in presence of concomitant oncogenic events. Cooperation of oncogenic alterations, including mitotic malfunction, might accordingly constitute a common pathway in generation of aneuploidy and tumor development.

CENTROSOME DISORDERS

Aneuploidy is often accompanied by centrosome disorders [58-60]. Centrosomes are structures within the cell, which are involved in the regulation of critical events like cell cycle progression, mitosis and cell polarity [61, 62]. A centrosome consists of a pair of centriols, which is embedded in pericentriolar matrix. The complex microtubuli network that extends throughout the entire cell derives from γ-tubulin rings surrounded by pericentriolar material. This structure forms the origin of microtubules, which defines the role of the centrosome as a microtubule-organizing center. During a normal cell cycle, centrosomes are duplicated synchronously with the genetic material indicating that centrosome cycle and cell cycle are tightly connected. Current knowledge points to a special responsibility for Rb-supervised activation of cyclin-dependent kinase 2 (Cdk2) as the cell cycle and centrosome coordinating event [63, 64]. In addition, a growing body of knowledge suggests a role for centrosomal binding of particular regulators such as cyclin E in order to facilitate entry into S-phase. Similarly, active complexes of cyclin B1 and CDK1 were first detected at the centrosomes, which underlines their potential role as integrative sites for proper cell cycle transition signaling [65]. Thus, centrosomes may represent protein interaction platforms that are essential to bring cell cycle regulators into close proximity, thereby contributing to accurate surveillance signaling.

The distribution of the two centrosomes during the centrosome cycle to the daughter cells is conducted during mitosis. After cell division, each cell inherits one centrosome which is reduplicated during the following S-phase. The two centrosomes are destined to form the poles of the mitotic spindle which guarantees equal distribution of chromosomes to the developing daughter cells [62]. Due to this important role, particular monitoring is needed to ensure that centrosome duplication takes place only once per cell cycle. Aberrant centrosome number were shown to result in multipolar mitotic spindles which are associated with generation of aneuploidy due to defects in chromosome distribution [60].

It has long been debated whether centrosomes have a direct role in the development of malignant cells. Centrosome abnormalities are observed among most types of cancer (brain, breast, bile duct, colon, head and neck, lung, prostate and cervical cancers [60]) and are often accompanied by structural and numerical chromosome aberrations. A striking correlation between centrosomal aberrations and chromosomal instability was found in chronic myelogenous leukemia [66]. Aberrant centrosome numbers could be found at early stages of tumor development. However, significant rise in karyotype instability was observed at later stages and appeared in the background of even more extensive centrosomal abnormalities. Similarly, centrosome aberrations were found to constitute an early event in the evolution of malignancy as shown for a number of lesions comprising low-grade tumors of the uterine, cervix, breast and prostate and even in situ carcinomas [60]. These findings clearly contradict the notion that centrosomal disorders simply reflect an unspecific phenomenon at later stages of tumor development.

Aberrant numbers of centrosomes were shown to arise by various different mechanisms. Uncoupling of cell division cycle and centrosome cycle allows multiple cycles of centrosome duplication within a single cell division cycle [60]. This process usually originates from prolonged arrest at cell cycle checkpoints in response to internal (e.g. DNA damage) or external (e.g. hydroxyurea, which blocks the cell at the G1-S transition) stress [67, 68].

Failure in cell division as a mechanism to give rise to supernumerous centrosomes reflects a direct link between mitotic control and centrosomal disorders. Aspects such as cleavage furrow regression and checkpoint adaptation (progression from metaphase arrest into G1 without chromosomal segregation and cell separation in presence of an active SAC [69]) lead to tetraploidization. This causes not only a doubled DNA content but also twice the number of centrosomes per cell. As outlined below, subsequent entry of tetraploid cells into the cell division cycle depends on inactivation of p53. Tetraploid cells induce a p53-mediated checkpoint response in G1, which prevents tetraploid cells from entering S-phase [70]. This represents a restriction not only for DNA synthesis but also for irregular centrosome duplication. Complementary to the tetraploidization effect that is discussed below, expression of the HPV oncoproteins E6 and E7, which are known to target the retinoblastoma protein (Rb) and p53, led to an increasing amount of cells harbouring supernumerous centrosomes [71]. As tetraploidization per se proved to directly increase the rate of loss and gain of chromosomes, this phenomenon could be enhanced by chromosomal missegregation resulting from the role of supernumerous centrosomes in segregation of a tetraploid set of chromosomes. As an underlying mechanism, overexpression of mitotic kinase Aurora A appeared to give rise to polyploidy as well as centrosome duplication by interfering with proper completion of mitosis [72]. Strikingly, the genomic location of the human Aurora A gene (20q13) is located in a region of the genome that is often amplified in tumors and might subsequently favor overexpression of the encoded gene products [73]. Overexpression of this kinase was reported for malignancies like breast, colorectal and gastric cancer [74]. Functions of Aurora A kinase in cell physiology include recruitment of regulatory proteins to the centrosome and modulation of p53-dependent signaling. Aurora A gain of function mutations cause loss of function of p53 by phosphorylation-specific enhancement of Mdm2-mediated degradation [75]. This might provide an explanation for unaffected entry of tetraploid cells into a subsequent cell cycle instead of elimination by the tetraploidy checkpoint. Emphasizing the strong interdependence of centrosomal regulation and mitotic spindle attachment, Aurora A was identified as a factor to coordinate localization of Aurora B to the inner centromere in early mitosis [76].

Figure 3. Multipolar mitotic spindles lead to chromosomal missegregation. The two poles of the mitotic spindle are usually formed by the two centrosomes which guarantees formation of a bipolar spindle. Supernumerous centrosomes favor generation of multipolar spindles. Equal distribution of chromosomes to the developing daughter cells by multipolar anaphases is severely compromised and might facilitate the generation of aneuploidy.

Another probably less important mechanism in tumorigenesis is doubling the centrosome number by cell fusion which might be favored by a number of stimuli [60]. De novo synthesis of centrioles entered the scene as another possible mechanism as recent findings suggest this mechanism to be a common theme among most kinds of cells [77].

Abnormal numbers of centrosomes can lead to multipolar mitotic spindles which affect proper chromosome distribution (Figure 3). Multipolar mitoses are observed in a large number of human malignancies [59]. In any case, it is believed that most of the malignant cells that segregate their chromosomes by multipolar anaphases are destined to die from apoptosis. However, as a single event certain cells are thought to benefit from inappropriate chromosomal distribution and might become subject of clonal expansion [59]. Importantly, regain of mitotic stability by reestablishment of bipolar cell division is regarded as a central aspect of successful expansion [59]. This might provide an explanation for the frequent observation of aneuploidy in cells that exhibit centrosomal abnormalities even though a direct role in the establishment of aneuploidy remains to be shown.

Defective Tetraploidy Checkpoint

Most of the aspects already discussed above for centrosomal disorders hold true for development of numerical chromosomal aberrations by tetraploidy, which is referred to as a subclass of polyploidy. Actually it is still not clear whether the process of polyploidization constitutes a physiological event as it is frequently observed among eukaryotes caused by a multitude of underlying conditions. For various cell types such as skeletal muscle cells, osteoclasts and megakaryocytes, adoption of polyploidy is considered as an integral part of the regular development and terminal differentiation. Generation of polyploidy is facilitated by either cell fusion, endoreduplication of the DNA or as the result of an abortive mitosis [78]. The resulting cell harbours a doubled set of chromosomes and twice the number of centrosomes. Apart from its not further clarified role in differentiation, several studies support the notion that polyploidy/tetraploidy arises under adverse conditions such as oxidative damage, mechanical or metabolic stress [79-82]. Accordingly, the degree of polyploidy in liver tissue was found to increase with the process of ageing and under high metabolic load. Nevertheless, several arguments suggest tetraploidy to constitute a pathological event as it is also commonly detected among certain tumors and their corresponding precursor lesions [83]. Furthermore, the existence of a specialized tetraploidy checkpoint (see below) that arrests and eliminates tetraploid cells strongly contradicts the assumption that tetraploidy might be a programmed response upon particular stimuli.

It was observed that these tetraploid cells mostly get stuck in G1 following mitotic abortion [70]. This particular G1-arrest is thought to be facilitated by the above-mentioned tetraploidy checkpoint. Functional p53 was shown to be essential as lack or malfunction provided a chance for tetraploid cells to escape from arrest and proceed into a subsequent cell division cycle. Importantly, a dramatic increase in chromosomal missegregation resulted from divisions of tetraploid cells [84, 85]. This led to massive accumulation of genetic aberrations that were subject to replication during a subsequent S-phase in case of permissive G1-checkpoint signaling. Targeted inactivation of crucial checkpoint proteins such as p53, p21 and Rb was shown to facilitate entry into a new cell division cycle [78]. The risk of chromosomal missegregation was shown to be up to 166fold increased when comparing a tetraploid cell to a compareable diploid one [85]. It seems unlikely that aberrant centrosome numbers are the sole factor to account for the elevated number of missegregation events during mitosis as recovery from supernumerous centrosomes by clustering was shown to guarantee bipolar anaphases. Besides mitotic failure, different mechanisms are thought to play an important role in generation of aneuploidy from tetraploidy. Recent results showed an increased vulnerability towards DNA damage in tetraploid cells. Double strand breakage as well as mitotic failure might therefore provide mechanisms that lead to gain and loss of chromosomes and genome instability [84].

Recent models suggest that generation of tetraploidy is just the initiating step of a process that is characterized by ongoing loss of single chromosomes [78]. This might finally result in a hyperdiploid karyotype. It is believed that this mechanism is responsible at least in part for rise of esophageal [83] and cervical cancer [86], as tetraploid cells are frequently found in the corresponding precursor lesions. Papilloma virus mediated inactivation of p53 and RB was furthermore shown to account for generation of tetraploid cells in cervical cancer

[87, 88]. Moreover, presence of tetraploid cells in esophageal lesions strongly correlates with loss of functional p53 [83].

TELOMERE DYSFUNCTION

Telomeres are specific repetitive TTAGGG sequences at the end of each chromosome. In the human genome these particular sequence elements are known to cover a length of more than 10.000 base pairs [89]. A short single stranded 3`-overhang at the end of the telomere sequence forms a circular structure that associates with a large set of different proteins. This structure is known as the T-loop [90]. The circular shape is generated by base pairing between the overhang and the repetitive sequences of the inner telomer. Base pairing is facilitated by the complementarity of this highly conserved nucleotide sequence. This loop protects the chromosome from exonucleolytic activity. Intact telomeres are a powerful mechanism to prevent uncontrolled fusion of genetic material. Loss of telomere protein binding was shown to give rise to uncontrolled end-to-end fusion of the chromosomes which underscores the complexity of this highly sophisticated system.

Loss of telomere sequence contributes to generation of genetic instability [91, 92]. Failure of DNA-dependent DNA polymerases to replicate the entire telomeric DNA causes shortening of these particular sequences with every replication process. As telomere sequences significantly shorten during a lifetime, one might expect that these changes contribute to the increasing risk of tumor development [93]. In contrast, a tumor suppressive role might also be imagineable, as telomer shortening and replicative senescence are thought to limit the number of cell divisions. They might protect the cell from uncontrolled proliferation [94, 95]. Strinking support for this notion is provided by the finding that short telomeres in telomerase-depleted mice turned out to reduce tumor incidence [96].

In immortalized cells telomere sequences retain a constant length due to reactivation of the enzyme telomerase. Telomerase can confer the capability of unlimited proliferation as it provides an RNA template for the generation of TTAGGG nucleotides at the end of the telomere. Nevertheless, telomere sequences in tumor tissue are frequently maximally shortened [97, 98]. Efficient telomerase expression and activity are limited to embryonic tissues, activated lymphocytes, hematopoietic and epithelial stem cells. Even though detected in a variety of somatic tissues, usually telomerase cannot compensate for loss of telomeric sequences [99]. Additional findings such as telomerase induction upon antigen challenge in lymphoid tissue further complicate our understanding of telomerase regulation. Some of the listed features such as telomerase activation are also expected to account for the development of malignant cells. Consistent with this, activation of telomerase was shown in 85% of all tumors [95]. Strikingly, reintroduction of telomerase into the telomerase-depleted mice mentioned above proved to increase tumor formation [96].

Telomere shortening is a measure of the number of cell divisions which limits the proliferation of cells at the senescence checkpoint. This limitation was shown to be circumvented by inactivation of the p53 and Rb mediated cell cycle inhibitory pathways. Bypass of senescence arrest that provides the possibility to further divide, ultimately drives the cells into telomer-induced crisis. As several findings confirmed a central role for p53 in damage signaling, cell cycle arrest and onset of apoptosis in response to telomere loss [100], inactivation of p53 represented a key to facilitate ongoing genomic destabilization in crisis. In rare cases single cells manage to recover from final arrest and regain the ability to indefinite proliferation. Extensive telomere arrosion favors uncontrolled fusion of genetic material during crisis [101]. Chromosome ends that lack the telomere sequences are subject of unspecific DNA ligase activity as a possible consequence of not fully understood double strand repair pathways (see below). This leads to generation of dicentric chromosomes which are characterized by their tendency to undergo multiple breakage and fusion events (Figure 4) [102]. Accordingly, a peak number of dicentric chromosomes is recognized during telomere crisis accompanied by enhanced genomic instability and rapid changes in chromosome number [101]. The process of ongoing genomic breakage and fusion of dicentric chromosomes is thought to be driven by the dividing force of opposite movement of single centromeres to the spindle poles during mitosis. The morphologic correlate of these processes are anaphase bridges. Breakage and subsequent fusion of chromosomes contribute to genomic instability and thus cause structural and numerical chromosome aberrations [92]. These processes are known as breakage-fusion-bridge (BFB) cycles. As these double strand breakages are considered as the origin of deletions and amplifications [103, 104], these changes are accompanied by a number of gene dosage effects that enable the cell to develop into a more malignant stage. After crisis the generated abnormal karyotype was found to be stabilized. This stabilization process is considered to be the result of the maintenance of telomere length in general by reactivated telomerase and might also be due to a decreased number of dicentric chromosomes [105, 106]. In addition to telomerase activation, alternative mechanisms for telomere lengthening such as homologous recombination are discussed to account for a small population of tumors [107]. Reactivation of telomerase was shown to directly suppress chromosome fusions and thereby to limit BFB cycles by addition of telomere sequences to broken chromosome ends [95]. Recent models suggest reactivation of telomerase during crisis as a possible result of genetic rearrangement involving the telomerase gene locus [101, 108]. These findings are consistent with clonal selection mechanisms within tumor tissue. It has been shown that cells characterized by large numbers of chromosome fusions in combination with regained telomere extension activity were positively selected after injection of highly proliferative transformed cells into SCID mice [95].

In humans several studies implicate activation of telomerase as a late event in tumor development. Cellular crisis and genomic instability instead are considered as earlier stages that contribute to the tumor phenotype and drive accumulation of cancer-associated chromosomal aberrations [108]. Thus, aneuploidy develops as a result of chromosomal fusion, mitotic breakage and missegregation of particular translocations with subsequent irregular DNA content within the two daughter cells.

Apart from its already mentioned role as an important factor to drive genomic instability which enables the cell to proceed into malignancy, telomere dysfunction appeared to be responsible for a variety of human disorders. Dyskeratosis congenita is known as a severe disease due to primary dysfunction in telomere maintenance which finally leads to progressive bone marrow failure. Mutationally induced compromised function of a particular protein called dyskerin causes a decrease in RNA templates, which are needed for proper telomerase function [109]. This disorder thereby restricts telomerase activity and is responsible for premature telomere shortening. Additional disorders are characterized by premature telomere shortening due to an increased proliferative demand. These conditions comprise chronic infections such as chronic hepatitis, liver cirrhosis [110] and enhanced apoptosis due to DNA damage repair syndromes [99]. Remarkably, all of these disorders are accompanied by an increased risk to develop a malignant tumor [111] which correlates well with the degree of telomere shortening in many cases [112-114]. The enhanced risk might be attributable at least in part to the increased genetic instability that is found during telomere-induced crisis.

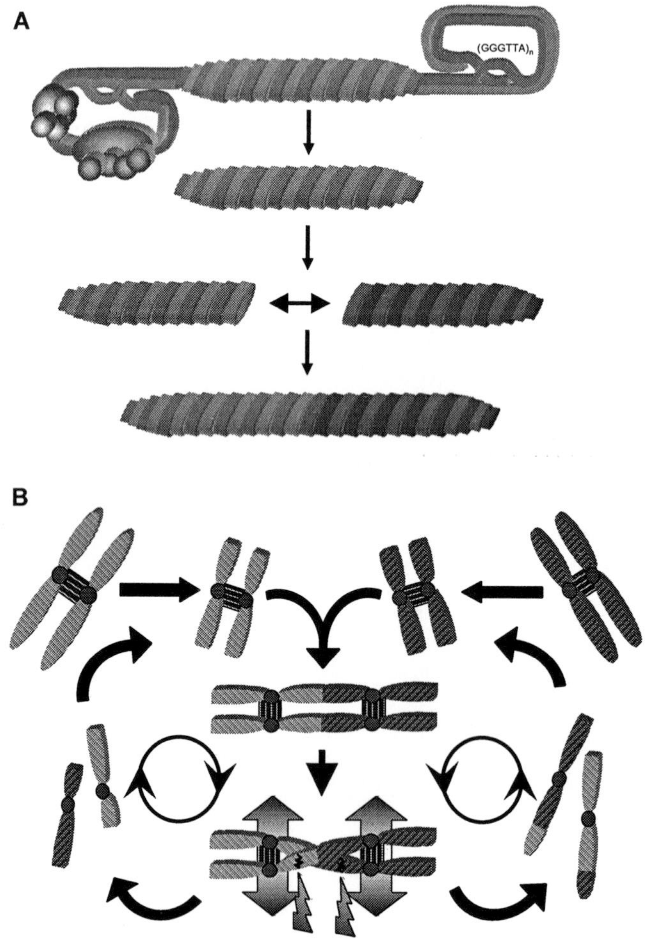

Figure 4. Genetic instability by cycles of chromosomal breakage and fusion. **A.** The telomere sequence at the end of each chromosome protects the DNA from nucleolytic activity and makes the linear chromosome ends distinguishable from DNA double-strand breaks. Loss of telomere sequences promote genetic instability. Unprotected chromosome ends that lack their telomere sequences are subject of uncontrolled ectopic fusion to compareable chromosomal lesions. **B.** Uncapping of chromosome ends and subsequent fusion events are the origin of dicentric chromosomes. These particular chromosomal structures are highly vulnerable towards breakage during mitosis as a result of the opposite pulling forces. Ongoing breakage and fusion of whole chromosomes with concomitant extensive amplifications and deletions are also known as breakage-fusion-bridge (BFB) cycles. DNA double-strand breaks that are not recognized and processed in time due to deficient DNA damage signaling and improper repair pathways are also subject of BFB cycles.

DEFECTS IN DNA DAMAGE SIGNALING

Proper DNA damage repair mechanisms are essential for the integrity of the genome. A highly sophisticated network that consists of different recognition and repair pathways guarantees genomic stability. Defects in recognition and repair of DNA damage are associated with an increased incidence of tumors [103, 115-119]. While point mutations arise from deficient repair of single misincorporated DNA bases, chromosomal lesions such as extensive genome instability are generally attributed to inappropriate repair of double strand DNA breaks (DSB) [103]. Essential players that mediate these signal transduction processes in response to DSBs are the phosphatidylinositol 3-kinase related kinases (PIKKs), namely ataxia telangiectasia mutated (ATM), which is considered as the main protein to respond to DSBs [116], and ataxia telangiectasia and RAD3-related (ATR). The ATM signaling pathway involves particular effector proteins that on the one hand are capable to induce cell cycle arrest to provide sufficient time for repair mechanisms and on the other can lead to apoptosis. Due to this general role, it is not surprising that ATM modulates a variety of pivotal cell cycle regulators as well. Accordingly, inhibitory influence on cell cycle progression is guaranteed as ATM enhances recruitment and activation/stabilization of particular mediator proteins such as p53 and BRCA-1 [115, 116]. Furthermore, ATM-mediated destruction of Cdc25A as well as sequestration/inactivation of Cdc25C via Chk1 and/or Chk2 interferes with CDK activation which is necessary to satisfy checkpoint signaling both at G1-S- and G2-M transition [116]. This implicates that ATM signaling has powerful impact on G1-S-, S-phase- and G2-M-checkpoints [116]. Recent data indicate that molecules like ATM might be important not only in damage recognition but also in DNA repair pathways themselves [116, 120]. The two major mechanisms of DSB repair comprise non-homologous end-joining (NHEJ) and homologous recombination (HR) [115]. Defects in damage sensation (e.g. ATM) and processing of the concerned lesions (e.g. LIG4 [103]) are generally thought to strongly compromise DNA rejoining fidelity and might therefore constitute a major cause of karyotype destabilizing events. The significance of malfunctions in DNA damage and recognition is further underscored by analyses of DSB repair deficiency syndromes and knockout mice, that lost essential proteins involved in repair pathways [117-119]. A growing body of evidence suggests that unrepaired DSBs form highly fusogenic sites that undergo ectopic fusion to different breakage sites or compareable lesions and thus contribute to genomic recombination. This mechanism might further promote generation of dicentric chromosomes and thereby favor onset of breakage-fusion-bridge (BFB) cycles (Figure 4). As outlined above, this process is considered to be responsible for the subsequent generation of extensive genome instability [102] and may cause nonreciprocal translocations which are known as the most common chromosomal abnormalities in carcinoma cells [121, 122]. Precious further information was gained from examination of human DNA repair deficiency syndromes. These comprise Ataxia telangiectasia (AT), Nijmegen breakage syndrome (NBS), Ataxia telangiectasia-like disorder (ATLD), Bloom syndrome and Werner syndrome all of which are considered as cancer predisposition syndromes [119]. Studies based on AT cell lines showed that about 10% of the DSBs fail to rapidly rejoin and thus might be subject of uncontrolled fusion events [115]. In normal cells ectopic fusion of genetic material following DNA breakage would suddenly lead to checkpoint activation and apoptosis. In specific DNA damage syndromes such as AT or Nijmegen breakage syndrome (NBS) checkpoint signaling is deficient [115, 119, 123]. This might provide an explanation for disease progression and bypass mechanisms in order to prevent apoptosis. AT as well as NBS are accompanied by an increased risk for development of malignancy, especially arising from the lymphoreticular tissues. AT is known as a human genetic disorder that is characterized by loss of function or inactivation of the ATM protein. AT phenotypically leads to cerebellar degeneration, thymic and gonadal atrophy, immunodeficiency and high-grade sensitivity towards ionizing radiation [116]. Of importance in terms of oncogenic potential, analyses based on heterozygous carriers of ATM mutations, that are estimated at about 1-2% of the overall population, revealed much about the underlying pathogenic mechanisms. People inheriting ATM mutations are divided into two groups. The first group exhibited an elevated risk of malignancy whereas people of the second group did not show any cancer predisposition. It is believed that these differences are due to two different kinds of mutations. Sequence alterations that lead to particular protein truncations, which constitute the major part of mutations affecting the ATM gene, are not considered to cause cancer predisposition. In contrast, missense mutations that are characterized by substitution of single bases within the sequence are thought to act in a dominant negative manner thereby affecting the remaining functional ATM allele. These lesions might therefore account for generation of aneuploidy and might provide a useful explanation for the elevated risk of malignancy among heterozygous carriers of ATM missense mutations [116, 124]. Importantly, recent studies highlight the essential role of the ATM-Chk2-p53 pathway in tumor suppression [125, 126]. Activated DNA damage responses were detected in a variety of precursor lesions of high-incidence malignancies most likely as a result of oncogenic effects (e.g. DNA replication stress) that interfere with genome stability. As this phenomenon is shared by several major types of human tumors one can estimate the importance of improper DNA damage signaling in generation of genetic instability and rise of malignancy. Therefore, tumor progression from these lesions seems to rely on inactivation of components such as ATM or p53 to facilitate onset of genomic instability and aneuploidy in order to generate karyotypes resembling those typically found among human tumors.

Chromosomal Translocations and their Role in Maligant Transformation

Telomere dysfunction as well as improper DSB repair are implicated in generation of chromosomal translocations. It is accepted that this section of chromosomal abnormalities divides into two major subgroups [127].

The first group of translocations is usually referred to as simple translocations. Cells mostly exhibit a single reciprocal (balanced) translocation. These abnormalities are commonly observed among hematological malignancies, childhood sarcomas, germ-cell tumors and a small number of epithelial tumors [128]. As epithelial cancers much more frequently display complex karyotypes, balanced translocations are difficult to distinguish from the multitude of chromosomal abnormalities and are therefore rarely known from these malignancies. However, a recent study suggests a common role for balanced translocation in rise of epithelial cancers even in presence of complex concomitant rearrangements [129]. Several mechanisms are thought to be responsible for generation of these particular translocations. Impaired V(D)J- and class switch recombinations, homologous- and non-homologous end-joining as well as genome fragile sites might favor establishment of genetic fusion events [128]. Consistently, numerous DNA damage repair deficiency syndromes such as AT and NBS are associated with an increased risk of chromosomal translocations and lympoid tumors [128]. Moreover, analysis of both translocation-forming chromosome fragments in patients suffering from leukemia and sarcomas clearly exhibited features characteristic for DNA processing by non-homologous end-joining mechanisms [130, 131]. The idea that these lesions might represent causal events in rise of malignancy is underlined by the enormous specificity for a certain tumor entity and the well-established role of translocations as precious guides during diagnosis. The tight association between lesion and disease suggests common functional pathways which involve these fusion events in transformation. Translocations of the simple type were demonstrated either to encode for chimeric transcription factors that interfere with differentiation (e.g. AML1-ETO, PML-RARα [132-135]. 90-95% of the transformed cells in chronic myelogenous leukemia harbour a reciprocal translocation between chromosomes 9 and 22. This chromosomal fusion gives rise to a chimeric tyrosine kinase that is characterized by deregulated phosphorylation activity (e.g. BCR-ABL) [136]. Cells harbouring a reciprocal translocation involving the immunoglubin heavy locus (IgH) exhibit abundant expression of important cellular regulator proteins due to massive transcriptional enhancement (e.g. BCL2-IgH, MYC-IgH) [137, 138]. In conclusion, these fusion events strengthen promitotic signaling, confer resistance towards apoptosis and lead to a differentiation block, which strikingly fits with its suggested role in malignant transformation. However, the finding, that some common leukemogenic translocations were also found among healthy individuals strongly emphasized the notion that additional lesions are needed for onset of disease [139].

In contrast, the second group of translocations (nonreciprocal translocations) is typically found in the background of complex karyotype abnormalities. These cells exhibit a complex pattern of numerical and structural chromosomal aberrations. Nonreciprocal translocations are often found in epithelium-derived solid tumors and are the result of random fusion events [122]. Accordingly, they are not specific for a particular kind of tumor. Nevertheless, common reciprocal translocations are also present in cells that predominantly harbour nonreciprocal translocations in the context of a complex pattern of chromosomal rearrangements (as mentioned above). Random nonreciprocal translocations are usually accompanied by extensive deletions and amplifications and lead to losses of heterozygosity and gene dosage effects due to gain or loss of the encoding genes [91, 122]. The exact molecular pathway that leads to these complex translocations is still a matter of debate. It is widely accepted that simultaneous breaks give rise to translocations as well as dicentric chromosomes. Whereas dicentrics undergo subsequent BFB cycles, translocations are stably transmitted through cell division [122]. The simultaneous occurrence of dicentric chromosomes provides an explanation for the observed complex cytogenetic background [140]. In accordance with the knowledge that telomere dysfunction is associated with end-to-end chromosome fusion, a recent study demonstrated that late-generation telomerase/p53 negative mice exhibit a high frequency of chromosome fusions, anaphase bridges and nonreciprocal translocations [91]. In contrast to reciprocal translocations, nonreciprocal translocations not only give rise to oncogenic fusions but importantly lead to loss and gain of genetic information. This results in gene dosage alterations which interfere with basic signal transduction networks [91].

Reciprocal (balanced) and nonreciprocal translocations share the characterisic that both are capable to confer a selective advantage to the concerned cell. This favors outgrowth of specific clones, that are able to benefit from the acquired genetic alteration in their contemporary environmental context [122].

Aneuploidy and Tumorigenesis – Functional Implications

It is still a matter of debate whether aneuploidy simply reflects an epiphenomenon during rise of malignant diseases or if it constitutes a key to rapid progression of tumors and increased proliferation. Several aspects underscore the assumption that aneuploidy provides an important tool to confer an advantage by means of clonal selection to the concerned cell [141]. Aneuploidy is part of the adenoma to carcinoma sequence in colorectal carcinogenesis [142]. Thus, in addition to various oncogenic hits, loss or gain of chromosomes is a regular phenomenon in the development of colorectal cancer. Furthermore, a large number of preneoplastic lesions harbour an aneuploid karyotype. Chromosome numbers of more or less than 46 were frequently observed in premalignant lesions such as hepatocellular dysplasia [143], prostatic intraepithelial neoplasia [144] and in Barrett's esophagitis [145]. All of these lesions share the characteristic of an enhanced risk for the development of malignancy. A lot of different tumor tissues exhibit a deregulated expression of spindle checkpoint genes or in a few cases show mutations in such class of genes. As a consequence functional impairment in SAC signaling might lead to chromosomal missegregation and possibly aneuploidy. In contrast some tumors show a highly conserved diploid karyotype. Nevertheless, due to the high prevalence of SAC defects in tumors, aneuploidy seems to be an important factor in the pathogenesis of different tumors.

Several recently established models aim to explain the effects of aneuploidy on the carrier cell. An additional chromosome may lead to expression of supernumerous oncogenes. Loss of whole chromosomes may lead to loss of tumor suppressor genes and can cause a decrease in the total gene copy number. The resulting gene dosage variation may ultimately favor absolute or relative gain of promitotic genes and lead to enhanced proliferation. As losses of heterozygosity, losses of maternal or paternal alleles, are frequently observed in tumors, chromosome number instabilities may constitute a tool to favor this state, which rises the probability for gene inactivating events [127]. Gain or loss of single chromosomes may contribute to an altered expression pattern of synergistic genes and thereby have a detrimental impact on highly sophisticated signaling networks. Impairment of complex signaling networks as a result of imbalance of the interacting proteins might compromise crucial cell functions such as cell cycle checkpoints or damage recognition and repair.

A large number of questions concerning genetic variability remain unclear. There are no data on the degree of genetic variability being needed to confer an advantage in terms of clonal expansion to a particular cell without causing substantial damage. A characteristic of malignant cells is the ability to deal with different challenges due to enhanced adaptational capacity. This feature is provided by an increase in genetic instability and thus by a concomitant gain of intrinsic variation. Instability and variation further lead to inhomogeneity within the concerned cell population. This dramatically increases the chance that a particular subset of cells is capable to deal with newly evolving challenges and thus perfectly adapts to a changing environment. However, these genetic changes can lead to a failure of basic cellular functions that are essential for the housekeeping metabolism of the cell. That means that rapid changes bear the danger of basic functional disruption suggesting a threshold level of intrinsic stability being essential even for malignant cells [59].

Mutations in SAC genes are rarely found and loss of an entire allele by truncation or a possibly resulting dominant negative interference might create more instability than the cell is capable to deal with. The degree of instability resulting from such defects seems to be a crucial factor. Recent experiments support this view, as SAC gene knockdown, that led to a complete loss of checkpoint function, proved to result in early cell death for instance during embryogenesis [146, 147]. Likewise, complete loss of the spindle checkpoint protein Mad2 by RNA interference caused mitotic failure and early death in human somatic cells [43]. In contrast haploinsufficiency of various spindle checkpoint genes in mice revealed that these animals had a high risk to develop tumors either after long latencies or after exposition to particular carcinogens [44, 148]. This leads to the idea that genetic changes in the development of malignancy have to occur in a more gentle way. Gene dosage changes by epigenetic mechanisms might be a possible underlying cause. The notion that extensive genome instability is preceded by gentle changes is underscored by findings in chronic myeloid leukemia, when centrosome aberrations were shown to appear only to a very moderate degree among early disease stages but rise to a tremendous degree during disease progression. Strikingly, phenotypical changes such as an abnormal karyotype were not detected in the early disease stages but were regularly present in progressive cells [66].

Genetic instability is undoubtedly associated with a high risk for the cell to die from failure in basic housekeeping functions. But there is no defined degree of instability being needed to confer the best possible adaptation capacity to the particular carrier cell.

Cell cycle checkpoints are known as special restrictional mechanisms constituting an important link between instability and cell cycle progression. They serve as control mechanisms responsible for damage recognition and subsequent cell cycle delay in order to provide sufficient time for repair mechanisms. If the recognized damage cannot be reverted to the original proper state in time, the cell is destined to commit suicide to prevent further events that might possibly affect the entire organism (such as rise of malignancy). That means that increased instability has to be considered as an aspect of negative selection due to checkpoint activation, block in cell cycle progression and severe damage in essential housekeeping functions. It is conceivable that loss of key components that are involved in damage recognition pathways might have enormous impact on the cell`s destiny. Loss of p53, for instance, was shown to improve cell survival in presence of genome destabilization [91, 104, 122]. This permissive effect might be due to checkpoint inactivation (remember tetraploidy checkpoint) and loss of proper damage recognition. Thus, instability-permissive constellations might favor acquisition of genetic damage and instability by reduction of adverse effects such as delay in cell cycle progression and subsequent onset of apoptosis. Nevertheless, these cells are faced with an elevated risk to loose particular housekeeping functions which might provide a limit to genetic instability by ultimately increasing the fraction of cells that undergo apoptosis. Whether reactivation of telomerase constitutes a factor to restabilize the karyotype at this point and thereby limits the dimension of genetic instability to subtoxic levels is a hypothesis that remains to be proven [91, 149].

As residual spindle checkpoint activity is observed in most cases of malfunction of mitotic signaling, such defects should more precisely be referred to as spindle checkpoint insufficiency. This state is found when more than a single unattached kinetochore is needed to induce metaphase delay. Insufficiency in contrast to complete checkpoint absence might still enable the cell to recover from vast chromosomal missegregation by induction of a transient cell cycle arrest. Importantly, the gentle changes of the genome might be generated by facing this particular kind of control deficiency during an aberrant mitosis [150]. According to this idea, tumorigenesis might originate from changes in gene transcription or might be due to posttranscriptional effects that affect control genes which possibly lead to spindle checkpoint insufficiency. Transcriptional and posttranscriptional repression of particular control genes might lead to checkpoint insufficiency and might thereby cause rapid recovery from mitotic arrest with subsequent missegregation in presence of moderate misalignment or impaired spindle attachment. As insufficiency still guarantees recognition of deleterious disorders, missegregation is thought to affect only single chromosomes in a relatively small subset of cells. This might prevent the severe consequences of massive chromosomal instability even though providing a precious tool for the generation of enhanced environmental adaptation and increased proliferative capacity. This could therefore be a basic mechanism to facilitate slow but continuous progression of an early premalignant tissue to a malignant tumor.

A direct correlation between the degree of aneuploidy and the dimension of genetic instability was recently shown [151]. Cells of chinese hamster embryo that harboured a hyperdiploid set of chromosomes were faced to much more extensive chromosomal instability compared to diploid cells. This hypothesis was retrospectively confirmed when analysing different human colon carcinoma cell lines [152]. According to these data the authors suggested a two step mechanisms to account for tumor development [151, 153]. The first step describes the generation of an aneuploid set of chromosomes that leads to destabilization of the karyotype and thus to a heterogenous tumor, providing enhanced adaptational capacity in order to deal with a broad variety of challenges. Roads to aneuploidy might comprise epigenetic downregulation of control genes or more rarely mutations of particular genes that functionally compromise chromosome segregation. Particular changes affecting mitotic regulators as well as chemical and physical effects are factors to favor spontaneous arising chromosomal missegregation. The second step is defined by an imbalance of interacting proteins that is directly caused by aneuploidy and the subsequent changes in the gene copy number. Gene interactions in sophisticated networks such as in mitotic control or DNA damage recognition are thought to be frequently targeted during this process. This step leads to further destabilization of the karyotype and an increasing degree of chromosomal instability.

Conclusion

More than a century after scientists first suggested a link between malignancy and an aberrant number of chromosomes, a large set of data underscores the notion that aneuploidy is an important factor in tumorigenesis. As genetic instability entered the scene as one of the major aspects of tumors, aneuploidy was rather considered as a central driving force of genetic instability than being a nonspecific byproduct during transformation. This idea was supported by the finding that tumors that did not exhibit an aneuploid karyotype frequently took different ways to generate genetic instability such as defective DNA mismatch repair (MMR) [127, 150]. Aneuploidy thus appears as a state to favor loss of heterozygosity and particular gene amplification whereas in contrast MMR-deficient non-aneuploid tumors found different ways to imbalance oncogene regulation and facilitate inactivation of tumor suppressor alleles. This leads to the conclusion that aneuploidy must be considered as an essential contributor in the generation of most cancers.

References

[1] Bharadwaj R and Yu H. The spindle checkpoint, aneuploidy, and cancer. *Oncogene* 2004;23(11):2016-27.

[2] Rieder CL, Schultz A, Cole R and Sluder G. Anaphase onset in vertebrate somatic cells is controlled by a checkpoint that monitors sister kinetochore attachment to the spindle. *J Cell Biol* 1994;127(5):1301-10.

[3] Stern BM and Murray AW. Lack of tension at kinetochores activates the spindle checkpoint in budding yeast. *Curr Biol* 2001;11(18):1462-7.

[4] Tanaka TU, Stark MJ and Tanaka K. Kinetochore capture and bi-orientation on the mitotic spindle. *Nat Rev Mol Cell Biol* 2005;6(12):929-42.

[5] Biggins S, Severin FF, Bhalla N, Sassoon I, Hyman AA and Murray AW. The conserved protein kinase Ipl1 regulates microtubule binding to kinetochores in budding yeast. *Genes Dev* 1999;13(5):532-44.

[6] Cheeseman IM, Anderson S, Jwa M, Green EM, Kang J, Yates JR, 3rd, et al. Phospho-regulation of kinetochore-microtubule attachments by the Aurora kinase Ipl1p. *Cell* 2002;111(2):163-72.

[7] Westermann S, Cheeseman IM, Anderson S, Yates JR, 3rd, Drubin DG and Barnes G. Architecture of the budding yeast kinetochore reveals a conserved molecular core. *J Cell Biol* 2003;163(2):215-22.

[8] Lampson MA, Renduchitala K, Khodjakov A and Kapoor TM. Correcting improper chromosome-spindle attachments during cell division. Nat Cell Biol 2004;6(3):232-7.

[9] Adams RR, Carmena M and Earnshaw WC. Chromosomal passengers and the (aurora) ABCs of mitosis. *Trends Cell Biol* 2001;11(2):49-54.

[10] Gassmann R, Carvalho A, Henzing AJ, Ruchaud S, Hudson DF, Honda R, et al. Borealin: a novel chromosomal passenger required for stability of the bipolar mitotic spindle. *J Cell Biol* 2004;166(2):179-91.

[11] Ditchfield C, Johnson VL, Tighe A, Ellston R, Haworth C, Johnson T, et al. Aurora B couples chromosome alignment with anaphase by targeting BubR1, Mad2, and Cenp-E to kinetochores. *J Cell Biol* 2003;161(2):267-80.

[12] Hauf S, Cole RW, LaTerra S, Zimmer C, Schnapp G, Walter R, et al. The small molecule Hesperadin reveals a role for Aurora B in correcting kinetochore-microtubule attachment and in maintaining the spindle assembly checkpoint. *J Cell Biol* 2003;161(2):281-94.

[13] Norden C, Mendoza M, Dobbelaere J, Kotwaliwale C, Biggins S and Barral Y. The NoCut pathway links completion of cytokinesis to spindle midzone function to prevent chromosome breakage. *Cell* 2006;125:85-98.

[14] Ota T, Suto S, Katayama H, Han ZB, Suzuki F, Maeda M, et al. Increased mitotic phosphorylation of histone H3 attributable to AIM-1/Aurora-B overexpression contributes to chromosome number instability. *Cancer Res* 2002;62(18):5168-77.

[15] Sudakin V, Chan GK and Yen TJ. Checkpoint inhibition of the APC/C in HeLa cells is mediated by a complex of BUBR1, BUB3, CDC20, and MAD2. *J Cell Biol* 2001;154(5):925-36.

[16] Yu H. Regulation of APC-Cdc20 by the spindle checkpoint. *Curr Opin Cell Biol* 2002;14(6):706-14.

[17] Zou H, McGarry TJ, Bernal T and Kirschner MW. Identification of a vertebrate sister-chromatid separation inhibitor involved in transformation and tumorigenesis. *Science* 1999;285(5426):418-22.

[18] Uhlmann F, Lottspeich F and Nasmyth K. Sister-chromatid separation at anaphase onset is promoted by cleavage of the cohesin subunit Scc1. *Nature* 1999;400(6739):37-42.

[19] McGrew JT, Goetsch L, Byers B and Baum P. Requirement for ESP1 in the nuclear division of Saccharomyces cerevisiae. *Mol.Biol.Cell* 1992;3:1443-1454.

[20] Yamamoto A, Guacci V and Koshland D. Pds1p is required for faithful execution of anaphase in the yeast, Saccharomyces cerevisiae. *J.Cell Biol.* 1996;133:85-97.

[21] Jallepalli PV, Waizenegger IC, Bunz F, Langer S, Speicher MR, Peters JM, et al. Securin is required for chromosomal stability in human cells. *Cell* 2001;105(4):445-57.

[22] Zhang X, Horwitz GA, Prezant TR, Valentini A, Nakashima M, Bronstein MD, et al. Structure, expression, and function of human pituitary tumor-transforming gene (PTTG). *Mol Endocrinol* 1999;13(1):156-66.

[23] Wäsch R and Cross FR. APC-dependent proteolysis of the mitotic cyclin Clb2 is essential for mitotic exit. *Nature* 2002;418(6897):556-62.

[24] Wäsch R and Engelbert D. Anaphase-promoting complex-dependent proteolysis of cell cycle regulators and genomic instability of cancer cells. *Oncogene* 2005;24(1):1-10.

[25] Harper JW, Burton JL and Solomon MJ. The anaphase-promoting complex: it's not just for mitosis any more. *Genes Dev* 2002;16:2179-206.

[26] Reimann JD, Freed E, Hsu JY, Kramer ER, Peters JM and Jackson PK. Emi1 is a mitotic regulator that interacts with Cdc20 and inhibits the anaphase promoting complex. *Cell* 2001a;105(5):645-55.

[27] Song MS, Song SJ, Ayad NG, Chang JS, Lee JH, Hong HK, et al. The tumour suppressor RASSF1A regulates mitosis by inhibiting the APC-Cdc20 complex. *Nat Cell Biol* 2004;6(2):129-37.

[28] Potapova TA, Daum JR, Pittman BD, Hudson JR, Jones TN, Satinover DL, et al. The reversibility of mitotic exit in vertebrate cells. *Nature* 2006;440(7086):954-8.

[29] Carrano AC, Eytan E, Hershko A and Pagano M. SKP2 is required for ubiquitin-mediated degradation of the CDK inhibitor p27. *Nat Cell Biol* 1999;1(4):193-9.

[30] Tsvetkov LM, Yeh KH, Lee SJ, Sun H and Zhang H. p27(Kip1) ubiquitination and degradation is regulated by the SCF(Skp2) complex through phosphorylated Thr187 in p27. *Curr Biol* 1999;9(12):661-4.

[31] Massague J. G1 cell-cycle control and cancer. *Nature* 2004;432(7015):298-306.

[32] Hanks S, Coleman K, Reid S, Plaja A, Firth H, Fitzpatrick D, et al. Constitutional aneuploidy and cancer predisposition caused by biallelic mutations in BUB1B. *Nat Genet* 2004;36(11):1159-61.

[33] Percy MJ, Myrie KA, Neeley CK, Azim JN, Ethier SP and Petty EM. Expression and mutational analyses of the human MAD2L1 gene in breast cancer cells. *Genes Chromosomes Cancer* 2000;29(4):356-62.

[34] Hernando E, Orlow I, Liberal V, Nohales G, Benezra R and Cordon-Cardo C. Molecular analyses of the mitotic checkpoint components hsMAD2, hBUB1 and hBUB3 in human cancer. *Int J Cancer* 2001;95(4):223-7.

[35] Nomoto S, Haruki N, Takahashi T, Masuda A, Koshikawa T, Fujii Y, et al. Search for in vivo somatic mutations in the mitotic checkpoint gene, hMAD1, in human lung cancers. *Oncogene* 1999;18(50):7180-3.

[36] Tsukasaki K, Miller CW, Greenspun E, Eshaghian S, Kawabata H, Fujimoto T, et al. Mutations in the mitotic check point gene, MAD1L1, in human cancers. *Oncogene* 2001;20(25):3301-5.

[37] Gemma A, Seike M, Seike Y, Uematsu K, Hibino S, Kurimoto F, et al. Somatic mutation of the hBUB1 mitotic checkpoint gene in primary lung cancer. *Genes Chromosomes Cancer* 2000;29(3):213-8.

[38] Ohshima K, Haraoka S, Yoshioka S, Hamasaki M, Fujiki T, Suzumiya J, et al. Mutation analysis of mitotic checkpoint genes (hBUB1 and hBUBR1) and microsatellite instability in adult T-cell leukemia/lymphoma. *Cancer Lett* 2000;158(2):141-50.

[39] Ru HY, Chen RL, Lu WC, Chen JH and . hBUB1 defects in leukemia and lymphoma cells. *Oncogene* 2002;21:4673-9.

[40] Hempen PM, Kurpad H, Calhoun ES, Abraham S and Kern SE. A double missense variation of the BUB1 gene and a defective mitotic spindle checkpoint in the pancreatic cancer cell line Hs766T. *Hum Mutat* 2003;21(4):445.

[41] Cahill DP, Lengauer C, Yu J, Riggins GJ, Willson JK, Markowitz SD, et al. Mutations of mitotic checkpoint genes in human cancers. *Nature* 1998;392:300-3.

[42] Shin HJ, Baek KH, Jeon AH, Park MT, Lee SJ, Kang CM, et al. Dual roles of human BubR1, a mitotic checkpoint kinase, in the monitoring of chromosomal instability. *Cancer Cell* 2003;4(6):483-97.

[43] Michel L, Diaz-Rodriguez E, Narayan G, Hernando E, Murty VV and Benezra R. Complete loss of the tumor suppressor MAD2 causes premature cyclin B degradation and mitotic failure in human somatic cells. *Proc Natl Acad Sci U S A* 2004;101(13):4459-64.

[44] Dai W, Wang Q, Liu T, Swamy M, Fang Y, Xie S, et al. Slippage of mitotic arrest and enhanced tumor development in mice with BubR1 haploinsufficiency. *Cancer Res* 2004;64(2):440-5.

[45] Li Y and Benezra R. Identification of a human mitotic checkpoint gene: hsMAD2. *Science* 1996;274(5285):246-8.

[46] Wang X, Jin DY, Wong YC, Cheung AL, Chun AC, Lo AK, et al. Correlation of defective mitotic checkpoint with aberrantly reduced expression of MAD2 protein in nasopharyngeal carcinoma cells. *Carcinogenesis* 2000;21(12):2293-7.

[47] Wang X, Jin DY, Ng RW, Feng H, Wong YC, Cheung AL, et al. Significance of MAD2 expression to mitotic checkpoint control in ovarian cancer cells. *Cancer Res* 2002;62(6):1662-8.

[48] Hernando E, Nahle Z, Juan G, Diaz-Rodriguez E, Alaminos M, Hemann M, et al. Rb inactivation promotes genomic instability by uncoupling cell cycle progression from mitotic control. *Nature* 2004;430(7001):797-802.

[49] Lee H, Trainer AH, Friedman LS, Thistlethwaite FC, Evans MJ, Ponder BA, et al. Mitotic checkpoint inactivation fosters transformation in cells lacking the breast cancer susceptibility gene, Brca2. *Mol Cell 1999*;4(1):1-10.

[50] Daniels MJ, Wang Y, Lee M and Venkitaraman AR. Abnormal cytokinesis in cells deficient in the breast cancer susceptibility protein BRCA2. *Science* 2004;306(5697):876-9.

[51] Wang RH, Yu H and Deng CX. A requirement for breast-cancer-associated gene 1 (BRCA1) in the spindle checkpoint. *Proc Natl Acad Sci U S A* 2004;101(49):17108-13.

[52] Kops GJ, Weaver BA and Cleveland DW. On the road to cancer: aneuploidy and the mitotic checkpoint. *Nat Rev Cancer* 2005;5(10):773-85.

[53] Iwanaga Y and Jeang KT. Expression of mitotic spindle checkpoint protein hsMAD1 correlates with cellular proliferation and is activated by a gain-of-function p53 mutant. *Cancer Res* 2002;62(9):2618-24.

[54] Chun AC and Jin DY. Transcriptional regulation of mitotic checkpoint gene MAD1 by p53. *J Biol Chem* 2003;278(39):37439-50.

[55] Gupta A, Inaba S, Wong OK, Fang G and Liu J. Breast cancer-specific gene 1 interacts with the mitotic checkpoint kinase BubR1. *Oncogene* 2003;22(48):7593-9.

[56] Jin DY, Spencer F and Jeang KT. Human T cell leukemia virus type 1 oncoprotein Tax targets the human mitotic checkpoint protein MAD1. *Cell* 1998;93(1):81-91.

[57] Cotsiki M, Lock RL, Cheng Y, Williams GL, Zhao J, Perera D, et al. Simian virus 40 large T antigen targets the spindle assembly checkpoint protein Bub1. *Proc Natl Acad Sci U S A* 2004;101(4):947-52.

[58] Albertson DG, Collins C, McCormick F and Gray JW. Chromosome aberrations in solid tumors. *Nat Genet* 2003;34(4):369-76.

[59] Brinkley BR. Managing the centrosome numbers game: from chaos to stability in cancer cell division. *Trends Cell Biol* 2001;11(1):18-21.

[60] Nigg EA. Centrosome aberrations: cause or consequence of cancer progression? *Nat Rev Cancer* 2002;2(11):815-25.

[61] Doxsey SJ. Centrosomes as command centres for cellular control. *Nat Cell Biol* 2001;3(5):E105-8.

[62] Meraldi P and Nigg EA. The centrosome cycle. *FEBS Lett* 2002;521(1-3):9-13.

[63] Lacey KR, Jackson PK and Stearns T. Cyclin-dependent kinase control of centrosome duplication. *Proc Natl Acad Sci U S A* 1999;96(6):2817-22.

[64] Matsumoto Y, Hayashi K and Nishida E. Cyclin-dependent kinase 2 (Cdk2) is required for centrosome duplication in mammalian cells. *Curr Biol* 1999;9(8):429-32.

[65] Doxsey S, Zimmerman W and Mikule K. Centrosome control of the cell cycle. *Trends Cell Biol* 2005;15(6):303-11.

[66] Giehl M, Fabarius A, Frank O, Hochhaus A, Hafner M, Hehlmann R, et al. Centrosome aberrations in chronic myeloid leukemia correlate with stage of disease and chromosomal instability. *Leukemia* 2005;19(7):1192-7.

[67] Balczon R, Bao L, Zimmer WE, Brown K, Zinkowski RP and Brinkley BR. Dissociation of centrosome replication events from cycles of DNA synthesis and mitotic division in hydroxyurea-arrested Chinese hamster ovary cells. *J Cell Biol* 1995;130(1):105-15.

[68] Wong C and Stearns T. Centrosome number is controlled by a centrosome-intrinsic block to reduplication. *Nat Cell Biol* 2003;5(6):539-44.

[69] Rieder CL and Maiato H. Stuck in division or passing through: what happens when cells cannot satisfy the spindle assembly checkpoint. *Dev Cell* 2004;7(5):637-51.

[70] Andreassen PR, Lohez OD, Lacroix FB and Margolis RL. Tetraploid state induces p53-dependent arrest of nontransformed mammalian cells in G1. *Mol Biol Cell* 2001;12(5):1315-28.

[71] Duensing S and Munger K. Centrosome abnormalities and genomic instability induced by human papillomavirus oncoproteins. *Prog Cell Cycle Res* 2003;5:383-91.

[72] Zhou H, Kuang J, Zhong L, Kuo WL, Gray JW, Sahin A, et al. Tumour amplified kinase STK15/BTAK induces centrosome amplification, aneuploidy and transformation. *Nat Genet* 1998;20(2):189-93.

[73] Bar-Shira A, Pinthus JH, Rozovsky U, Goldstein M, Sellers WR, Yaron Y, et al. Multiple genes in human 20q13 chromosomal region are involved in an advanced prostate cancer xenograft. *Cancer Res* 2002;62(23):6803-7.

[74] Andrews PD. Aurora kinases: shining lights on the therapeutic horizon? *Oncogene* 2005;24(32):5005-15.

[75] Katayama H, Sasai K, Kawai H, Yuan ZM, Bondaruk J, Suzuki F, et al. Phosphorylation by aurora kinase A induces Mdm2-mediated destabilization and inhibition of p53. *Nat Genet* 2004;36(1):55-62.

[76] Kunitoku N, Sasayama T, Marumoto T, Zhang D, Honda S, Kobayashi O, et al. CENP-A phosphorylation by Aurora-A in prophase is required for enrichment of Aurora-B at inner centromeres and for kinetochore function. *Dev Cell* 2003;5(6):853-64.

[77] Khodjakov A, Rieder CL, Sluder G, Cassels G, Sibon O and Wang CL. De novo formation of centrosomes in vertebrate cells arrested during S phase. *J Cell Biol* 2002;158(7):1171-81.

[78] Storchova Z and Pellman D. From polyploidy to aneuploidy, genome instability and cancer. *Nat Rev Mol Cell Biol* 2004;5(1):45-54.

[79] Vliegen HW, Eulderink F, Bruschke AV, van der Laarse A and Cornelisse CJ. Polyploidy of myocyte nuclei in pressure overloaded human hearts: a flow cytometric study in left and right ventricular myocardium. *Am J Cardiovasc Pathol* 1995;5(1):27-31.

[80] Gorla GR, Malhi H and Gupta S. Polyploidy associated with oxidative injury attenuates proliferative potential of cells. *J Cell Sci* 2001;114(Pt 16):2943-51.

[81] Hixon ML and Gualberto A. Vascular smooth muscle polyploidization--from mitotic checkpoints to hypertension. *Cell Cycle* 2003;2(2):105-11.

[82] Oberringer M, Lothschutz D, Jennewein M, Koschnick M, Mutschler W and Hanselmann RG. Centrosome multiplication accompanies a transient clustering of polyploid cells during tissue repair. *Mol Cell Biol Res Commun* 1999;2(3):190-6.

[83] Galipeau PC, Cowan DS, Sanchez CA, Barrett MT, Emond MJ, Levine DS, et al. 17p (p53) allelic losses, 4N (G2/tetraploid) populations, and progression to aneuploidy in Barrett's esophagus. *Proc Natl Acad Sci U S A* 1996;93(14):7081-4.

[84] Fujiwara T, Bandi M, Nitta M, Ivanova EV, Bronson RT and Pellman D. Cytokinesis failure generating tetraploids promotes tumorigenesis in p53-null cells. *Nature* 2005;437(7061):1043-7.

[85] Shi Q and King RW. Chromosome nondisjunction yields tetraploid rather than aneuploid cells in human cell lines. *Nature* 2005;437(7061):1038-42.

[86] Olaharski AJ, Sotelo R, Solorza-Luna G, Gonsebatt ME, Guzman P, Mohar A, et al. *Tetraploidy and chromosomal instability are early events during cervical carcinogenesis.* Carcinogenesis 2005.

[87] Duensing S and Munger K. Centrosome abnormalities, genomic instability and carcinogenic progression. *Biochim Biophys Acta* 2001;1471(2):M81-8.

[88] Olaharski AJ and Eastmond DA. Elevated levels of tetraploid cervical cells in human papillomavirus-positive Papanicolaou smears diagnosed as atypical squamous cells of undetermined significance. *Cancer* 2004;102(3):192-9.

[89] Alberts B, Johnson A., Lewis J., Raff M., Roberts K., Walter P. *Molecular biology of the cell.* fourth edition ed. New York: Garland Science; 2002.

[90] de Lange T. T-loops and the origin of telomeres. *Nat Rev Mol Cell Biol* 2004;5(4):323-9.

[91] Artandi SE, Chang S, Lee SL, Alson S, Gottlieb GJ, Chin L, et al. Telomere dysfunction promotes non-reciprocal translocations and epithelial cancers in mice. *Nature* 2000;406(6796):641-5.

[92] O'Hagan RC, Chang S, Maser RS, Mohan R, Artandi SE, Chin L, et al. Telomere dysfunction provokes regional amplification and deletion in cancer genomes. *Cancer* Cell 2002;2(2):149-55.

[93] DePinho RA. The age of cancer. *Nature* 2000;408(6809):248-54.

[94] Engelhardt M and Finke J. Does telomere shortening count? Blood 2001;98(3):888-90.

[95] Hackett JA and Greider CW. Balancing instability: dual roles for telomerase and telomere dysfunction in tumorigenesis. *Oncogene* 2002;21(4):619-26.

[96] Greenberg RA, Chin L, Femino A, Lee KH, Gottlieb GJ, Singer RH, et al. Short dysfunctional telomeres impair tumorigenesis in the INK4a(delta2/3) cancer-prone mouse. *Cell* 1999;97(4):515-25.

[97] Engelhardt M, Kumar R, Albanell J, Pettengell R, Han W and Moore MA. Telomerase regulation, cell cycle, and telomere stability in primitive hematopoietic cells. *Blood* 1997;90(1):182-93.

[98] Engelhardt M, Wäsch R and Guo Y. Telomeres and telomerase in normal and leukemic hematopoietic cells. *Leuk Res* 2004;28(10):1001-4.

[99] Wong JM and Collins K. Telomere maintenance and disease. *Lancet* 2003;362(9388):983-8.

[100] Chin L, Artandi SE, Shen Q, Tam A, Lee SL, Gottlieb GJ, et al. p53 deficiency rescues the adverse effects of telomere loss and cooperates with telomere dysfunction to accelerate carcinogenesis. *Cell* 1999;97(4):527-38.

[101] DePinho RA and Polyak K. Cancer chromosomes in crisis. *Nat Genet* 2004;36(9):932-4.

[102] Pihan G and Doxsey SJ. Mutations and aneuploidy: co-conspirators in cancer? *Cancer* Cell 2003;4(2):89-94.

[103] Sharpless NE, Ferguson DO, O'Hagan RC, Castrillon DH, Lee C, Farazi PA, et al. Impaired nonhomologous end-joining provokes soft tissue sarcomas harboring chromosomal translocations, amplifications, and deletions. *Mol Cell* 2001;8(6):1187-96.

[104] Zhu C, Mills KD, Ferguson DO, Lee C, Manis J, Fleming J, et al. Unrepaired DNA breaks in p53-deficient cells lead to oncogenic gene amplification subsequent to translocations. *Cell* 2002;109(7):811-21.

[105] Counter CM, Avilion AA, LeFeuvre CE, Stewart NG, Greider CW, Harley CB, et al. Telomere shortening associated with chromosome instability is arrested in immortal cells which express telomerase activity. *Embo J* 1992;11(5):1921-9.

[106] Ducray C, Pommier JP, Martins L, Boussin FD and Sabatier L. Telomere dynamics, end-to-end fusions and telomerase activation during the human fibroblast immortalization process. *Oncogene* 1999;18(29):4211-23.

[107] Henson JD, Neumann AA, Yeager TR and Reddel RR. Alternative lengthening of telomeres in mammalian cells. *Oncogene* 2002;21(4):598-610.

[108] Maser RS and DePinho RA. Connecting chromosomes, crisis, and cancer. *Science* 2002;297(5581):565-9.

[109] Mitchell JR, Wood E and Collins K. A telomerase component is defective in the human disease dyskeratosis congenita. *Nature* 1999;402(6761):551-5.

[110] Aikata H, Takaishi H, Kawakami Y, Takahashi S, Kitamoto M, Nakanishi T, et al. Telomere reduction in human liver tissues with age and chronic inflammation. *Exp Cell Res* 2000;256(2):578-82.

[111] Feldser DM, Hackett JA and Greider CW. Telomere dysfunction and the initiation of genome instability. *Nat Rev Cancer* 2003;3(8):623-7.

[112] Engelhardt M, Drullinsky P, Guillem J and Moore MA. Telomerase and telomere length in the development and progression of premalignant lesions to colorectal cancer. *Clin Cancer Res* 1997;3(11):1931-41.

[113] Engelhardt M, Mackenzie K, Drullinsky P, Silver RT and Moore MA. Telomerase activity and telomere length in acute and chronic leukemia, pre- and post-ex vivo culture. *Cancer Res.* 2000;60:610-7.

[114] Engelhardt M, Ozkaynak MF, Drullinsky P, Sandoval C, Tugal O, Jayabose S, et al. Telomerase activity and telomere length in pediatric patients with malignancies undergoing chemotherapy. *Leukemia* 1998;12(1):13-24.

[115] O'Driscoll M and Jeggo PA. The role of double-strand break repair - insights from human genetics. *Nat Rev Genet* 2006;7(1):45-54.

[116] Shiloh Y. ATM and related protein kinases: safeguarding genome integrity. *Nat Rev Cancer* 2003;3(3):155-68.

[117] Maser RS and DePinho RA. Take care of your chromosomes lest cancer take care of you. *Cancer Cell* 2003;3(1):4-6.

[118] Mills KD, Ferguson DO and Alt FW. The role of DNA breaks in genomic instability and tumorigenesis. *Immunol Rev* 2003;194:77-95.

[119] van Gent DC, Hoeijmakers JH and Kanaar R. Chromosomal stability and the DNA double-stranded break connection. *Nat Rev Genet* 2001;2(3):196-206.

[120] Riballo E, Kuhne M, Rief N, Doherty A, Smith GC, Recio MJ, et al. A pathway of double-strand break rejoining dependent upon ATM, Artemis, and proteins locating to gamma-H2AX foci. *Mol Cell* 2004;16(5):715-24.

[121] Gisselsson D. Chromosome instability in cancer: how, when, and why? *Adv Cancer Res* 2003;87:1-29.

[122] Lengauer C. How do tumors make ends meet? *Proc Natl Acad Sci U S A* 2001;98(22):12331-3.

[123] Lobrich M and Jeggo PA. Harmonising the response to DSBs: a new string in the ATM bow. *DNA Repair (Amst)* 2005;4(7):749-59.

[124] Gatti RA, Tward A and Concannon P. Cancer risk in ATM heterozygotes: a model of phenotypic and mechanistic differences between missense and truncating mutations. *Mol Genet Metab* 1999;68(4):419-23.

[125] Gorgoulis VG, Vassiliou LV, Karakaidos P, Zacharatos P, Kotsinas A, Liloglou T, et al. Activation of the DNA damage checkpoint and genomic instability in human precancerous lesions. *Nature* 2005;434(7035):907-13.

[126] Bartkova J, Horejsi Z, Koed K, Kramer A, Tort F, Zieger K, et al. DNA damage response as a candidate anti-cancer barrier in early human tumorigenesis. *Nature* 2005;434(7035):864-70.

[127] Lengauer C, Kinzler KW and Vogelstein B. Genetic instabilities in human cancers. *Nature* 1998;396(6712):643-9.

[128] Aplan PD. Causes of oncogenic chromosomal translocation. *Trends Genet* 2006;22(1):46-55.

[129] Mitelman F, Mertens F and Johansson B. Prevalence estimates of recurrent balanced cytogenetic aberrations and gene fusions in unselected patients with neoplastic disorders. *Genes Chromosomes Cancer* 2005;43(4):350-66.

[130] Reichel M, Gillert E, Nilson I, Siegler G, Greil J, Fey GH, et al. Fine structure of translocation breakpoints in leukemic blasts with chromosomal translocation t(4;11): the DNA damage-repair model of translocation. *Oncogene* 1998;17(23):3035-44.

[131] Zucman-Rossi J, Legoix P, Victor JM, Lopez B and Thomas G. Chromosome translocation based on illegitimate recombination in human tumors. *Proc Natl Acad Sci U S A* 1998;95(20):11786-91.

[132] Licht JD. AML1 and the AML1-ETO fusion protein in the pathogenesis of t(8;21) AML. *Oncogene* 2001;20(40):5660-79.

[133] Tenen DG. Disruption of differentiation in human cancer: AML shows the way. *Nat Rev Cancer* 2003;3(2):89-101.

[134] Alcalay M, Orleth A, Sebastiani C, Meani N, Chiaradonna F, Casciari C, et al. Common themes in the pathogenesis of acute myeloid leukemia. *Oncogene* 2001;20(40):5680-94.

[135] Speck NA and Gilliland DG. Core-binding factors in haematopoiesis and leukaemia. *Nat Rev Cancer* 2002;2(7):502-13.

[136] Steelman LS, Pohnert SC, Shelton JG, Franklin RA, Bertrand FE and McCubrey JA. JAK/STAT, Raf/MEK/ERK, PI3K/Akt and BCR-ABL in cell cycle progression and leukemogenesis. *Leukemia* 2004;18(2):189-218.

[137] Kuppers R. Mechanisms of B-cell lymphoma pathogenesis. *Nat Rev Cancer* 2005;5(4):251-62.

[138] Sanchez-Beato M, Sanchez-Aguilera A and Piris MA. Cell cycle deregulation in B-cell lymphomas. *Blood* 2003;101(4):1220-35.

[139] Greaves MF and Wiemels J. Origins of chromosome translocations in childhood leukaemia. *Nat Rev Cancer* 2003;3(9):639-49.

[140] Gisselsson D, Pettersson L, Hoglund M, Heidenblad M, Gorunova L, Wiegant J, et al. Chromosomal breakage-fusion-bridge events cause genetic intratumor heterogeneity. *Proc Natl Acad Sci U S A* 2000;97(10):5357-62.

[141] Hanahan D and Weinberg RA. The hallmarks of cancer. *Cell* 2000;100(1):57-70.

[142] Kinzler KW and Vogelstein B. Lessons from hereditary colorectal cancer. *Cell* 1996;87(2):159-70.

[143] Rubin EM, DeRose PB and Cohen C. Comparative image cytometric DNA ploidy of liver cell dysplasia and hepatocellular carcinoma. *Mod Pathol* 1994;7(6):677-80.

[144] Baretton G, Vogt T, Valina C, Schneiderbanger K and Lohrs U. [Prostate cancers and potential precancerous conditions: DNA cytometric investigations and interphase cytogenetics]. *Verh Dtsch Ges Pathol* 1993;77:86-92.

[145] Persons DL, Croughan WS, Borelli KA and Cherian R. Interphase cytogenetics of esophageal adenocarcinoma and precursor lesions. *Cancer Genet Cytogenet* 1998;106(1):11-7.

[146] Dobles M, Liberal V, Scott ML, Benezra R and Sorger PK. Chromosome missegregation and apoptosis in mice lacking the mitotic checkpoint protein Mad2. *Cell* 2000;101(6):635-45.

[147] Kalitsis P, Earle E, Fowler KJ and Choo KH. Bub3 gene disruption in mice reveals essential mitotic spindle checkpoint function during early embryogenesis. *Genes Dev* 2000;14(18):2277-82.

[148] Michel LS, Liberal V, Chatterjee A, Kirchwegger R, Pasche B, Gerald W, et al. MAD2 haplo-insufficiency causes premature anaphase and chromosome instability in mammalian cells. *Nature* 2001;409(6818):355-9.

[149] Gisselsson D, Jonson T, Petersen A, Strombeck B, Dal Cin P, Hoglund M, et al. Telomere dysfunction triggers extensive DNA fragmentation and evolution of complex chromosome abnormalities in human malignant tumors. *Proc Natl Acad Sci U S A* 2001;98(22):12683-8.

[150] Rajagopalan H and Lengauer C. Aneuploidy and cancer. *Nature* 2004;432(7015):338-41.

[151] Duesberg P, Rausch C, Rasnick D and Hehlmann R. Genetic instability of cancer cells is proportional to their degree of aneuploidy. *Proc Natl Acad Sci U S A* 1998;95(23):13692-7.

[152] Lengauer C, Kinzler KW and Vogelstein B. Genetic instability in colorectal cancers. *Nature* 1997;386:623-7.

[153] Li R, Sonik A, Stindl R, Rasnick D and Duesberg P. Aneuploidy vs. gene mutation hypothesis of cancer: recent study claims mutation but is found to support aneuploidy. *Proc Natl Acad Sci U S A* 2000;97(7):3236-41.

In: New Research on Genomic Instability
Editor: Eleanor J. Glascow, pp. 77-98
ISBN: 1-60021-320-0

Chapter III

Spontaneous and Induced Chromosomal Instability in Patients with Carcinoma of Cervix Uteri

Eva Bozsaky[1]*, Ladislava Wsólová[2], Jozef Kállay[3] and Ivan Chalupa[1]

[1] Cancer Research Institute, Slovak Academy of Sciences, Bratislava, Slovak Republic
[2] Slovak Medical University, Bratislava, Slovak Republic
[3] St. Elisabeth's Oncological Institute, Department of Gynecologic Oncology, Bratislava, Slovak Republic

ABSTRACT

A characteristic feature of neoplastic cells is the increased occurrence of various genetic changes ranging from point mutations to chromosomal alterations. Based upon this general observation, it has been suggested that cancer cells are genetically unstable, and that acquisition of genomic instability may be an important prerequisite of tumorigenesis. Defects in basic cellular mechanisms maintaining genomic integrity and guarding against genomic mutations might lead to accumulation of multiple stable mutations in the genome of affected cells. Consequently, genomic instability can be regarded as any condition that predisposes a cell to the accumulation of stable genetic changes at a greater rate than normal. This relationship is well documented in the chromosomal instability syndromes, often characterized by increased incidence of different neoplasias.

Besides the spontaneous occurrence, genomic instability can also be induced or significantly increased by exposure to cytotoxic agents such as ionizing radiation. Recently, evidence has accumulated supporting a relationship between genetic

* Corresponding author: Address: Cancer Research Institute of Slovak Academy of Sciences, Vlárska 7, 83391 Bratislava, Slovak Republic. Telephone number: 421-2-59327510; Fax number: 421-2-59327250; E-mail address: bozsakye@freemail.hu

background and the radiation induced genomic instability phenotype. Therefore, the monitoring of genomic instability is useful from both the oncological and radiobiological points of view. Detection of genomic instability can contribute to the effective screening of early clinical stages of neoplasia and the identification of persons with increased cancer risk.

In the present study, we compared stability of the genome between untreated cervical cancer patients and healthy women according to the spontaneous and γ-ray induced cytogenetic changes in their non-cancer cells (peripheral lymphocytes). Lymphocyte cultures from whole venous blood of patients with cervical neoplasia (stage T1 according to the TNM classification) and healthy female volunteers were cultivated and irradiated *in vitro* with different doses (2, 4 a 6 Gy) using ^{60}Co-gamma source. Slides were prepared using the standard air-drying procedure. The metaphases were analysed for the following structural chromosomal aberrations (CA): chromatid gaps and breaks, isochromatid gaps and breaks, and exchanges (dicentrics, rings, quadriradials, triradials and double minutes).

Generally, a dose-dependent increase of structural changes was observed in healthy women as well as in patients, while statistical analysis has shown significantly higher CA frequencies in patients as compared with healthy women only after γ-ray treatment. As the lymphocytes were cultivated in presence of 5-bromo-2'-deoxyuridine (BrdU) and further stained by the fluorescence-plus Giemsa (FPG) technique, it allowed the identification of metaphases in first, second and third mitotic divisions. Accordingly, the progression of structural chromosomal changes has been also monitored. Moreover, the simultaneous evaluation and quantification of both, CA and sister chromatid exchanges (SCE) in the same metaphases can provide information about the relationship of these two cytogenetic parameters generally accepted as the indicators of chromosomal instability (CIN).

The obtained results indicate differences in the chromosomal stability between healthy women and patients, and suggest the necessity of the mutagen treatment because of the intensification of the differences between individuals with normal and decreased genomic stability.

Keywords: genomic instability, chromosomal aberration, sister chromatid exchanges, γ-irradiation, cervical cancer

INTRODUCTION

The basic prerequisits for the survival and proper functioning of the living organisms are the accurate maintenance and transmission of the genetic instructions coded in DNA, with the highest possible fidelity. The faithful course of these processes is guaranteed by the genomic stability which ensures both cellular homeostasis and genetic continuity in multicellular organisms. The integrity of genetic material is under permanent attack by numerous exogenous and endogenous sources. Consequently, formation of DNA damages is a common event in living cells. However, how cells negotiate or respond to such alterations is not uniform and seems to be genetically determined.

Epidemiological data strongly indicate that certain chemical, physical and biological factors can induce a pronounced increase in the incidence of certain cancers compared with that in the human population generally, but not all persons exposed to the same agents in similar environments develop cancer. The reasons of various susceptibility or resistance to carcinogenesis are not fully understood, but the genetic makeup of individuals conceivably plays a role here. Further fact supporting this suggestion is the existence of childhood cancers. If exposure to carcinogens alone would be the principal cause of carcinogenesis, the children should constitute the group least affected. However, the pediatric oncology is still an active field of medicine (Hsu, 1983).

Evidence of genetic predisposition to neoplasia in human population was described by Lynch et al. (1977) who termed this phenomenon as *cancer family syndromes*. Furthermore, Wurster-Hill et al. (1974, 1979) have observed an increased frequency of structural chromosomal aberrations in families with a high incidence of neoplasias affecting both, cancer patients and their blood relations without evidence of malignancy. The relationship between the high frequency of chromosomal aberrations and the predisposition to cancer is well known in the *chromosomal instability syndromes*. These disorders initially were described as clinical syndromes, independent of the basic research into their mechanisms of action. In the chromosomal instability syndromes both the chromosomal instabilities and the neoplastic outcomes are mechanistically related to abnormalities of DNA metabolism, DNA repair, cell cycle governance, or control of apoptosis (Duker, 2002).

It is now generally accepted that malignant transformation is the result of the stepwise accumulation of multiple genetic alterations and/or epigenetic changes influencing the gene expression patterns (Jones and Baylin, 2002). Neoplastic cells typically possess numerous genomic lesions ranging from point mutations, affecting a specific gene, to chromosomal aberrations, affecting numerous genes within a chromosomal segment. Such relative instability of cancer cells compared to normal cells led Nowell (1976) to suggest that genomic instability may itself contribute to that stepwise progression. He argued that if multiple genomic changes are required for the full development of the cancer phenotype, then an increased rate of acquired alterations would accelerate the development of cancer. As acquisition of genomic instability represents an early step in the process of carcinogenesis, neoplasia may represent a delayed phenotypic manifestation of genomic instability (Cheng and Loeb, 1993).

In normal cells, multiple overlapping metabolic pathways control the overall accuracy of a variety of cellular functions maintaining the cellular homeostasis, including DNA replication, DNA repair, cell cycle progression and cell differentiation. Mutational alteration of the genes encoding proteins involved in these cellular processes could contribute to genomic instability. Loss of normal cellular mechanisms that guard against genomic mutation and the ensuing genomic instability might lead to accumulation of multiple stable mutations in the genome of affected cells, resulting in neoplastic transformation. Consequently, genomic instability can be regarded as any condition that predisposes a cell to the accumulation of stable genetic changes at a greater rate than normal (Coleman and Tsongalis, 1995).

In the last decades the efforts to identify cancer susceptibility genes were made. Genes involved in maintaining the integrity of the genome – thereby the genomic stability – were shown to be responsible for inherited predisposition to cancer by indirect mechanism. More recently, the concept of 'gatekeepers and caretakers' have added complexity to the understanding of genetic basis of cancer and the neoplastic pathway(s), completing the classical Knudson's 'two hit' hypothesis (1971). According to Kinzler and Vogelstein's model (1997) gatekeepers are genes that directly regulate the growth of tumors by inhibiting growth or promoting death, whereas caretaker genes are involved in the maintenance of genome stability. Each cell type has only one (or a few) gatekeepers, and inactivation of a given gatekeeper leads to a very specific tissue distribution of cancer. In contrast, inactivation of a caretaker gene does not promote tumor initiation directly. Rather, it leads to genetic instabilities which in turn result in increased mutation rates of all genes, including gatekeepers. Therefore, the chance of a mutation at a gatekeeper gene is increased by the presence of a mutation in a caretaker gene. Once tumor initiation occurs due to inactivation of a gatekeeper gene, the tumor may progress rapidly because of the accelerated rate of mutations in other genes (Kinzler and Vogelstein, 1997).

Manifestation Levels of Genomic Instability

Two main forms of genomic instability are associated with tumor cells. The mutational instability (MIN) phenotype is characterized by mutations at the nucleotide level resulting in base substitutions or deletions or insertions of a few nucleotides. This type of instability is usually connected to mismatch repair defects leading to the accumulation of point mutations in the DNA sequence (e.g. microsatellite instability in hereditary nonpolyposis colorectal cancer). These are not detectable through cytogenetic analysis as they display an apparently normal karyotype (i.e. normal number and structure of chromosomes). The vast majority of tumors exhibit the instability at the chromosome level – chromosomal instability (CIN) – manifesting as the gross structural rearrangement(s) and/or numerical changes of chromosomes. Common chromosomal aberrations include the loss or gain of whole chromosomes or chromosome fragments, and the amplification of chromosome segments (Lengauer et al., 1998; Coleman and Tsongalis, 1999; van Gent et al., 2001; Charames and Bapat, 2003; Sieber et al., 2003). These changes can lead to the inactivation of tumor suppressor genes (for example, by loss of heterozygosity) or to the promotion of tumorigenesis by the activation of protooncogenes (Lengauer et al., 1998) or by the induction of multidrug resistance after cytostatic drug treatment (Ramachandran and Melnick, 1999). A different type of gross chromosomal aberration that is often observed in tumors is the translocation, in which chromosome segments are exchanged. The complex type of translocation can result in gains and losses of chromosomal material, as well as in the deregulation of gene expression or the generation of new fusion gene products that then acquire oncogenic potential (Lengauer et al., 1998; van Gent et al., 2001). The mechanism by which these translocations are produced has remained obscure for many years, although the prevailing idea is that double-strand breaks (DSBs) on the two participating chromosomes might be the initiating lesions (van Gent et al., 2001).

Possible Molecular Mechanisms Governing DNA Damages and Genomic Instability

As described above, the genomic instability driving tumorigenesis is 'supplied' by DNA damage and by errors made by the DNA machinery. The fragile and vulnerable structure of DNA molecule does not guarantee life-long stability or proper function of the genes. Cellular DNA is continually subjected to a plethora of environmental insults or intrinsic cellular processes causing a wide variety of lesions. Environmental agents such as the ultraviolet (UV) light, ionizing radiation and numerous genotoxic chemicals induce alterations in DNA structure, which if left unrepaired, may lead to mutations that enhance cancer risk. Endogenous sources include oxidative free radicals generated during essential metabolic processes, inappropriate DNA methylation, and aberrant telomeres. Some chemical bonds in DNA tend to spontaneously disintegrate under physiological conditions. In addition, DNA breaks can be induced during the process of mobile element transposition (Friedberg et al., 1995; Hoeijmakers, 2001; Duker, 2002). Accordingly, it is not surprising that during the evolution an intricate network of sophisticated, interwoven DNA-repair systems has developed. The various features of DNA damages require different ways to counteract their deleterious effects. Repair mechanisms splicing out the damaged regions and inserting new bases require the intact complementary strand as a template to fill the gap, followed by ligation (i.e. nucleotide excision repair, NER; base excision repair, BER; mismatch repair, MMR). In another group of DNA lesions both DNA strands are damaged (e.g. DSBs), which prevents to use of the complementary DNA strand as a template for repair. DSBs are a very serious form of DNA damage which must be efficiently repaired for cell survival. In eukaryotic cells, two main DSB repair pathways have been identified that differ in their requirements for DNA homology: homologous recombination (HR) and non-homologous end joining (NHEJ). HR seems to be preferred after replication (S or G2 phase), when a second identical DNA copy (e.g. a homologous chromosome or a sister chromatid) is available to align the breaks. Otherwise, the cells rely on less-accurate and more error-prone NHEJ, which is more relevant in G1 phase of the cell cycle, when the second copy of DNA is not available (Takata et al., 1998).

Global DNA repair capability might be influenced by the time that is allowed for repair during the cell cycle. This is coordinated by several checkpoints which prevent cells from starting DNA replication (G1/S checkpoint), from progressing with replication (intra S checkpoint) or from going into mitosis (G2/M checkpoint), if they contain damaged DNA (Skorski, 2002). These checkpoints control both, the length of the cell cycle phase and the activation of DNA repair pathways and movement of DNA repair proteins to sites of DNA damage. For the maintenance of genomic stability, the connection and faithful cooperation between DNA repair pathways and cell cycle progression is crucial. Molecular alteration(s) of genes that encode proteins important in these processes results in dysfunctional repair or/and checkpoint pathway(s), consequently causing genetic disorders known as 'DNA repair deficiency syndromes'. As such situation can predispose cells to the accumulation of mutations, patients exhibit increased sensitivity to certain deleterious factors. This phenomenon is associated with an increased incidence of cancer and elevated occurrence of certain chromosomal abnormalities (therefore termed also as 'chromosomal instability syndromes', e.g. xeroderma pigmentosum, ataxia telangiectasia, Bloom syndrome, Fanconi anemia, etc.).

Regarding the cytogenetic manifestation of genomic instability (CIN), DSBs are of special importance, being the potent inducers of chromosomal aberrations (CA). If left unrepaired, DSBs may lead to broken chromosomes. Unrepaired DSBs and CAs (that arise by misrepair of DSBs) may be cell lethal at interphase or mitosis. If repaired improperly, DSBs may lead to mutations, chromosome rearrangements, and oncogenic transformation (Obe et al., 2002). DSB, a particularly genotoxic form of DNA damage can be caused by many different factors. Naturally occurring reactive oxygen molecules, ionizing radiation and certain chemicals are prevalent sources of such damages. They can also arise as a result of replication of single-strand breaks. Finally, DSBs can be formed during several essential cellular processes, such as recombination in meiosis, recombination in the T cell receptor and immunoglobulin genes in T and B cells (van Gent et al., 2001; Eyfjord and Bodvarsdottir, 2005).

Cytogenetic Manifestation of Spontaneous and Induced Genomic Instability

The most characteristic cytogenetic endpoints reflecting damaged stability of the genome are chromosomal aberrations (CA), micronuclei (MN) and sister chromatid exchanges (SCE). Accordingly, these parameters are widely used as reliable indicators of chromosomal instability.

Spontaneous and Radiation Induced Chromosomal Aberrations (CA)

Epidemiological studies have shown that people with elevated frequencies of CA in their peripheral blood lymphocytes have a significantly elevated risk of developing cancer (Hagmar, 1998; Bonassi, 2000). Many types of cancers are associated with specific types of CA which are etiologic for the cancer in question (Mitelman et al., 1997). In addition, CAs have long been recognised to be an important biomarker of human exposure to ionizing radiation and genotoxic chemicals (Ramalho et al., 1988; Natarajan et al, 1996; Maffei et al, 2004).

Regarding the induction of chromosomal aberrations, the most efficient inducers are ionizing radiation (IR) and certain chemical agents. Different mechanisms of CA formation induced by these two different deleterious sources were observed, which led Evans (1968) to suggest, that 'aberrations are something that the cell makes and not something that the radiation makes'. IR induces chromosome type aberrations when affecting the cells in G1 phase of cell cycle and chromatid type aberrations in S and G2 phases. Radiomimetic chemicals, such as bleomycin have a similar effect. However, chemical mutagens, which do not induce directly DNA strand breaks but cause other lesions, were shown to induce only chromatid type aberrations irrespective of the DNA synthetic stage treated. An intervening S phase is necessary to visualise the aberrations (Evans and Scott, 1969). Accordingly, chromosome breaking agents were classified as S-dependent (alkylating agents, UV) and – independent (IR, radiomimetic chemicals) agents.

Cellular exposure to IR results in a variety of directly and indirectly induced DNA lesions, including DNA base alterations, DNA-DNA and DNA-protein crosslinks, and single- and double-strand breaks (Ward, 1988). The cells react rapidly to irradiation, evoking a stress response that includes a plethora of biological responses (e.g. cell cycle specific growth arrest, repair of damaged DNA). These early events are likely to be preconditions and determinants of the later fate of the irradiated cells (i.e. whether a cell will necrose, senesce, apoptose or ultimately survive and proliferate). If a cell does survive, the initial biological response to the radiation induced insult may influence whether the cell participates in normal differentiation, exhibits a limited lifespan or proliferates and begins to acquire – via genomic instability – characteristics associated with the neoplastic transformation of a normal cell to a cancerous cell (Morgan et al., 1996).

Ionizing radiation can induce two types of chromosomal aberrations in mammalian cells. Unstable aberrations include dicentric chromosomes, ring chromosomes, large deletions, and fragments, and are usually lethal to dividing cells. These types of aberrations do not allow the equal distribution of genetic material into daughter cells. In many cases, the frequency of such aberrations correlates well with the cytotoxic effects of radiation. Stable aberrations include changes such as small deletions, reciprocal translocations, and aneuploidy. These aberrations do not preclude the cell from dividing and proliferating (Little, 2003).

Nowadays, there is much excitement about the non-targeted effects of IR, the radiation-induced genomic instability (RIGI) and the radiation-induced bystander effects (BSE). There is considerable evidence that cells that themselves are not exposed to ionizing radiation but are the progeny of cells irradiated many cell divisions previously may express a high frequency of gene mutations, chromosomal aberrations and cell death (these effects are known as RIGI). As these effects can be observed days, weeks, months, even years after exposure of the original progenitor cell to radiation, it has been suggested that radiation initiates a process within a cell that is perpetuated in the progeny of that irradiated cell (Morgan et al., 1996; Kaplan et al., 1997; Morgan et al., 2002). The radiation-induced BSE appears in non-irradiated cells exhibiting cellular responses typically associated with direct radiation exposure. This effect occurs as a consequence of contact with irradiated cells or by receiving soluble signals from irradiated cells (Morgan et al., 2002; Lorimore and Wright, 2003).

The biological effect of IR depends on several factors, e.g. type of radiation, temporal period of exposure (single dose, fractionated dose, or chronic exposure), energy and quality of absorbed radiation, size of irradiated area, radiosensitivity of irradiated cells, etc. Final effect reflecting the biological cell response can be influenced by individual genetic background, genomic stability status. It is well established that there are wide variations in the inherent radiation sensitivity (Limoli et al., 2001). In particular, chromosomal destabilization could have a dramatic effect on the likelihood that tumor cells will develop radioresistance. It can arise by several possible mechanisms: differential cellular responses to DSBs, especially the regulation of cell cycle checkpoints and/or differences in DSB repair can contribute to radioresistant phenotype (Morgan and Murnane, 1995). Cytogenetic and DNA DSB rejoining assays suggest that radiation sensitivity is a function of DNA repair efficiency: radioresistant cells rejoin DSBs at a faster rate than the more sensitive cells (Schwartz et al., 1988; 1991). Variations in radiosensitivity are an important phenomenon in point of view of response to radiation therapy (West, 1994).

Structural or numerical CA – lagging chromosome fragments and whole chromosomes – can be manifested after the division of aberrant cell in form of *micronuclei (MN)*. MN derived from acentric chromosome fragment can be distinguished from those derived from whole chromosome by fluorescent in situ hybridization (FISH) using a pancentromeric probe, which will detect the centromere in the MN. Thus, clastogenic and aneugenic events can be detected by the MN technique. It is used as a routine method in *in vitro* studies in genotoxicology as well as in population monitoring (e.g. to estimate radiation effects) (Natarajan, 2002).

Sister Chromtid Exchanges (SCEs)

The second cytogenetic parameter reflecting genomic instability is SCE. It is a cytological manifestation of DNA breakage and rejoining at apparently homologous sites on the two chromatids of a single chromosome (Natarajan, 1993). Since the early study of McClintock (1938), describing the evidence for formation of SCEs on ring chromosomes in maize, there were several milestones in history of SCE. Methods of SCE visualization were continually modernized according to the just available technical possibilities (Taylor, 1958; Latt, 1974; Kato, 1974; Perry and Wolf, 1974; Becher and Sandberg, 1982). Regarding the mechanism of SCE origin, more theories were formulated (Kato, 1977; Shafer, 1977; Ishii and Bender, 1980; Painter, 1980; Cleaver, 1981; Lindenhahn and Schubert, 1983). Although, the exact molecular mechanisms of SCE formation are still not completely understood, SCE analysis is one of the most sensitive cytogenotoxic assays and has been widely used for the detection of mutagenic and carcinogenic potential of many environmental agents on the chromosomal level (Tucker et al., 1993). Though, there is obviously a role for DNA damage in the induction of SCE, attempts to correlate their production with known DNA-repair systems have failed (Palitti and Becchetti, 1977; De Weerd-Kastelein et al., 1977) or are highly controversial (Sasaki, 1982).

Altered levels of SCE were reported in patients with various types of neoplasias (Mitra et al., 1982; Livingston et al., 1983; Dosaka et al., 1987; Murty et al., 1987; Adhvaryu et al., 1985, 1988a, b, c; Li et al., 1989; Lukovic and Milasin, 1992; Cheng et al., 1995; Dhillon et al., 1996; Dhillon and Dhillon, 1998; Cortés-Gutiérrez et al., 2000; Bozsakyová et al., 2005). The close association between initial steps in cell transformation and induction of DNA alterations was described to make SCE a relevant indicator of the neoplastic transformation processes, illustrated by the capacity of several carcinogens to affect the SCE rate (Popescu et al., 1981). Furthermore, the SCE frequency was suggested to serve as a preclinical marker for early detection of cancer (Skolnick et al., 1980; Mitra et al., 1982).

THE AIM OF STUDY

The aim of present study was to investigate

- whether there is any difference between the genomic stability status of patients with carcinoma of cervix uteri (the second most common malignant neoplasia in women worldwide) and healthy persons on the chromosomal level,
- whether there is any difference in cellular response ('measured' by cytogenetic parameters) to mutagen (ionizing radiation) treatment between cancer patients and healthy individuals.

In order to answer these questions, the spontaneous and γ-ray induced chromosomal changes were evaluated in non-cancer cells (peripheral lymphocytes) focussing on the two main cytogenetic endpoints – chromosomal aberrations and sister chromatid exchanges – widely used as reliable indicators of genomic instability.

MATERIAL AND METHODS

Subjects

The subjects of this study were 10 patients with carcinoma of the cervix uteri classified according to the UICC's TNM classification criteria (Sobin and Wittekind, 1997) two cases belonging to stage Tis (tumor *in situ*) and eight cases to stage T1. All of them were untreated at the time of examination.

For comparison, 10 healthy female nonsmoking volunteers were studied under the same experimental conditions. Controls and patients had no history of chronic infection, drug intake, or irradiation. Healthy women and cancer cases were from 33 to 50 and 27 to 59 years old with a mean age of 39 and 39.6 years, respectively.

In order to examine non-cancer cells, heparinized blood samples were obtained by venipuncture from patients (before any anticancer therapy) and healthy women, as well.

Cell Cultures

Lymphocyte cultures were set up from each sample by adding 0.4ml of whole heparinized venous blood to 4.6ml of Eagle's Minimum Essential Medium containing 20% foetal bovine serum (FBS, SEBAK GmbH, Aidenbach, Germany), 2% phytohaemagglutinin (Murex Diagnostics Ltd., Dartford, England), 294.1 μg/ml L-glutamine (DIFCO, Detroid, USA) and antibiotics (100 μg/ml streptomycin and 80 μg/ml kanamycin). Lymphocytes were incubated at 37°C in a humidified atmosphere of 95% air and 5% CO_2.

After 24h cultivation the cultures were exposed to the different doses of γ-radiation. To achieve differential staining of sister chromatids in order to distinguish mitosis of three consecutive cell divisions, immediately after irradiation, 5-bromo-2'-deoxyuridine (BrdU) (Serva, Heidelberg, Germany) at a final concentration of 5μg/ml was added. The cultures were cultivated further 48h in the dark and at 2h prior to harvesting, colchicine (0.75μg/ml) was added . Duplicate cultures were prepared from each treated samples.

Irradiation Conditions

Samples were irradiated using a ^{60}Co gamma source (Chisobalt, Chirana, Praha, Czech Republic) located at St. Elizabeth's Oncological Institute in Bratislava. The doses of γ-irradiation used – 2, 4 and 6 Gy – were selected on the basis of results of a preliminary cytotoxicity study. Dose determinations were made by the experts of the Department of Radiophysics of the hospital after the calibration of the apparatus.

Cytogenetic Analyses

Slides were prepared using the standard air drying procedure. In order to visualize differential staining of chromosomes, slides were stained according to the modification of the fluorescence-plus Giemsa technique (Blaško and Chalupa, 1984) of Becher and Sandberg (1982). Briefly, five days old preparations were stained in 0.5 μg/ml Hoechst 33258 for 15 min. Then slides mounted with Sorensen's solution (pH 8.0) heated to 50°C were exposed to UV light (2x500W) for 30 min. Slides were well rinsed with distilled water, dried, and subsequently stained with 2% Giemsa stain.

Whenever possible, 100 well-spread metaphases per dose were analysed in each sample. The stage of division (first, second and third cell division; M1, M2 and M3, respectively) was determined in evaluated metaphases according to the staining pattern of chromosomes. In the metaphases the number of chromosomes and the presence of structural chromosomal aberrations were assessed. The classification of aberrations was carried out as described by Venitt and Parry (1984) and in the International System for Cytogenetic Nomenclature (ISCN) (1995). The metaphases were analysed for the following structural aberrations: chromatid gaps and breaks; isochromatid gaps and breaks; and exchanges (dicentrics, rings, quadriradials, triradials and double minutes). In cells of the second mitotic division, the number of SCEs was evaluated simultaneously. Also, the number of SCE per chromosome were calculated.

Statistical Analysis

The statistical analyses were realized using the SPSS software (versions 6.0.1 and 8.0).

Differences in frequency of aberrant metaphases between the group of patients and the group of healthy women as well as between the applied radiation doses (0, 2, 4 and 6 Gy) were analysed by the chi-square test. Total numbers of chromosomal aberrations between the group of patients and healthy women as well as between applied radiation doses were compared by Mann-Whitney test. The overall difference in frequencies of chromosomal aberrations between healthy women and patients, between untreated and variously irradiated samples, between different cell divisions, and the interaction between these factors, were tested by multifactor analysis of variance (MANOVA).

Differences in mean SCE frequencies (SCE/cell and SCE/chromosome) between aberrant and non-aberrant metaphases were analysed in group of healthy women as well as in patients by the two-tailed Student's *t*-test at each different exposure, separately.

RESULTS

We investigated spontaneous and γ-ray induced chromosomal instability in low-stage cervical neoplasia cases. Cytogenetic endpoints were evaluated in their *in vitro* cultivated and irradiated normal cells – peripheral blood lymphocytes, and compared to those in group of age-matched healthy donors.

The frequencies of metaphases with structural chromosomal aberrations (CA) (i.e. aberrant metaphases) and the total numbers of structural CA are presented in Figures 1 and 2, respectively. Generally, the dose-dependent increase of both parameters was observed in healthy women as well as in patients. Frequencies of both, aberrant metaphases and total numbers of CA were compared between each applied radiation dose. The statistical analyses of both parameters have shown highly significant differences ($p<0.001$) between doses in group of healthy women as well as of patients. This corresponds with known clastogenic effect of ionizing radiation (IR), which is a clear consequence of its DSB causing effect. The phenomenon of dose-related increase of IR-induced CA provides to use the CA assay as biological dosimeter in the case of radiation accidents (Ramalho et al., 1988; Natarajan et al., 1996). Comparing frequencies of aberrant metaphases and total numbers of CA between healthy women and patients, the higher γ-ray induced levels were found always in group of patients (statistical significances are indicated in Figures 1 and 2). These findings suggest elevated predisposition to chromosomal destabilization in response to IR in case of patients in comparison to healthy women. As the non-cancer cells were subjected to the analyses, the results reflect the genetic makeup, degree of genomic stability resp. instability, generally. Differences between healthy donors and patients indicate their various, genetically determined radiation sensitivity. This is further supported by the frequency distributions of the CA ranges listed in Table 1. The data show that in case of patients lymphocytes with higher induced CA frequencies prevailed, whilst in case of healthy women, lymphocytes with lower CA frequencies were seen. Increased chromosomal radiosensitivity in *in vitro* cultivated non-cancer cells was observed in different cancers (Scott et al., 1999; Terzoudi et al., 2000; Baeyens et al., 2002), moreover in cells of blood relations of cancer patients (Knight et al., 1993; Roberts et al., 1999). As a cause of higher degree of radiosensitivity, mutations in genes important in cellular response to radiation induced damages were suggested. Genetic susceptibility, which could result from defective repair mechanisms (Müller-Vogt et al., 2003), cell cycle control, or apoptosis was shown to be associated with multiple cancers (Keller et al., 2005). An accumulation of defective repair or control processes is finally manifested as a high degree of cytogenetic instability.

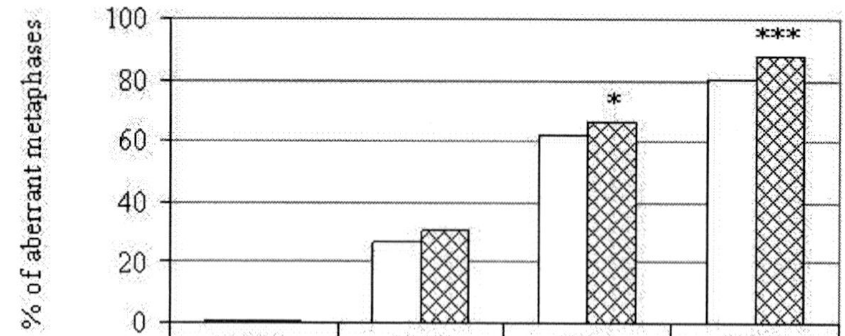

Dose	0 Gy	2 Gy	4 Gy	6 Gy
□ healthy women	0,7	26,5	62,1	80,6
⊠ patients	0,4	30,3	66,7	87,9

* - statistically different from the healthy women ($0.01 < p < 0.05$)
*** - statistically different from the healthy women ($p < 0.001$)

Figure 1. % of metaphases with structural chromosomal aberration(s) in lymphocytes of healthy women and patients with carcinoma of cervix uteri after in vitro exposure to different doses of g-irradiation.

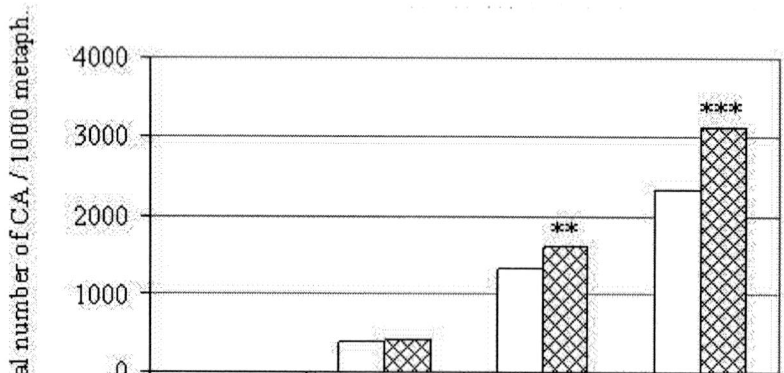

Dose	0 Gy	2 Gy	4 Gy	6 Gy
□ healthy women	8	389	1321	2335,4
⊠ patients	4	419	1592,8	3120,6

** - statistically different from the healthy women ($0.001 < p < 0.01$)
*** - statistically different from the healthy women ($p < 0.001$)

Figure 2. Total number of structural chromosomal aberrations (CA) in lymphocytes of healthy women and patients with carcinoma of cervix uteri after in vitro exposure to different doses of g-irradiation.

Table 1. Frequency distribution of the range of chromosomal aberrations in untreated and in vitro γ-irradiated lymphocytes of healthy women and patients with carcinoma of cervix uteri

Dose [Gy]	Group of women	% of metaphases with the indicated range of CA / cell					
		0	1 - 3	4 - 6	7 - 9	10 - 12	> 13
0	H	99.3	0.7				
	P	99.6	0.4				
2	H	73.5	25.3	1.0	0.2		
	P	69.7	29.7	0.6			
4	H	37.9	53.3	8.4	0.4		
	P	33.3	54.7	10.0	1.4	0.1	0.3
6	H	19.4	55.5	19.7	4.1	0.4	
	P	12.1	51.7	27.0	7.0	1.6	0.4

CA, chromosomal aberration; H, healthy women; P, patients

Table 2. Spontaneous and γ-ray induced chromosomal aberrations in *in vitro* cultuvated lymphocytes of healthy women and cervical cancer patients

Dose [Gy]	Group of women	No. of evaluated metaph.	Number of structural chromosomal aberrations								Total no. of CA	*p*
			Chromatid		Isochromatid		Exchange					
			g	b/f	g	b/f	dic	r	qr	tr		
0	H	1000	18	3	14	3	2	-	-	-	8	0.364
	P	1000	11	4	2	-	-	-	-	-	4	
2	H	1000	60	58	31	101	190	37	1	2	389	0.074
	P	1000	9	24	6	118	238	37	1	1	419	
4	H	1000	49	146	61	342	680	139	2	12	1321	0.002
	P	975	23	77	15	583	790	86	8	9	1553**	
6	H	981	69	152	91	668	1242	201	15	13	2291	< 0.001
	P	929	25	196	21	1049	1445	185	8	16	2899***	

H, healthy women; P, patients
g, gap; b/f, break and/or fragment; dic, dicentric; r, ring; qr, quadriradial; tr, triradial
** significantly different from the healthy women ($0.001 < p < 0.01$)
*** significantly different from the healthy women ($p < 0.001$)

Frequencies of various type of structural CAs – chromatid and isochromatid gaps and breaks and exchange-type aberrations (dicentrics, rings, quadriradials and triradials) – are presented in Table 2 in detail. From the table is obvious the prevalence of dicentrics and isochromatid breaks in all irradiated cases. This effect was described in lymphocytes of hospital workers occupationally exposed to low doses of IR (Paz-y-Mino et al., 1995; Maffei et al., 2004). The higher occurrence of isochromatid aberrations in comparison to chromatid aberrations generally, refers to the role of timing factor at the cell treatment. The stage of cell cycle at the treatment and the duration of further cultivation are of crucial importance regarding the final cytogenetic manifestation of IR effect. In our experiment, the aim of 24h pre-cultivation of phytohaemagglutinin-stimulated lymphocytes prior to irradiation was to achieve more or less synchronous start of cell division. The prevalence of isochromatid aberrations indicates the formation of most damages in G1 phase of cell cycle arising as aberrations affecting both chromatids after cell division. The lower portion of chromatid aberrations reflects presence of small cell fraction in S- or G2 phase at the treatment.

Table 3. Frequencies of chromosomal aberrations in three consecutive cell divisions in untreated and in vitro γ-irradiated lymphocytes of healthy women and patients with carcinoma of cervix uteri

Dose [Gy]	Group of women	Total number of CA / 100 metaphases in the indicated cell division		
		M1	M2	M3
0	H	0.8	0.9	0.6
	P	0	0.5	0.6
2	H	49.6	29.0	46.1
	P	51.1	38.6	34.9
4	H	147.4	103.6	141.1
	P	174.4	145.0	149.2
6	H	246.0	175.7	259.3
	P	313.1	302.0	331.6

CA, chromosomal aberration; H, healthy women; P, patients
M1, metaphases in first cell division; M2, metaphases in second cell division, M3, metaphases in third cell division

The incorporation of 5-bromo-2'-deoxyuridine (BrdU) during *in vitro* cultivation and its visualization by FPG technique providing differential staining pattern of chromosomes enabled to identify three consecutive cell divisions. In the cells of first mitotic division (M1) both chromatids were uniformly stained with Giemsa. In the second cell division (M2), one of the two chromatids stained lightly with Giemsa because of bifiliar substitution with BrdU in one chromatid and unifiliar substitution in the other. In the third division (M3) three-quarters of the chromatids remained unstained as a result of complete substitution with BrdU (Perry, 1983; Adhvaryu et al., 1985). According to this pattern, the analysis of structural CA in certain cell divisions (M1, M2, M3) was performed. Data in Table 3 display the dynamics of structural chromosomal changes through three consecutive cell divisions. Concerning the overall statistical characterization of CA frequencies in different mitotic divisions after different irradiation doses, MANOVA analysis has shown significant differences between healthy women and patients, between various exposures (0, 2, 4 and 6 Gy), and between certain cell divisions (M1, M2 and M3). Moreover, the interactions between these factors (group of women, radiation dose and cell division) significantly influenced the differences in CA frequencies. The data indicate that the analysis of first-division metaphases is important for determination of IR effect as it was described also at the monitoring of individuals exposed to radiation (Paz-y-Mino et al., 1995; Maffei et al., 2004). Accordingly, international cytogenetic protocols for testing the chromosome damaging effects of potential mutagens and carcinogens call for analysis of metaphase chromosomes from the first mitosis after exposure. Slightly higher levels of CA in M3 cells in comparison to M1 cells after 6 Gy irradiation can be explained by dramatically affected cell cycle checkpoints and abortive cell divisions as a consequence of a high radiation dose, and/or by the accumulation of unrepaired stably persisting aberrations.

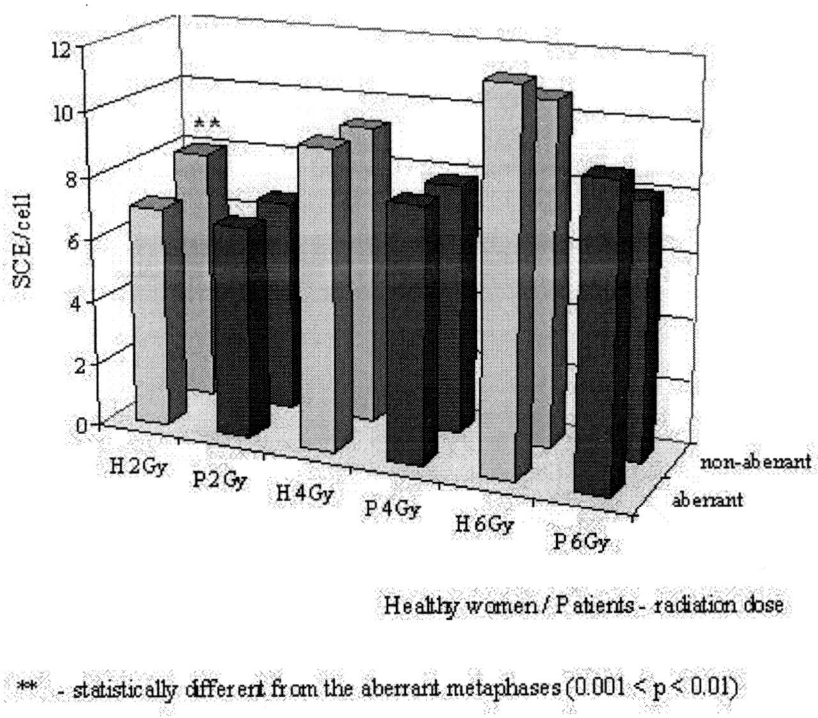

Figure 3. SCE/cell in metaphases with and without chromosomal aberrations in in vitro cultivated and irradiated lymphocytes of healthy women and patients with carcinoma of cervix uteri.

The differential staining pattern of chromosomes further allowed us the simultaneous evaluation of CA and SCE in metaphases of second cell division (M2). As the exact mechanisms of origin of both cytogenetic endpoints are still not clarified, and the possible connection between them remained doubtful, the comparison of SCE levels between aberrant and non-aberrant metaphases could provide useful information. In Figures 3 and 4 are illustrated frequencies of SCE/cell and SCE/chromosome in cells with and without structural CA. Because of the low number of spontaneously occurring aberrant metaphases, non-irradiated cases were not included into the statistical analysis. As indicated in the figures, statistically significant differences were found only in several cases. After exposure to 2 Gy dose of γ-radiation, the SCE/cell were higher in metaphases without CA, while in case of 4 Gy and 6 Gy irradiation doses these values were higher in aberrant metaphases. Such discrepancies could be caused by the varying chromosome numbers observed mainly after irradiation with higher doses (data not shown). Therefore, to obtain an reliable parameter, rather the chromosome than the cell seems to be useful as a proportional unit of cytogenetic event. Frequencies of SCE/chromosome were higher in non-aberrant cells than in aberrant ones (except the group of patients after 6 Gy irradiation). Despite the absence of statistical significance in some cases, the higher SCE/chromosome values in non-aberrant cells can suggest the relationship between CA and SCE on the molecular level of cellular response. As described above, the phenomenon of genomic stability and instability is a complex network of properly functioning or damaged cellular processes, respectively. One of these processes is the repair machinery. Probable relationship between repair and SCE has already been discussed in earlier studies (Swift, 1977; Latt and Shreck, 1980), though there is no proof of its connection. Recently, the relationship between mechanisms of SCE formation and homologous recombination (a major DSB repair pathway) was described (Sonoda et al., 1999). Furthermore, Johnson and Jasin (2000) indicated that a common mechanism for both chromatid and chromosome mitotic recombination may provide a safeguard against genomic alterations. Our present results together with earlier findings regarding SCE (Bozsakyová et al., 2005) seem to support this assumption, though being aware that the results obtained only by classical cytogenetic analyses are not sufficient to draw definite conclusion.

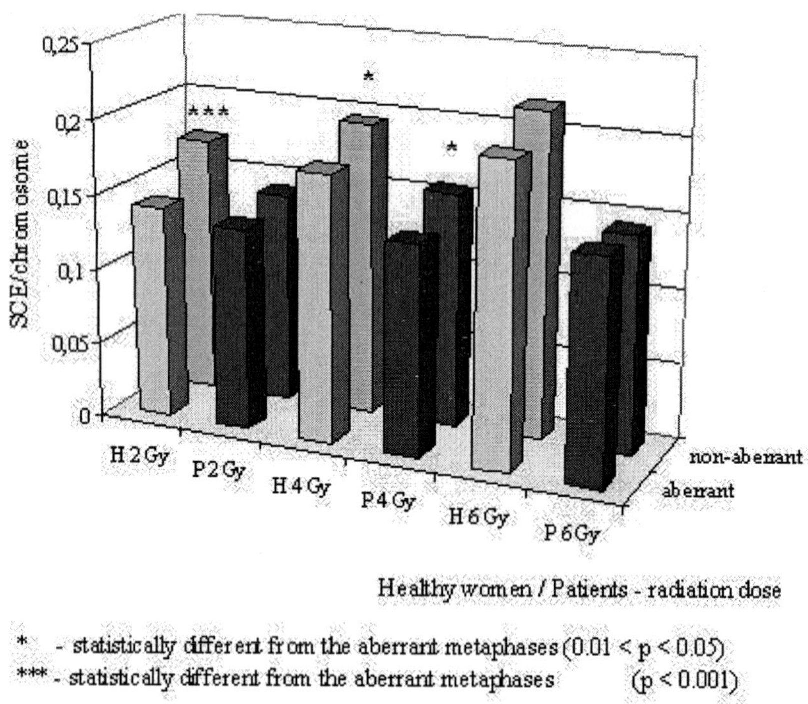

Figure 4. SCE/chromosome in metaphases with and without chromosomal aberrations in in vitro cultivated and irradiated lymphocytes of healthy women and patients with carcinoma of cervix uteri.

CONCLUSION

The most important aim of the present study and related works was to investigate the degree of genomic stability respectively instability in cervical cancer patients in comparison to healthy women. We observed the cytogenetic manifestation of spontaneous and γ-ray induced changes in *in vitro* cultivated and irradiated lymphocytes. The higher values of chromosomal aberrations in cells of patients – especially after mutagen treatment – suggest their higher predisposed vulnerability. As it is described by several authors, such differences can be caused by decreased efficiency of any process of cellular response pathways ensuring the maintenance of cellular homeostasis. According to our results, the cytogenetic evaluation of easily obtainable cells, peripheral blood lymphocytes, may serve as a useful parameter at the identification of persons with increased cancer risk.

ACKNOWLEDGEMENTS

The authors thank to Dr. E. Bolješíková (Department of Radiophysics, St. Elizabeth's Oncological Institute, Bratislava) for providing the use of Chisobalt equipment, and Dr. G. Králik and the staff of the Department of Radiophysics for help with dosimetry and radiation treatment. The authors are grateful to Miss L. Šebová for invaluable technical assistance.
This work was supported by the grant of the Slovak Grant Agency of SAS VEGA No.2/7137/27.

REFERENCES

Adhvaryu, SG; Vyas, RC; Dave, BJ; Trivedi, AH; Parikh, BN. Spontaneous and induced sister chromatid exchange frequencies and cell cycle progression in lymphocytes of patients with carcinoma of the uterine cervix. *Cancer Genet., Cytogenet.,* 1985 14, 67-72.

Adhvaryu, SG; Rawal, UM; Patel, JV; Patel, DD; Balar, DB. Increased frequency of sister chromatid exchanges in lymphocytes of breast cancer patients. *Int. J. Cancer,* 1988a 41, 394-398.

Adhvaryu, SG; Rawal, UM; Patel, JV. Elevated sister chromatid exchange frequencies in the lymphocytes of esophageal cancer patients. *Cancer,* 1988b 61, 1867-1871.

Adhvaryu, SG; Rawal, UM; Patel, JV. Enhancement of lymphocytic SCE frequencies in patients with ovarian cancer. *Neoplasma,* 1988c 35, 103-108.

Baeyens, A; Thierens, H; Claes, K; Poppe, B; Messiaen, L; De Ridder, L; Vral, A. Chromosomal radiosensitivity in breast cancer patients with a known or putative genetic predisposition. *Br. J. Cancer,* 2002 87, 1379-1385.

Becher, R; Sandberg, AA. Rapid method for Giemsa staining of sister chromatids. *Cancer Genet. Cytogenet.,* 1982 7, 223-230.

Blaško, M; Chalupa, I. A modified method for differential staining of sister chromatid exchanges by the Giemsa stain. (in Slovak). *Biológia,* 1984 39, 1115-1119.

Bonassi, S; Hagmar, L; Stromberg, U; Montagud, AH; Tinnerberg, H; Forni, A; Heikkila, P; Wanders, S; Wilhardt, P; Hansteen, IL; Knudsen, LE; Norppa, H. Chromosomal aberrations in lymphocytes predict human cancer independently of exposure to carcinogens. European Study Group on Cytogenetic Biomarkers and Health. *Cancer Res.,* 2000 60, 1619-1625.

Bozsakyová, E; Wsólová, L; Chalupa I. Spontaneous and γ-ray-induced sister chromatid exchanges in patients with carcinoma of cervix uteri. *Int. J. Radiat. Biol.,* 2005 81, 177-185.

Charames, GS; Bapat, B. Genomic instability in cancer. *Curr. Mol. Med.,* 2003 3, 589-596.

Cheng, KC; Loeb, LA. Genomic instability and tumor progression: mechanistic considerations. *Advances in Cancer Research,* 1993, 60, 121-156.

Cheng, TJ; Christiani, DC; Wiencke, JK; Wain, JC; Xu, X; Kelsey, KT. Comparison of sister chromatid exchange frequency in peripheral lymphocytes in lung cancer cases and controls. *Mutat. Res.,* 1995 348, 75-82.

Cleaver, JE. Correlations between sister chromatid exchange frequencies and replicon sizes. *Exp. Cell Res.,* 1981 136, 27-30.

Coleman, BC; Tsongalis, GJ. Multiple mechanisms account for genomic instability and molecular mutation in neoplastic transformation. *Clin.Chem.,* 1995 41, 644-657.

Coleman, WB; Tsongalis, GJ. The role of genomic instability in human carcinogenesis. *Anticancer Res.,* 1999 19, 4645-4664.

Cortés-Gutiérrez, EI; Cerda-Flores, RM; Leal-Garza, CH. Sister chromatid exchanges in peripheral lymphocytes from women with carcinoma of the uterine cervix. *Cancer Genet. Cytogenet.,* 2000 122, 121-123.

De Weerd-Kastelein, EA; Keijzer, W; Rainaldi, G; Bootsma, D. Induction of sister-chromatid exchanges in xeroderma pigmentosum cells after exposure to ultraviolet light. *Mutat Res,* 1977 45, 253-261.

Dhillon, VS; Kler, RS; Dhillon, IK. Chromosome instability and sister chromatid exchange (SCE) studies in patients with carcinoma of cervix uteri. *Cancer. Genet. Cytogenet.,* 1996 86, 54-57.

Dhillon, VS; Dhillon, IK. Chromosome aberrations and sister chromatid exchange studies in patients with prostate cancer: possible evidence of chromosome instability. *Cancer Genet.* Cytogenet., 1998 100, 143-147.

Dosaka, H; Abe, S; Sasaki, M; Miyamoto, H; Kawakami, Y. Sister chromatid exchange induction by benzo(a)pyrene in cultured peripheral blood lymphocytes of lung cancer patients and healthy individuals with or without familial history of neoplasms. *Int. J. Cancer,* 1987 39, 329-332.

Duker, NJ. Chromosome breakage syndromes and cancer. *Am. J. Med. Genet.,* 2002 115, 125-129.

Evans, HJ. Repair and recovery at chromosome and cellular levels: similarities and differences. *Brookhaven Symp. Biol.,* 1968 20, 111-133.

Evans, HJ; Scott, D. The induction of chromosome aberrations by nitrogen mustard and its dependence on DNA synthesis. *Proc. R. Soc. B Biol. Sci.,* 1969 173, 419-512.

Eyfjord, JE; Bodvarsdottir, SK. Genomic instability and cancer: Networks involved in response to DNA damage. *Mutat. Res.,* 2005 592, 18-28.

Friedberg, EC; Walker, GC; Siede, W. *DNA Repair and Mutagenesis.* Washington DC: American Society of Microbiology Press; 1995.

Hagmar, L; Bonassi, S; Stromberg, U; Brogger, A; Knudsen, LE; Norppa, H; Reuterwall, C. Chromosomal aberrations in lymphocytes predict cancer: a report from the European

Study Group on Cytogenetic Biomarkers and Health (ESCH). *Cancer Res.,* 1998 58, 4117–4121.

Hoeijmakers, JHJ. Genome maintenance mechanisms for preventing cancer. *Nature*, 2001 411, 366-374.

Hsu, TC. Genetic instability in the human populations: A working hypothesis. *Hereditas,* 1983 98, 1-9.

ISCN, *An International System for Human Cytogenetic Nomenclature,* Ed.: Felix Mitelman, Cytogenet. Cell Genet., 1995.

Ishii, Y; Bender, MA. Effects of inhibitors of DNA synthesis on spontaneous and ultraviolet light-induced sister-chromatid exchanges in Chinese hamster cells. *Mutat. Res.,* 1980 79, 19-32.

Johnson, RD; Jasin, M. Sister chromatid gene conversion is a prominent double-strand break repair pathway in mammalian cells. *EMBO J,* 2000 19, 3398-3407.

Jones, PA; Baylin, SB. The fundamental role of epigenetic events in cancer. *Nature Rev. Genet.,* 2002 3, 415-428.

Kaplan, MI; Limoli, CL; Morgan, WF. Perpetuating radiation-induced chromosomal instability. *Radiat. Oncol. Invest.,* 1997 5, 124-128.

Kato, H. Spontaneous sister chromatid exchanges detected by BudR-labelling method. *Nature,* 1974 251, 70-72.

Kato, H. Mechanism of sister chromatid exchanges and the relation to the production of chromosomal aberrations. *Chromosoma,* 1977 59, 179-191.

Keller, U; Grabenbauer, G; Kuechler, A; Sprung, C.N; Muller, E; Sauer, R; Distel, L. Cytogenetic instability in young patients with multiple primary cancers. *Cancer Genet. Cytogenet.,* 2005 157, 25-32.

Kinzler, KW; Vogelstein, B. Cancer-susceptibility genes. Gatekeepers and caretakers. *Nature*, 1997 386, 761-763.

Knight, RD; Parshad, R; Price, FM; Tarone, RE; Sanford, KK. X-ray induced chromatid damage in relation to DNA repair and cancer incidence in family members. *Int. J. Cancer,* 1993 54, 589-593.

Knudson, AG. Mutation and cancer: statistical study of retinoblastoma. *Proc. Natl. Acad. Sci.* USA, 1971 68, 820-823.

Latt, SA. Localization of sister chromatid exchanges in human chromosomes. *Science* 1974 185, 74-76.

Latt, SA; Schreck, RR. Sister chromatid exchange analysis. *Am. J. Hum. Genet.,* 1980 32, 297-313.

Lengauer, C; Kinzler, KW; Vogelstein, B. Genetic instabilities in human cancers. *Nature,* 1998 396, 643-649.

Li, G-Y; Yao, K-T; Glaser, R. Sister chromatid exchange and nasopharyngeal carcinoma. *Int. J. Cancer,* 1989 43, 613-618.

Limoli, CL; Corcoran, JJ; Jordan, R; Morgan, WF; Schwartz, JL. A role for chromosomal instability in the development of and selection for radioresistant cell variants. *Br. J. Cancer,* 2001 84, 489-492.

Lindenhahn, M; Schubert, I. On the origin of hydroxyurea-induced chromatid aberrations in G_2 chromosomes with BrdUrd in only one of the sister chromatids. *Mutat. Res.*, 1983 108, 301-316.

Little, JB. Ionizing radiation., In: Kufe, DW; Pollock, RE; Weichselbaum, RR; Bast, RC Jr; Gansler, TS; Holland, JF; Frei IIIE. Cancer medicine, Hamilton, Canada, 2003.

Livingston, GK; Cannon, LA; Bishop, DT; Johnson, P; Fineman, RM. Sister chromatid exchange: variation by age, sex, smoking and breast cancer status. *Cancer Genet. Cytogenet.*, 1983 9, 289-299.

Lorimore, SA; Wright, EG. Radiation-induced genomic instability and bystander effects: related inflammatory-type responses to radiation-induced stress and injury? *A review. Int. J. Radiat. Biol.*, 2003 79, 15-25.

Lukovic, L; Milasin, J. Sister chromatid exchanges in patients with carcinoma in situ of cervix uteri. *Cancer Genet. Cytogenet.*, 1992 59, 84-85.

Lynch, HT; Guirgis, HA; Lynch, PM; Lynch, JF; Harris, RE. Familial cancer syndromes: a survey. *Cancer,* 1977 38, 1867-1881.

Maffei, F; Angelini, S; Forti, GC; Violante FS; Lodi, V; Mattioli, S; Hrelia, P. Spectrum of chromosomal aberrations in peripheral lymphocytes of hospital workers occupationally exposed to low doses of ionizing radiation. *Mutat. Res.*, 2004 547, 91-99.

McClintock, B. The production of homozygous deficient tissues with mutant characteristics by means of aberrant mitotic behaviour of ring-shaped chromosomes. *Genetics*, 1938 23, 315–376.

Mitelman, F; Martens, F; Johansson, FA. A breakpoint map of recurrent chromosomal rearrangements in human neoplasia. *Nature Genetics,* 1997 15, 317–474.

Mitra, AB; Murty, VS; Luthra, UK. Sister chromatid exchanges in leukocytes of patients with cancer of cervix uteri. *Hum.Genet.*, 1982 60, 214-215.

Morgan, WF; Murnane, JP. A role for genomic instability in cellular radioresistance? *Cancer Metastasis Rev.*, 1995 14, 49-58.

Morgan, WF; Day, JP; Kaplan, MI; McGhee, EM; Limoli, CL. Genomic instability induced by ionizing radiation. *Radiat. Res.*, 1996 146, 247-258.

Morgan, WF; Hartmann, A; Limoli, CL; Nagar, S; Ponnaiya, B. Bystander effects in radiation-induced genomic instability. *Mutat. Res.*, 2002 504, 91-100.

Murty, VVVS; Mitra, AB; Sharma, A; Das, BC; Luthra, UK. Mitomycin C induced chromosomal aberrations and sister chromatid exchanges (SCEs) in lymphocytes of patients with precancerous and cancerous lesions of the uterine cervix. *Neoplasma*, 1987 34, 101-105.

Müller-Vogt, US; Maurer, JE; Grabenbauer, GG; Distel, LV; Iro, H; Steinhart, H. Altered DNA repair capacity in young patients suffering from multiple cancers. *Int. J. Mol. Med.*, 2003 11, 669-674.

Natarajan, A.T. Mechanisms for induction of mutations and chromosome alterations. Environment. *Health Persp. Suppl.*, 1993 101, 225-229.

Natarjan, AT. Chromosome aberrations: past, present and future. *Mutat. Res.*, 2002 504, 3-16.

Natarajan, AT; Boei, JJWA; Darroudi, F; van Diemen, PCM; Dulout, F; Hande, MP; Ramalho, AT. Current cytogenetic methods for detecting exposure and effects of mutagens and carcinogens. *Environ. Health Persp.,* 1996 104, 445-448.

Nowell, PC. The clonal evolution of tumor cell populations. Science, 1976 194, 23-28.

Obe, G; Pfeiffer, P; Savage, JRK; Johannes, C; Goedecke, W; Jeppesen, P; Natarajan, AT; Martínez-López, W; Folle, GA; Drets, ME. Chromosomal aberrations: formation, identification and distribution. *Mutat. Res.,* 2002, 504, 17-36.

Painter, RB. A replication model for sister-chromatid exchange. *Mutat. Res.,* 1980 70, 337-341.

Palitti, F; Becchetti, A. Effect of caffeine on sister chromatid exchanges and chromosomal aberrations induced by mutagens in Chinese hamster cells. *Mutat Res,* 1977 45, 157-159.

Paz-y-Mino, C; Leone, PE; Chavez, M; Bustamante, G; Cordova, A; Gutierrez, S; Penaherrera, MS; Sanchez, M.E. Follow up study of chromosome aberrations in lymphocytes in hospital workers occupationally exposed to low levels of ionizing radiation. *Mutat. Res.,* 1995 335, 245-251.

Perry, PE. Induction of sister chromatid exchange by chemical mutagens. In: Amleto Castellani. *The use of human cells for the evaluation of risk from physical and chemical agents.* Plenum Publishing Corporation; 1983

Perry, PE; Wolff, S. New Giemsa method for differential staining of sister chromatids. *Nature,* 1974 261, 156-158.

Popescu, NC; Amsbaugh, SC; DiPaolo, JA. Relationship of carcinogen-induced sister chromatid exchange and neoplastic cell transformation. *Int. J. Cancer,* 1981 28, 71-77.

Ramachandran, C; Melnick, SJ. Multidrug resistance in human tumors--molecular diagnosis and clinical significance. *Mol. Diagn.,* 1999 4, 81-94.

Ramalho, AT; Nascimento, ACH; Natarajan, AT. Dose assessments by cytogenetic analysis in the Goiania (Brazil) radiation accident. *Radiat. Protect. Dosim.,* 1988 25, 97-100.

Roberts, SA; Spreadborough, AR; Bulman, B; Barber, JBP; Evans, GDR; Scott, D. Heritability of cellular radiosensitivity: A marker of low-penetrance predisposition genes is breast cancer? *Am. J. Hum. Genet.,* 1999 65, 784-794.

Sasaki, MS. Sister chromatid exchange as a reflection of cellular DNA repair. In: Sandberg, AA. *Sister chromatid exchanges.* New York: Liss; 1982; 135-161.

Schwartz, JL; Rotmensch, J; Giovanazzi, S; Cohen, MB; Weichselbaum, RR. Faster repair of DNA double-strand breaks in radioresistant human tumor cells. *Int. J. Radiat. Oncol. Biol. Phys.,* 1988 15, 907-912.

Schwartz, JL; Mustafi, R; Beckett, MA; Czyzewski, EA; Farhangi, E; Grdina, DJ; Rotmensch, J; Weichselbaum, RR. Radiation-induced DNA double-strand break frequencies in human squamous cell carcinoma cell lines of different radiation sensitivities. *Int. J. Radiat. Biol.,* 1991 59, 1341-1352.

Scott, D; Barber, JBP; Spreadborough, AR; Burrill, W; Roberts, SA. Increased chromosomal radiosensitivity in breast cancer patients: a comparison of two assays. *Int. J. Radiat. Biol.,* 1999 75, 1-10.

Shafer, DA. Replication bypass of sister chromatid exchanges and implications for Bloom's syndrome and Fanconi anemia. *Hum. Genet.,* 1977 39, 177-190.

Sieber, OM; Heinimann, K; Tomlinson, PM. Genomic instability – the engine of tumorigenesis? *Nature Rev. Cancer,* 2003 3, 701-708.

Skolnick, M; Livingston, GK; Fineman, RM; Johnson, P; King, MC; McLellan, T; Bishop, DT. Genetics of breast cancer: geneological clusters, major genes, linkage and sister chromatid exchange as a preclinical marker. *Am J Hum Genet,* 1980 32, A151.

Skorski, T. Oncogenic tyrosine kinases and the DNA-damage response. *Nat. Rev. Cancer,* 2002 2, 351-360.

Sobin, LH; Wittekind, C. *TNM classification of malignant tumors.* New York: Wiley-Liss; 1997.

Sonoda, E; Sasaki, MS; Morrison, C; Yamaguchi-Iwai, Y; Takata, M; Takeda, S. Sister chromatid exchanges are mediated by homologous recombination in vertebrate cells. Molecular and Cellular Biology, 1999 19, 5166-5169.

Swift, M. Malignant neoplasms in heterozygous carriers of genes for certain autosomal recessive syndromes. In: Mulvihill JJ; Miller RW; Fraumeni JE Jr. *Genetics of human cancer,* New York: Raven Press; 1977; 200-215.

Takata, M; Sasaki, MS; Sonoda, E; Morrison, C; Hashimoto, M; Utsumi, H; Yamaguchi-Iwai, Y; Shinohara, A; Takeda, S. Homologous recombination and non-homologous end-joining pathways of DNA double-strand break repair have overlapping roles in the maintenance of chromosomal integrity in vertebrate cells. *EMBO J.,* 1998 17, 5497-5508.

Taylor, JH. Sister chromatid exchanges in tritium labelled chromosomes. *Genetics*, 1958 43, 515.

Terzoudi, GI; Jung, T; Hain, J; Vrouvas, J; Margaritis, K; Donta-Bakoyiannis, C; Makropoulos, V; Angelakis, P.H; Pantelias, G.E. Increased chromosomal radiosensitivity in cancer patients: the role of cdk1/cyclin-B activity level in the mechanisms involved. *Int. J. Radiat. Biol.,* 2000 76, 607-615.

Tucker, JD; Auletta, A; Cimino, MC; Dearfield, KL; Jacobson-Kram, D; Tice, RR; Carrano, AV. Sister-chromatid exchange: second report of the Gene-Tox Program. *Mutat Res,* 1993 297, 101-180.

Van Gent, DC; Hoeijmakers, JHJ; Kanaar, R. Chromosomal stability and the DNA double-stranded break connection. *Nature Rev. Genet.,* 2001 2, 196-206.

Venitt, S; Parry, JM. *Mutagenicity testing: a practical approach.* Oxford, Washington DC: IRL Press; 1984.

Ward, JF. DNA damage produced by ionizing radiation in mammalian cells: Identities, mechanisms of formation, and reparability. *Prog. Nucleic Acid Res. Mol. Biol.,* 1988 35, 95-125.

West, CML. Predictive assays in radiation therapy. *Adv. Radiat. Biol.* 1994 18, 149-180.

Wurster-Hill, DH; Cornwell, GC; McIntyre, OR. Chromosome aberrations and neoplasm - a family study. *Cancer,* 1974 33, 72-81.

Wurster-Hill, DH; Cornwell, GC; McIntyre, OR. Chromosome aberrations of myeloid and lymphoid cells in cancer patients and family members without evidence of cancer. *Cancer Detect. Prev.,* 1979 2, 125-126.

In: New Research on Genomic Instability
Editor: Eleanor J. Glascow, pp. 99-120

ISBN: 1-60021-320-0

Chapter IV

Tumors: Their Complex Morphologies and Genetic Instabilities

Leon P. Bignold*

Division of Tissue Pathology, Institute of Medical and Veterinary Science, Rundle Mall, Australia

Abstract

In addition to excess growth, the cells of most types of tumors exhibit abnormalities both of their nuclear morphology and of their specialised functions. "Malignant" tumors usually show the same changes to greater degrees, and in addition, tend to invade adjacent tissues, and spread by the blood- and lymphatic-vascular systems to distant parts of the body. However, according to tumor type, numerous exceptions to these general associations are known to occur. This chapter reviews these aspects of tumors and the theories of their possible genetic bases. The genomic effects of etiological factors and the mutations which are associated with inherited predispositions are mentioned. It is proposed that the co-mutational aspect of DNA-replicative-infidelity-type genetic instability might be involved in the morphological and behavioural abnormalities of tumor cells.

Introduction

From Graeco-Roman times, tumors have been classified as "benign" and "malignant". In the early nineteenth century, tumors were studied according to the concept of "tissues" (Bichat, 1802) and these lesions were found to be too complex in their structures and behaviours to be classified into only these two groups. Laennec in 1804 classified tumors as

* E-mail leon.bignold@adelaide.edu.au

either "homologous" or "heterologous" (Wolff, 1907). Lobstein (1829) on the other hand, classified swellings in general as "homeoplastic" (which were mainly reparative fibrotic processes), "heteroplastic" (which included some true tumors, as well as some inflammatory lesions) and "cacoplastic", which corresponded, more or less, to malignant tumors as they are recognised today (Bignold et al; 2007).

Microscopically-identifiable rounded "globules" in tumors were reported by Home (1820). Further developments in the histological period of cancer research began in the mid-1830s, with improvements in the achromatic lens for microscopes, along with better sectioning and staining of specimens (Bradbury, 1967; Clark and Kasten, 1983). The characteristic continuous variation ("pleomorphism") of the sizes and shapes of tumour cells and their nuclei were illustrated by Müller (1838). This amounted to the first documentation of "cellular instability" in tumors. In the 1840s, attempts to correlate the microscopic appearances of tumors with their clinical behaviour began in earnest (Rather, 1978). However, microscopes with achromatic lenses cannot optically resolve objects less than 1-2 microns in diameter, so that small intranuclear structures, such as chromosomes, could not be seen. Thus, although cell biological data of the time allowed the inference that the nucleus is the location of the significant hereditary material of the cell (Virchow, 1858), little progress could be made in the investigation of any possible somatic-genetic abnormality of tumor cells. It was, however, clearly established that, when comparing clinically "benign" with "malignant" tumors, there is a common, but only approximate, correlation between the degree of histological abnormalities of tumor cells and the aggressiveness of the clinical behaviour of the corresponding tumors (Rather, 1978).

With progress of these studies during the middle of the nineteenth century, greater and greater complexity of the morphology and behaviour of tumor cells was revealed. Virchow (1863) classified tumors according to structural complexity, *viz*: - "histiocytoid-", "organoid-" and "teratomatous" types. However subsequently, it was discovered that particular tumor types derive from particular cell types (due especially to the work of Thiersch, Waldeyer, Remak and others - Wolff, 1907; Rather, 1978). Because of this, "histogenetic" classifications (i.e. according to cell-type-of-origin) of tumors became dominant. Following this, studies of large numbers of cases of each tumor type showed that even within the "benign" and "malignant" groupings of each histogenetic category, tumors exhibit differing degrees of their deviations from their corresponding cell type of origin according to which particular cell-of-origin is involved. By the 1870s, studies of the histopathological features which distinguish benign from malignant tumors specifically for each cell-type-of-origin were well underway (Wolff, 1907; Rather, 1978).

Rapid changes in the study of biology and of tumors followed the work of Abbe (1840-1905) at the Zeiss works in Jena in the 1870s. Abbe invented the apoachromatic lens, and as well, improved substage condensers, and other components of microscopes (Bradbury, 1967). With these new microscopes, by the early 1880s, chromosomes and the events of mitosis were documented (Harris, 1999; Lima-de-Faria, 2003). Shortly afterwards, abnormal mitoses and chromosomes were observed in tumor cells, and descriptions were presented (without using the exact phrase) of mitotic and chromosomal instability as a feature of such cells (Wolff, 1907).

Beginning in 1890, David Paul Hansemann (1858-1920; "von", 1902) published a series of papers (Hansemann, 1890; 1891; 1892) and a book (1893) in which he asserted that asymmetric mitoses and/or other chromosomal abnormalities result in "abnormal balances of chromosomes" (i.e. "aneuploidy") in cells. He postulated that these in turn are the basis of tumor formation (see Bignold et al, 2006; 2007). However, other authors at the time found that asymmetric mitoses are difficult to identify early in tumor formation, and also, that abnormal mitoses can be seen in non-tumorous pathological conditions. Mainly for these reasons, Hansemann's chromosomal theory (and the modified version of it put forwards by Boveri, 1914; see Bignold et al, 2006; 2007) was rejected at the time (Israel, 1902).

In the same period, reliable histopathologic observations accumulated to the extent that identifications of pathological processes in tissues became sufficiently accurate as to be of diagnostic value. As large hospitals began to establish departments of histopathology to provide this diagnostic service, data were accumulated concerning the clinico-pathological complexity of tumors in larger and larger series of cases. It became appreciated that among some tumor types (such as tumors of the transitional cell epithelium of the lower urinary tract), there is no sharp division between "benign" and "malignant" tumors, but rather there exists a spectrum of clinico-pathological abnormalities. Further, among many carcinomas (malignant epithelial tumors), there are degrees of malignancy which can be "graded" by applying morphological criteria. Hansemann (1890; 1893) in fact invented this approach, using the terms "de-differentiation" (*Entdifferenzierung*) for the observed changes and "anaplasia" for a hypothetical fundamental process underlying it. According to Hansemann, "anaplasia" could be present in different degrees in different cases of malignant tumour. This terminology provided for grades of malignancy, which earlier classifications – especially "homeoplastic", "heteroplastic" etc – of tumors (see above) did not. Hansemann's grading of malignant changes continues to be the basis of the description of tumors for diagnostic purposes today.

We can also note that Hansemann's idea of "anaplasia" implied some kind of biological process. In Hansemann's early papers (1890; 1891; 1892), anaplasia was actually applied to normal oogenesis as if it was the process which might be abnormally induced in all tissue cell types to produce tumours (Bignold et al, 2006; 2007). However, this was discounted at the time, and oogenesis is currently not considered to be related to tumor formation. However, the general idea of tumor formation deriving from a deranged normal biological process is, at the present time, still considered valid. One current example is the suggestion that tumors form because of a mutation of a "master gene" of some developmental process, as for example, HOX genes (Carrio et al, 2005; Grier et al, 2005).

In the early twentieth century, somatic mutation (in the sense of an alteration of somatic cellular hereditary material which is analogous to germ-line alterations of the same material) was proposed as a possible basis of tumors (Murray, 1904; Tyzzer, 1916; Whitman, 1919; Bauer, 1928; Lockhart-Mummery, 1934). At the time however, this idea was rejected for several main reasons. First, somatic mutation was thought to be uncommon (e.g. Rous, 1959). Second, a simple, immediately-effective mutation could not explain the long latency (time delay) between the application of carcinogens and appearance of tumors either according to clinical observations (beginning with the chimney sweeps' scrotal cancers described by Pott – Wolff, 1907) or in the results of experimental studies (Willis, 1948; reviewed Bignold,

2003a; 2004a). Third, in classical germ-line genetics, one mutation produces one alteration of phenotype, and this change is stable in subsequent descendants of the original mutant individual. This last feature in particular is not consistent with the complex histopathological features of tumors, and especially inter-tumoral and intra-tumoral variations (see below). A good summary of many of these objections, and additional relevant issues was provided by Berenblum (1974) who considered that tumors have five important features: - irreversibility, diversity of types; multiplicity of exciting agents, multistep induction and rarity. He did not consider any genetic mechanism known at that time (including the "viral oncogene theory" or altered control of gene transcription), could account for all types of neoplasms. Thus, until the 1980s (and by some authors even later), "neoplasia" continued to be widely considered as some uncomprehended biological process, of which obvious genetic (in the form of mitotic, chromosomal or genetic changes) were the effects, not the cause (see Israel, 1902 and above).

Major Cell Biological Changes of Tumor Cells

For the purposes of the present discussion of possible explanations of the morphological abnormalities of tumors by genetic mechanisms, it is first necessary to mention the all biological abnormalities of tumor cells in detail. These changes are described in almost any text books of general pathology, and some aspects of this complexity are the subject of the next section of this chapter. Here however, the particular histopathological features of tumor cells of interest are grouped into three categories. The development of each of these may well require separate mutational and other pathogenetic mechanisms. From this, it is suggested that there are too many cellular abnormalities inherent in "dedifferentiation" or "anaplasia" for a single mutation to account for them, Because of this, many mutations (indicating genetic instability) are probably responsible for the histopathology of tumours (Bignold, 2002a; 2003b; 2003c; 2004b).

Nucleus-to-nucleus Variability of Appearances (Nuclear "Atypia", "Pleomorphism")

This change is appreciable microscopically in almost all cases of malignant solid tumors. The appearances of tumour cell nuclei after histological processing are governed by several major factors. The first is the "activation status" of the cell. This can be appreciated in the differences between "active" and "inactive" normal cells (e.g. active and "memory" lymphocyte nuclei). The biochemical bases of the morphological differences of "activation status" presumably include changes of content of water, of histone- and non-histone proteins and of RNA in the nuclei. The second major factor affecting the appearances of tumor cell nuclei is aneuploidy, which is often associated with abnormal chromosome-chromosome interactions. The latter are manifest in mitosis by chromosomal "stickiness" (Koller, 1957). The third group of factors relate to the fixative used in histological processing (Bignold, 2002b). Thus the "chromatin patterns" of the nuclei of cells relate to fixation: they are coarsest with coagulative fixatives, and virtually absent when air-drying and methanol are used for primary fixation (e.g. in the May-Grunwald-Giemsa stain). Although the mechanisms of variability of chromatin patterns of nuclei of tumor cells according to fixative have been little discussed in the literature, this variability may be associated with abnormalities of the non-histone protein composition of the nuclei (Bignold, 2002b; 2003c). Some basis for the validity of this idea is to be found in the abnormal translocation of cytoplasmic proteins into nuclei which has been documented as directly associated with mutant nuclear pore proteins (Martinez et al., 1999), which in turn could possibly occur in tumors as a result of genetic instability (see below).

Loss of Specialisation

This change (which can be termed loss of "tissue differentiation") is very common among the cells of solid tumors, and may well result from simple disabling mutations of the specific genes for the relevant metabolic process. Varying degrees of the change might be due to varying "hypo-morphisms" of the causative mutations. For example, markedly-effective disabling mutations of the enzymes for synthesis of keratin may result in a squamous cell carcinoma appearing "poorly-differentiated" compared to a "well-differentiated" tumor, in which the disabling mutations may be less severe, and keratin production is preserved to a greater extent. Alternatively, different degrees of loss of specialisation may arise from mutational "hits" on varying numbers of enzymes in the synthetic chain required for the particular specialised product of the cell. This last idea is effectively a "somatic polygenic" hypothesis for the variability of loss of specialisation in tumors.

Invasiveness and Colonisation of Remote Tissues (Metastasis)

These phenomena are essential aspects of malignancy. They are not part of the "normal" behaviour of most cell types (for example, epithelia), but are seen among embryonic cells, and among some adult cell types, such as lymphocytes. To explain their occurrence in tumor cells, it might be remembered that the genomes of cells of most types contain the genes for embryonic differentiation and for differentiation of other cell types (e.g. hemoglobin production in a keratinocyte) in repressed states. A common explanation, therefore, of the apparent "acquisition" of invasive and metastatic capacities by tumor cells is that "de-repression" of the relevant parts of the embryonic and/or other cell-type-genome has occurred in the particular cell type (Sell, 1993). The mechanism is attractive because cell invasiveness involves active movement on the part of the cell, which in turn requires several specialised cell structures, such as actin filaments to be synthesised and coordinated. Such complex and coordinated changes in a cell are more likely to be achieved by disabling of repressor gene(s) for a whole cell process that by a series of mutations affecting each cellular component.

Nevertheless, other theories of invasion and metastasis, such as loss of desmosomes, basement membrane or other "pericellular constraints" (Chidgey, 2002; Chrenedk et al, 2001; Wheelock and Johnson, 2003; Brinckerhoff et al, 2000; Montell, 2005) or alternatively, activation of latent motility in cells by autocrine factors (Levine et al, 1995; Silletti and Raz, 1996; Dobashi et al, 2006) are additional possibilities. It might be noted that there is no known reason for all tumor cells to have the same mechanism of invasion and metastasis; or indeed to have necessarily the same mechanism of any cell biological abnormality.

The Rate of Cell Accumulation cannot be Assessed Effectively in Histopathological Sections

Factors which result in cell accumulation include those affecting mitotic rate, cell senescence and local nutritional factors. Apart from causing altered numbers of mitotic figures and probably reduction of the manifestation of "tissue specialisation", this group of factors *per se* probably does not affect other histological abnormalities of tumors to any great degree.

It may be noted here that the numbers of mitotic figures are not a completely reliable indication of growth rate. This is because the numbers of mitotic figures in a histopathologically prepared specimen could also be increased if the average duration of mitosis in the cells of the particular tumor were to be increased. Histopathology provides only a “snapshot-in-time” of the lesions, and does not provided good indications of the rates of dynamic events.

Morphological Complexity of Tumors I: Independence of Nuclear Abnormalities, Loss of Specialisation and Invasiveness between Different Tumor Types

One of the problems of Hansemann’s concept of “de-differentiation”/“anaplasia” is that the grades of morphology and behaviour, and the same correlations with one another, do not occur in all tumor types. For example, keratinizing squamous cell carcinomas of the epidermis of the skin invade the dermis, but grow slowly and metastasise to lymph nodes in only a small percentage of cases. However, squamous cell carcinomas of the lung - showing the same degress of keratinisation to those of the epidemis – grow rapidly and metastasise very frequently to local lypmph nodes. In fact, among the various tumors, abnormalities of nuclear appearances, loss of specialisation and invasiveness can occur separately or together in any combination, according to tumor type (Bignold, 2003b), and is summarised in Figure 1. Here, invasive and metastatic behaviour is represented on the x-axis, altered nuclear morphology on the y-axis, and reduction of specialisation on the z-axis. In this, clinically “benign” tumors fall on the left hand side of the cube, while the most malignant tumors fall on the right hand side. Cytologically uniform tumors are at the lower surface of the cube, while pleomorphic tumors are at the upper surface. Specialising tumors occupy the front side, and non-specialising (“dedifferentiated”, “anaplastic”) tumors form the rear side. The histopathological features of all tumors fall inside the cube. The corners of the cube are occupied by those tumor types which exhibit the effects of each grouping in their extremes. These extreme examples, both purely and in combination, are as follows.

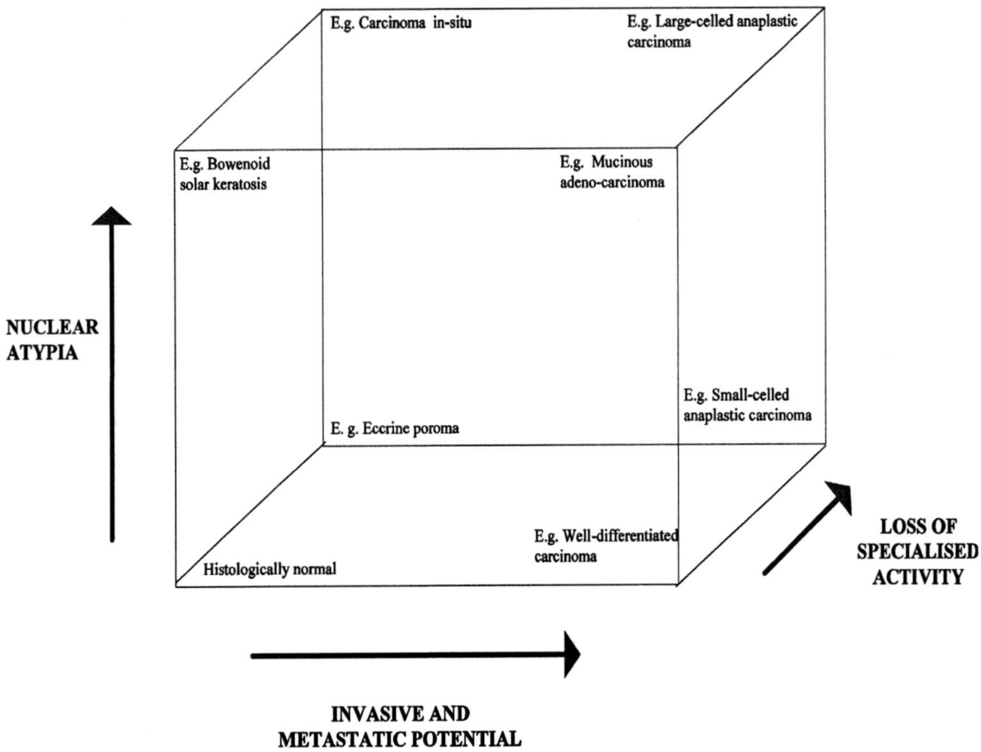

Figure 1. Diagrammatic arrangement of particular human tumor types to show that all possible combinations of degrees of nuclear atypia, tendency to invasion and metastasis (clinical aggressiveness) and loss of specialized activity occur in different tumor types.

Non-invasive Cells, Showing Nuclear Uniformity, with Normal Specialisation

Histologically normal cells may harbour neoplastic potentiality. Thus in patients with carcinoma of the bladder, the remaining transitional epithelium is more likely to develop tumors than is that of persons who have not had any tumor (this is part of the basis of "field theory" of tumors – Braakhuis et al, 2003). Several different, "histologically normal" epithelia adjacent to carcinomata have been shown to be genetically unstable (Hittelman, 2001). In classical two-stage chemical carcinogenesis, this is also the appearance of "initiated" tissue before "promotion" (Berenblum, 1974).

Non-invasive Cells which Have "Normal" Nuclei, but Show Loss of Specialisation

These changes are seen in eccrine poroma of skin (Figure 2a). This tumor is non-invasive, and the nuclei are within normal limits, but the cells exhibit no specialisation. Other skin lesions in this category could include some forms of seborrheic keratosis and hidradenoma simplex and basaloid adenomas of salivary glands. Several lesions currently categorised as (idiopathic) "hyperplasias" fulfil the criteria for this type of lesion. Examples are non-atypical ductal hyperplasias of the breast and "basal cell hyperplasia" of the prostate.

Non-invasive Tumors Showing Marked Nuclear Abnormalities but with Minimum Loss of Specialisation

This category is exemplified by some solar keratoses of the skin (Figure 2b). In these, the keratin is retained as a "horn" while the underlying epidermal cells are frequently bizarre, with abnormal mitoses.

Non-invasive Tumors Showing Marked Nuclear Abnormalities, and Loss of Specialisation

An example of these lesions is squamous cell carcinoma in-situ of the vagina (Figure 2c). These cells do not keratinise. The same grouping of abnormalities may be seen in in-situ carcinomas of the skin (Bowen's disease) and of the colon, stomach, ducts of breast etc.

Invasive Tumors Showing Minimal Nuclear Atypia, and Minimal Loss of Specialisation

These tumors are exemplified by some follicular carcinomas of the thyroid (Figure 2d), which can, in metastases, appear very similar to normal thyroid tissue (Rosai et al., 1992). Other examples of this combination of features may be demonstrated by "well-differentiated" adenocarcinomas of the biliary tree, which can metastasise to the lung and grow in a "lepidic pattern" and mimic "pulmonary adenomatosis". Such a case of pancreatic origin is illustrated in Colby et al (1995). Yet other examples of the category of tumor are "adenoma malignum" of the uterine cervix, some renal cell carcinomas etc.

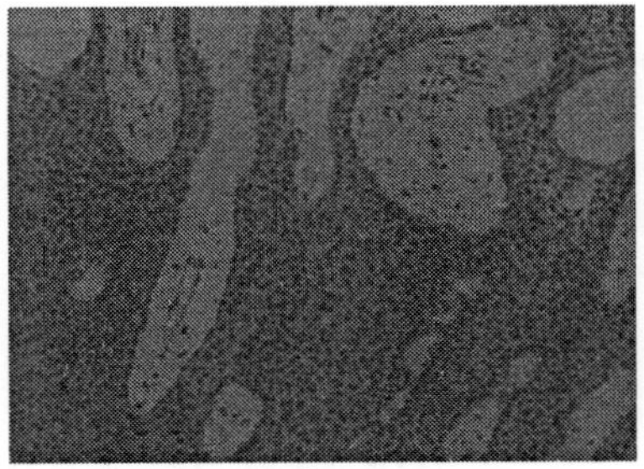

2a

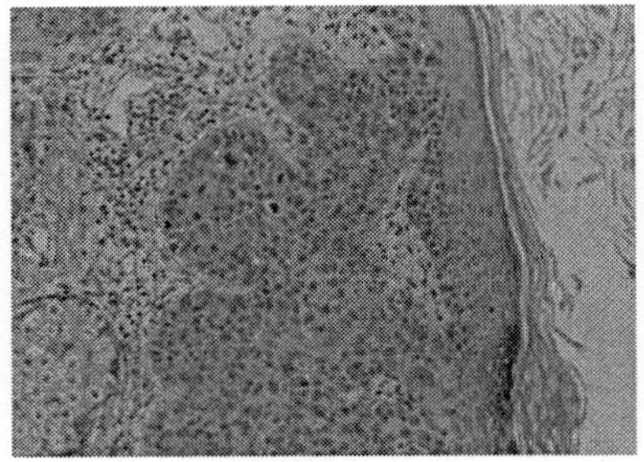

2b

2c

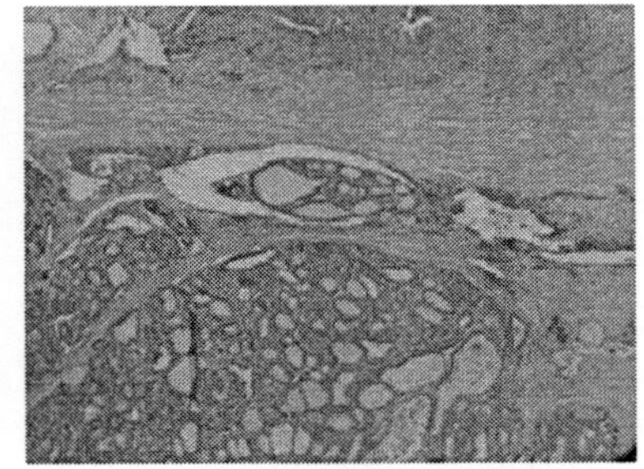

2d

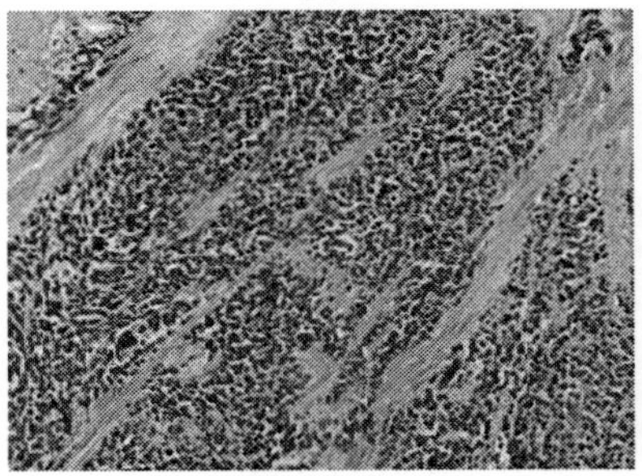

2e

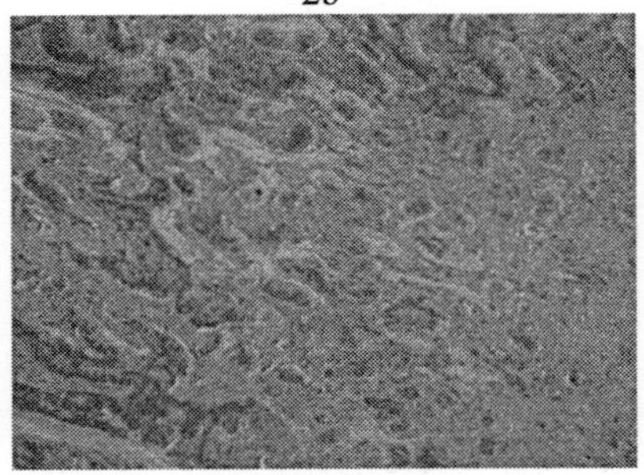

2f

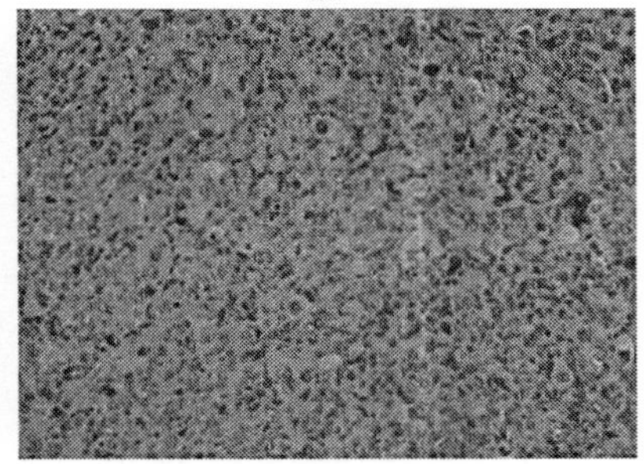

2g

Figure 2. Photomicrographs of tumors mentioned in Figure 1. **2a**, eccrine poroma (x 100); **2b**, solar keratosis (x 100); **2c**, squamous celled carcinoma in situ (x 40); **2d**, follicular carcinoma of thyroid (note tumour in blood vessel) (x 40); **2e**, anaplastic small-celled carcinoma of bronchus (x 100); **2f**, mucinous adenocarcinoma (x 40); **2g**, amelanotic malignant melanoma (x 100). All stained with hematoxylin and eosin.

Invasive Tumors Showing Minimal Nuclear Abnormalities, and Marked Loss of Specialisation

This category is typified by "anaplastic" small-celled carcinoma of the lung (Figure 2e). In this tumor, the nuclei are regular, and are similar to those of lymphoid cells. Other examples occur among carcinomas of the breast, stomach, uterine cervix (non-keratinising small cell type) etc.

Invasive Tumors with Marked Nuclear Abnormalities, but with Minimum Loss of Specialisation

Mucinous ("colloid") carcinomas, especially of the colon may exhibit these changes (Figure 2f). Lakes of mucus surround the invasive, pleomorphic tumor cells. Similar tumors may arise in the lung and breast.

Invasive Tumors with Marked Cytological Abnormalities, and Complete Loss of Specialisation

A typical example of this category is amelanotic melanoma (Figure 2g). Pleomorphic large-celled "anaplastic" carcinoma of the lung and similar tumors with these features can arise in many epithelia, including that of the bronchi.

From the above, it is obvious that most tumors fall in the middle of the cube (Figure 1) rather than at its corners. This is in keeping with the proposed model of randomness of large numbers of mutations and other genomic events which underlie the process, because it is expected that many mutations will be of partial effect.

Morphological Complexity of Tumors II: Occurrence of Non-cell-type-of-origin Features in Tumors: Metaplasias and Embryonic Reversions

One of the most puzzling aspects of certain tumor types has been the development in some or all cases of features not present in the cell type of origin. Early in his study of tumors, Virchow (1858) referred to such new features as "histological substitution". Later however, Virchow (1871; 1884) divided new features in cells into two types. The first was due to obvious extrinsic factors such as the flattening due to pressure, of previously columnar cells within a cyst ("adaptive" changes). The second type consisted of those altered features in which the cells of one type of specialised type in a certain place become replaced by cells of a different specialised type ("metaplasia"). The mechanisms of former are considered self-explanatory, but those of the latter - implying a "genomic switch" in specialization - are still under investigation, usually in terms of "switched master gene" activations of development or tissue differentiation (Frisch, 1997).

Also in the 1870s, Cohnheim (1882) noted that tumors can arise in adult connective tissues which have the histological appearances of embryonic mesenchyme (i.e. "myxomas"). Cohnheim (1882) used this observation to support his theory that all tumors arise from "embryonic rests", although such embryological "left overs" are identifiable in few tissues. Cohnheim's observation remains the prototype for all "embryonic reversive" phenomena in tumors (see Bignold, 2005), among which are now recognised such events as secretion of alpha-fetoprotein by hepatoma cells and "ectopic" antigen expression. Like other non-cell-type-of-origin features, embryonic reversions are seen to indicate some lesion of a developmental "master-gene" or "gene-activation-cascade" in the particular tissue in which the tumor arises.

Radiation and Alkylating Agents as Causes of Mutations, as well as Mitotic and Chromosomal Abnormalities in Living Cells

Immediately after WWII, the invention of the atomic bomb provoked greatly enhanced interest in radiobiology and pathology. In addition, the toxic effects of derivatives of mustard gas [*bis*(2-chloroethyl)sulphide], including the nitrogen analogues, had been studied in secret during WWII because of their possible military uses (which fortunately were never employed). The results of these war-time studies were only published in the late 1940s when the anti-mitotic - and hence possible anti-tumor - effects of the agents were established (Philips, 1950; Wheeler, 1962). In various animal and *in vitro* models, both irradiation (see Lea, 1962), and many alkylating compounds (Ross, 1962; Wheeler, 1962) were found to cause both mutations and mitotic and chromosomal abnormalities (see especially Symposia in the Annals of the New York Academy of Science vol 68, 1958; and vol 163, 1969). Nevertheless, despite continuing to the present time, such studies have not yet defined precisely the mechanisms of these effects by either irradiation (see Mothersill and Seymour, 2006) or alkylating agents (see Bignold, 2006). In relation to alkylating agents, early experimental studies showed that some such agents can inhibit DNA synthesis but not RNA synthesis in living cells (see Bignold, 2006). This observation, although little studied since, does not seem to support the theory that alkylation of DNA (i.e. "adduct-formation) is the mechanism of action of these substances (because the substrate is the same for both syntheses). These unresolved questions can be seen as a part of the continuing debate concerning the respective roles of aneuploidy and of mutation and of the possible mechanisms of both of these genetic changes in the pathogenesis of tumors (see articles in Mutation Research vol 504, issues 1-2, 2002; in Seminars in Cancer Biology vol 15, issue 2, 2005; as well as in Li et al, 2000; Duesberg, 2005; Duesberg et al, 2005; Pathak and Multani, 2006).

DNA-replicative-infidelity-type Genetic Instability

From all of the above, we can see that mitotic and chromosomal instability has been recognised (although not always described by that phrase) since the 1890s, both in human tumors and experimental studies of carcinogens. A completely new phase of consideration of genetic instability, however, began with the discovery of the structure of DNA. The history of impaired replicative fidelity of DNA as a type of genetic instability has been reviewed previously (Bignold, 2002a, 2003a; 2004b), but here it can be summarised as follows:

In his description of the properties of genes, H J Muller (1922) noted that genes must be reproduced accurately, and that if any error occurs in such replication, the changes would be transmitted exactly to daughter cells. Watson and Crick (1953) emphasised that their double helical model of DNA allowed for faithful replication of DNA sequence. In the 1960s, bacterial strains were identified which replicate their DNA less faithfully than do corresponding wild types. For these mutant strains, the term "mutator strain" was introduced (Cox and Yanofsky, 1967). Cleaver (1968) showed that a gene for repair of damaged DNA are mutant in the condition of xeroderma pigmentosum (a recessively-inherited disease characterised by multiple manifestations, including increased predisposition to cancers of the skin).

Although Speyer (1965) and Nelson and Mason (1972) suggested that replicative infidelity of DNA might occur in tumors, Loeb and co-workers (Loeb, 1996; 2001; Loeb et al, 1974; Loeb and Kunkel, 1982; Loeb and Loeb, 2000; Beckman and Loeb, 2005) has continuously championed such DNA-replicative infidelity as a relevant type of genetic instability for tumors, and (especially for the purposes of the next section), has shown that there are a variety of exogenous and endogenous sources of substances which can adversely interfere with fidelity of DNA replication cells (Loeb and Kunkel, 1982; Cheng and Loeb, 1997; Wogan et al, 2004). The occurrence of such a phenomenon has seemed the only available explanation of the high numbers (in the region of 10,000 per cell) of "genomic events" which occur in some cancers (Stoler et al., 1999; Gaasenbeek et al, 2006).

DNA-replicative Infidelity of DNA as a "One-off" in a Normal Cell, May Produce a Mutation Causing DNA-replicative Infidelity in Daughter Cells

Based on known aspects of the functions of the components of the DNA polymerase complexes, the present author has discussed previously (Bignold, 2003a; 2004a) the evidence that "one-off" impairment of DNA replicative infidelity during S phase in a cell might be the mechanism of effect of chemical carcinogens. Such a mechanism could explain the puzzling phenomenon of carcinogens which do not react with DNA ("non-genotoxic" carcinogens – Bignold, 2003a; 2004a). Further, a mutant half of a replication "bubble" (15,000- 150,000 nucleotides) would be long enough to include more than one genomic element, which in turn might cause multiple phenotypic abnormalities to appear in the one cell at the same time (Figure 3). For example, it could cause excessive growth and loss of specialisation which is the feature of lesions such as seborrheic keratosis - see above and see hypothesis below. As another feature, one molecule of carcinogen, if acting by impairing replicative fidelity would produce a focus of numerous changes of nucleotide sequence, rather than a single nucleotide change or a "frameshift" mutation (Figure 4). In this way, the extraordinary potencies of some carcinogens which have been revealed both in clinical and experimental studies, might be explained (Bignold, 2003a; 2004a).

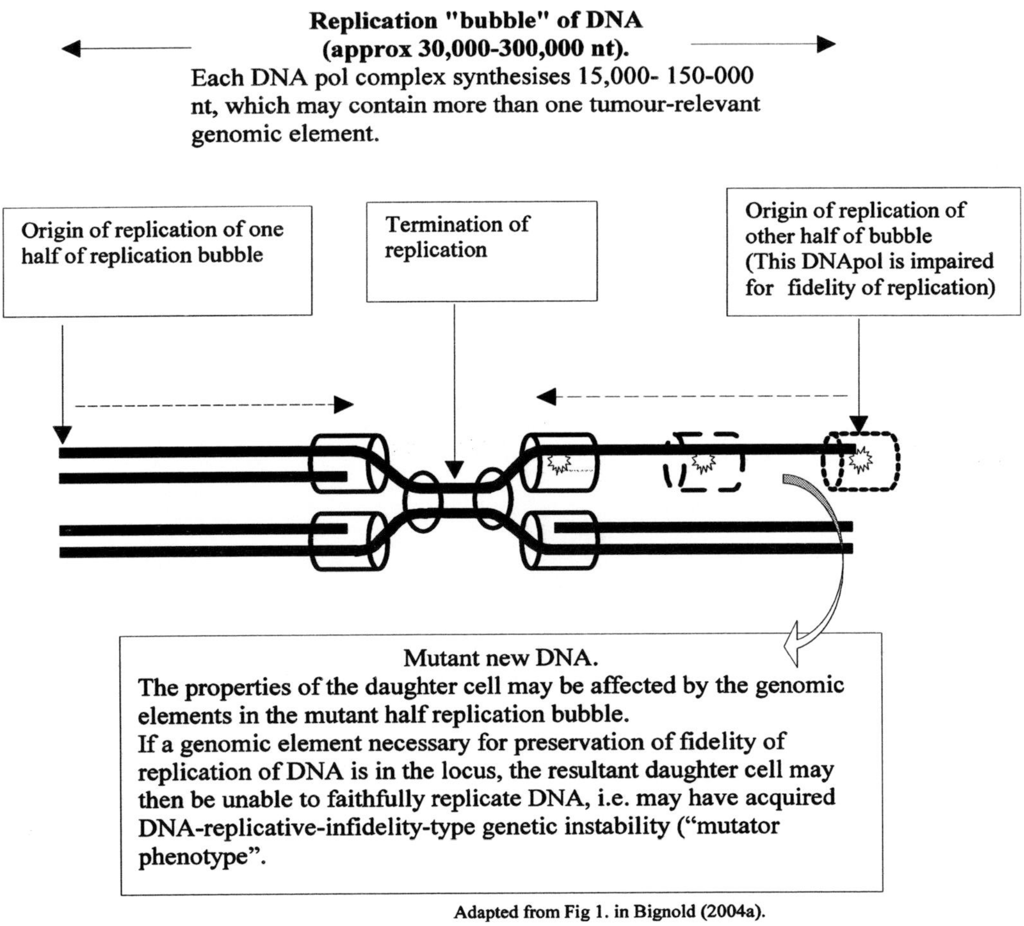

Figure 3. Diagram of DNA synthesis in "half replication bubbles", indicating effect of impairment of fidelity of replication of one DNA polymerase complex.

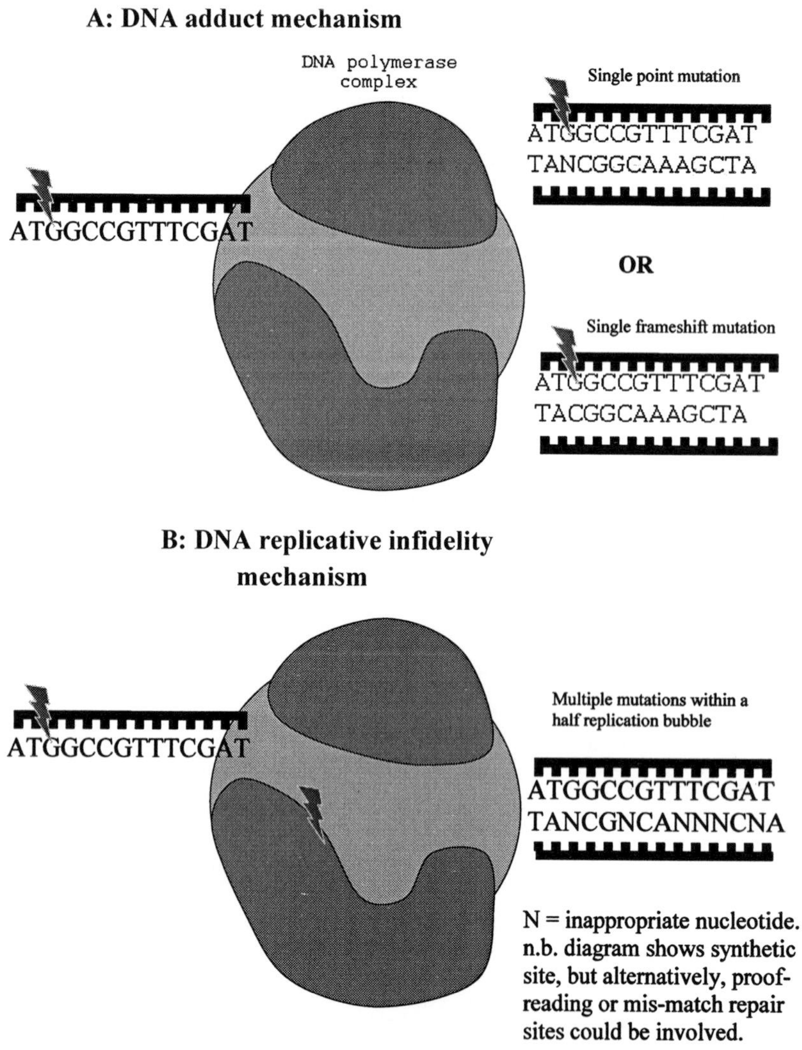

Figure 4. Comparison of effects on the nucleotide sequence in newly-synthesized DNA of chemical adducts versus impairment of DNA-replicative fidelity.

If this original mutation affected only genomic elements which are not relevant to fidelity of replication of DNA, then daughter cells would be likely to replicate DNA faithfully (i.e. not exhibit "mutator phenotype" - see above). However, DNA-replicative infidelity in daughter cells would occur if this "one-off" mutational event acted on genomic elements which are necessary for preservation of DNA replicative fidelity in daughter cells (Figure 5). Thus such an event of exogenous DNA-replicative infidelity during S phase in a somatic cell could then result in somatically-inherited DNA-replicative-infidelity-type genetic instability, i.e. "mutator phenotype" (Figure 5).

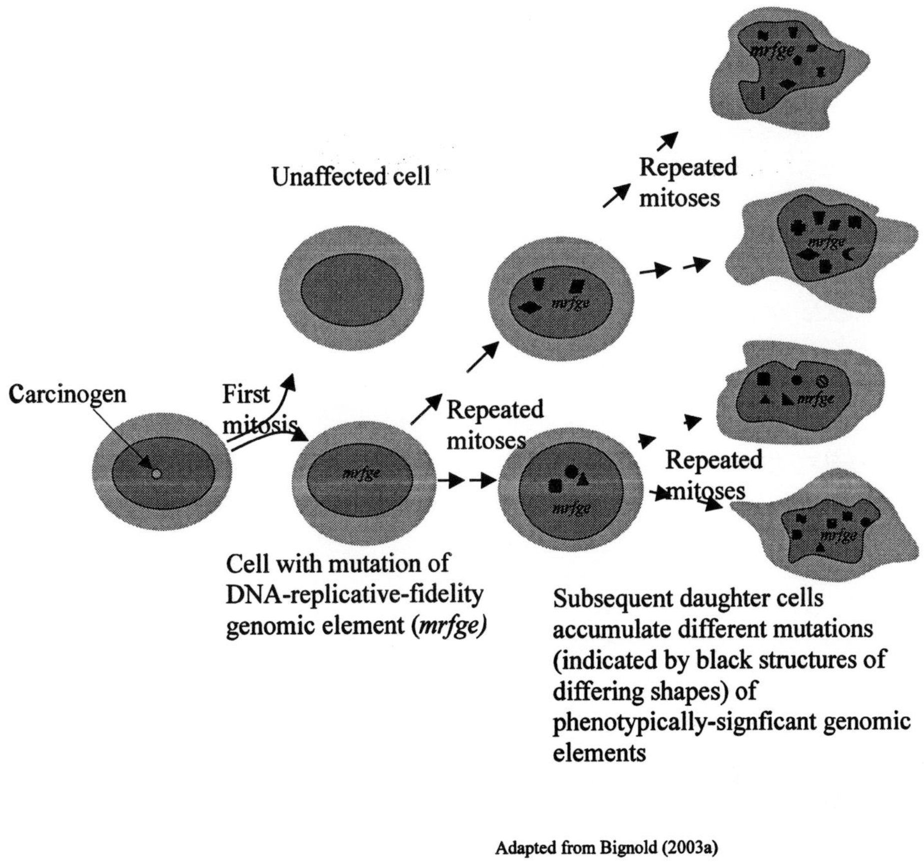

Figure 5. Effects on daughter cells if a disabling mutation occurs in a genomic element which is necessary for preservation of DNA replicative fidelity.

Evidence for "Co-mutation" in Tumor Formation: Inherited Predispositions

In searching for evidence of co-mutation as an event in human tumor pathogenesis, it may be useful to consider familial predispositions to particular tumors types. In the middle of the nineteenth century, the idea that certain families are predisposed to cancers of all types ("cancer diathesis") was widely considered (Wolff, 1907). However, by the early twentieth century, this had been disproved, especially by the careful work of Warthin (1913). At the same time, family studies had revealed a variety of disorders involving predispositions to tumors of particular types (as in xeroderma pigmentosum, neurofibromatosis, multiple enchondromas - Bauer, 1928). Subsequent clinical studies revealed more and more families with such tumor-type-specific inherited predispositions, and now over 100 inheritable tumor predispositions are recognised (Kinzler and Vogelstein, 1998; Marsh and Zori, 2002; Hino et al, 2006).

Of particular interest have been the predispositions which involve genes for replicative fidelity of DNA. These include xeroderma pigmentosum, in which the mutation is of a gene for a specific mechanism of repair of ultraviolet damage (see Pattinson and Davies, 2006), and the *BRCA* genes which are for mismatch repair (Marsh and Zori, 2002).

The present author (Bignold, 2004c) reviewed these predispositions from the point of view of the cell-type-of-origin-specificity of the tumor types which are involved. Further, the curious fact was discussed that in among different families, exactly the same nucleotide base changes which form the known mutation predispose to different tumor types with certain genes (*APC*, *RET* and *PTEN*). This seemed to indicate that additional genomic elements might be necessarily affected by the so-called second (somatic) mutational "hit" which is thought to underlie tumor formation in these specific predispositions (Vogelstein and Kinzler, 1998). Models were proposed suggesting details of these additional, albeit so far undiscovered, genomic elements which might be involved in such tumors. Further, it was proposed that these additional genomic elements are not necessarily mutant in the germ-line of the affected families, but just located close enough to the mutant genomic element to be involved in a co-mutational event with the sole "good copy" of the relevant gene in the affected patient's genome, and thus cause a tumor.

Hypothesis: Co-mutation Could be the Basis of the Cell-type-of-origin-dependency of the Features of Tumors

If we continue from the discussion of the possible effects of "one–off" DNA replicative infidelity of DNA as indicated in the section on that issue (above), it is clear that all the variations observable in tumor cells might be derived as follows.

More than one phenotypic abnormality could occur simultaneously in a cell if the relevant genes are sufficiently closely linked in the genome to be susceptible to co-mutation (Figure 6, sequences 1, 2 and 3). The cell-of-origin-dependency of some features seen in some tumor types might arise if the particular genomic sequence necessarily included a normally active, but not viability-essential, genomic element in the newly-mutant segment. An enormous range of possibilities arise if all forms of "genomic elements", i.e., insulator segments, transcription factors, microRNAs, gene-within-genes and other small, and/or compact regulatory elements are considered, in addition to the traditional protein-encoding "genes". Figure 6 gives only a few examples of the large number of possible combinations.

It can also be noted that none of the various possible mechanisms of mutations which are based on mitotic aberrations or lesions of single chromosomes appear to have sufficient consistency of association with tumor types to account for the histopathology of tumors (see Introduction). Similarly, smaller, but still long-segment mutation by frameshifts would provide very similar tumors, and little explanation of case-to-case continuous variability (i.e. degrees of "differentiation", see above). However, the case-to-case continuous variability of the characteristics of tumor cells could be explained by infidelity of replication of DNA being the mechanism of initiation of tumor formation in a somatic cell. This is because, in different cases of impaired replicative fidelity, the precise changes of nucleotide sequence, and their total number are unlikely to be the same.

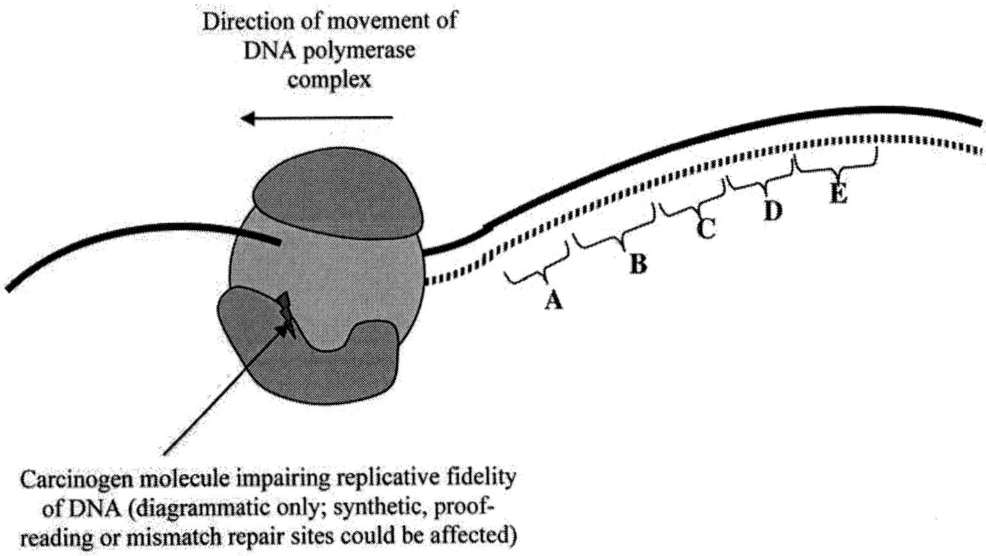

Figure 6. Some models by which co-mutation might cause different tumour types in different cell types.

1. If **A** is a tumour suppressor genomic element and **B**, C, **D**, and **E** are irrelevant, simple excess growth might occur.
2. If **A** is tumour suppressor and any of the others is a cell-type-specifically active genomic element for specialisation, then excess growth with loss of specialisation (consistent with a "benign" tumour) might occur.
3. If one element is a tumour suppressor, another is a cell-type-specifically active genomic element for specialisation and a third is for preservation of DNA replicative fidelity, then a genetically unstable and hence possibly "malignant" variant of tumour as in "2" might occur.
4. The order of the genomic elements in the segment might be important. For example, if **A** is a cell-type-specifically active genomic element for specialisation, **B** is an insulator and C is a repressed pro-growth genomic element, but not any other order, the result would be the same as in "2". If in addition at "**D**" another insulator was present, and at "**E**", a suppressed genomic element which might disturb DNA replicative fidelity in the particular cell type is included, the result might be the same as in "3".
5. If at any part of the sequence, there was a genomic element essential for the viability of the cell, then the cell would die and no tumour would result.
6. If one of the parts of the sequence was a genomic element for preservation of mitotic stability, then mitotic instability might be part of the phenotype of the resulting tumour cell from the start of the process.

n. b. Mechanisms "1", "2", and "3" are not dependent on the cell-type-of origin of the cell, but "4" does depend specifically on the cell type of origin because it requires a genomic element which is normally active in the cell, but is not essential for the viability of the cell, to be mutated.

Significance and Testing

The major significance of this possible mechanism of action of carcinogens is that it might form the basis of new assays of putative carcinogens. No perfect *in vitro* or *in vivo* method is currently available for assessment of the carcinogenic potency of chemicals in humans. Since new chemicals are being synthesised every year by the chemical industries around the world, the lack of such a test causes considerable problems for the introduction of new agents for agricultural, pharmaceutical and similar purposes. One possible method was described previously (Bignold, 2004a) but advances in methods of molecular biology are such that other techniques may well be feasible now or in the near future.

Conclusion

Since the late nineteenth century, alterations of the hereditary material somatic cells, either in the form of alterations of mitoses and chromosome numbers (aneuploidy), or in the form of germ-line-mutation-analogous somatic mutation, have been suggested as the basis of tumor formation. However, because of a variety of pathological and experimental observations, these ideas were largely rejected until the 1970s. Numerous discoveries have been made since then relating to the molecular basis of genes, the mechanisms of the faithful replication of these molecules during cell division, and in particular, that certain inherited predispositions to tumors are associated with impairment of mechanisms which ensure faithful replication of genes. These discoveries all indicate that some kind of genome-altering mechanism is the most likely basis of tumors. Here, the history of the possible role of genetic abnormalities in the morphology of tumors has been reviewed and schemes of tumor formation are described which involve infidelity of replication of DNA. The challenge may now be to establish reliable methods for assay of this possible effect of putative carcinogens for identification of environmental causes of cancer.

Acknowledgements

I am grateful to my colleagues, Emeritus Professor Brian Coghlan and Dr Hubertus Jersmann, for translations of the relevant texts in German and French. The staff members of the Barr-Smith Library, University of Adelaide, have been outstanding in the helpfulness and patience with my use of the contents and other resources of that Collection.

References

Bauer, K. H. (1928) *Mutationstheorie der Geschwülst-Entstehung/Mutation Theory in Tumour formation.* Julius Springer, Berlin.

Beckman, R. A. and Loeb, L. A. (2005) Genetic instability in cancer: theory and experiment. Semin. *Cancer Biol.* 15: 423-435.

Berenblum, I. (1974) Carcinogenesis as a biological problem. *Frontiers of Biology,* vol 34. North-Holland, Amsterdam, pp 275-321.

Bichat, X. (1802) *Anatomie generale, appliquee à la physiologie et à la medecine.* Nouvelle ed., avec des notes et additions par P.A. Beclard. Paris, J.-A. Brosson et J.-S. Chaude, 1821.

Bignold, L. P. (2002a) The mutator phenotype/clonal selection hypothesis can provide a basis for the complex and variable morphological and behavioural features of solid tumour malignancies. *Cell. Molec. Life Sci.* 59: 950-958.

Bignold, L. P. (2002b) Hypothesis for the influence of fixatives on the chromatin patterns of interphase nuclei, based on shrinkage and retraction of nuclear and perinuclear structures. Br. J. Biomed. Sci. 59: 105-13.

Bignold, L. P. (2003a) Initiation of genetic instability and tumour formation: a review and hypothesis of a nongenotoxic mechanism. *Cell. Molec. Life Sci.* 60: 1107-1117.

Bignold, L. P. (2003b) The mutator phenotype theory of carcinogenesis and the complex histopathology of tumours: support for the theory from the independent occurrence of nuclear abnormality, loss of specialisation and invasiveness among occasional neoplastic lesions. *Cell. Molec. Life Sci.* 60: 883-891.

Bignold, L. P. (2003c) Pathogenetic mechanisms of nuclear pleomorphism of tumour cells based on the mutator phenotype theory of carcinogenesis. *Histol. Histopathol.* 18: 657-664.

Bignold, L. P. (2004a) Carcinogen-induced impairment of enzymes for replicative fidelity of DNA and the initiation of tumours. *Carcinogenesis* 25: 299-307.

Bignold, L. P. (2004b) Chaotic genomes make chaotic cells: the mutator phenotype theory of carcinogenesis applied to clinicopathological relationships of solid tumors. *Cancer Invest.* 22: 338-343.

Bignold, L. P. (2004c) The cell-type-specificity of inherited predispositions to tumours: review and hypothesis. *Cancer Lett.* 216: 127-46.

Bignold, L. P. (2005) Embryonic reversions and lineage infidelities in tumour cells: genome-based models and role of genetic instability. *Int. J. Exp. Pathol.* 86: 67-79.

Bignold, L. P. (2006) Alkylating agents and DNA polymerases. *Anticancer Res.* 26 (2B): 1327-1336.

Bignold, L. P., Coghlan, B. D. L. and Jersmann, H. P. A. (2006) Cancer: Cell structures, carcinogens and genetic instability – a background. *EXS* 96: 1-24.

Bignold, L. P., Coghlan B. D. L. and Jersmann, H. P. A. (2007) *David Paul von Hansemann: Contributions to Oncology. Context, Comments and Translations.* Birkhäuser, Basel.

Boveri, T. (1914) *Zur Frage der Entstehung der Malignen Tumoren.* G Fischer, Jena. Trans M Boveri, and published as Origin of Malignant Tumors, 1929, Williams and Wilkins, Baltimore.

Braakhuis, B. J., Tabor, M. P., Kummer, J. A., Leemans, C. R. and Brakenhoff, R.H. (2003) A genetic explanation of Slaughter's concept of field cancerization: evidence and clinical implications. Cancer Res. 63:1727-1730.

Bradbury, S. (1967) *The Evolution of the Microscope*. Permagon Press, Oxford.

Brinckerhoff, C. E., Rutter, J. L. and Benbow, U. (2000) Interstitial collagenases as markers of tumor progression. *Clin. Cancer Res.* 6: 4823-4830

Carrio, M., Arderiu, G., Myers, C. and Boudreau, N. J. (2005) Homeobox D10 induces phenotypic reversion of breast tumor cells in a three-dimensional culture model. *Cancer Res.* 65: 7177-85.

Cheng, K.C. and Loeb, L.A. (1997) Genomic stability and instability: a working paradigm. *Curr. Top. Microbiol. Immunol.* 221: 5-18.

Chidgey, M. (2002) Desmosomes and disease: an update. *Histol. Histopathol.* 17: 1179-1192

Chrenek, M. A., Wong, P. and Weaver, V. M. (2001) Tumour-stromal interactions. Integrins and cell adhesions as modulators of mammary cell survival and transformation. *Breast Cancer Res.* 3: 224-229.

Clark, G. and Kasten, F. H. (1983) *History of Staining.* 3rd edn. Williams and Wilkins, Baltimore.

Cleaver, J. E. (1968) Defective repair replication of DNA in xeroderma pigmentosum. *Nature* 218: 652-656.

Cohnheim, J. (1882) *Lectures in General Pathology.* Trans A. B. Mckee, 1889, New Sydenham Society, London, section II, pp 746-821.

Colby, T. V., Koss, M. N. and Travis, W. D. (1995) *Tumors of the Lower Respiratory Tract. Atlas of Tumor Pathology,* Third Series, Fascicle 13, Armed Forces Institute of Pathology, Washington DC, p 533.

Cox, E. C. and Yanofsky, C. (1967) Altered base ratios in the DNA of an Escherichia coli mutator strain. *Proc. Natl Acad. Sci. U. S. A.* 58: 1895-1902.

Dobashi, Y., Watanabe, H., Matsubara, M., Yanagawa, T., Raz, A., Shimamiya, T. and Ooi, A. (2006) Autocrine motility factor/glucose-6-phosphate isomerase is a possible predictor of metastasis in bone and soft tissue tumours. *J. Pathol.* 208: 44-53.

Duesberg, P. (2005) Does aneuploidy or mutation start cancer? *Science.* 307: 41-2.

Duesberg, P., Li, R., Fabarius, A. and Hehlmann, R. (2005) The chromosomal basis of cancer. *Cell Oncol.* 27: 293-318.

Frisch, S. M. (1997) The epithelial cell default-phenotype hypothesis and its implications for cancer. *Bioessays* 19: 705-709

Gaasenbeek, M., Howarth, K., Rowan, A. J., Gorman, P. A., Jones, A., Chaplin, T., et al (2006) Combined array-comparative genomic hybridization and single-nucleotide polymorphism-loss of heterozygosity analysis reveals complex changes and multiple forms of chromosomal instability in colorectal cancers. *Cancer Res.* 66:3471-3479.

Grier, D. G., Thompson, A., Kwasniewska, A., McGonigle, G. J., Halliday, H. L. and Lappin, T. R. (2005) The pathophysiology of HOX genes and their role in cancer. *J. Pathol.* 205: 154-171.

Hansemann, D. P. (1890) Ueber asymmetrische Zelltheilung in epithelkrebsen und deren biologische Bedeutung/On the asymmetrical division of cells in epithelial carcinomata and their Biological Importance. *Virchow's Arch.* 119: 299-326. Translation in Bignold et al, 2007

Hansemann, D. P. (1891) Ueber pathologische Mitosen/On pathological mitoses. *Virchow's Arch.* 123: 356-370. Translation in Bignold et al, 2007

Hansemann, D. P. (1892) Ueber die Anaplasie der Geschwulstzellen und die asymmetrische Mitose. Anaplasia of cancer cells and asymmetrical mitosis. *Virchow's Arch.* 129: 436-449. Translation in Bignold et al, 2007.

Hansemann, D. P. (1893) *Studien über die Spezificität, den Altruismus und die Anaplasie der Zellen mit besonderer Berücksichtigung der Geschwülste/ Studies of the specificity, altruism and anaplasia of cells with special reference to tumours.* A Hirschwald, Berlin. Translation in Bignold et al, 2007.

Harris, H. (1999). *The Birth of the Cell.* Yale University Press, New Haven and London.

Hino, O., Kobayashi, T. and Okimoto, K.(2006) Genetic and environmental factors in hereditary predisposition to tumors: a conceptual overview. *EXS* 96: 269-292.

Hittelman, W. N. (2001) Genetic instability in epithelial tissues at risk of cancer. *Ann. N. Y. Acad. Sci.* 952: 1-12.

Home, E. (1820) The Croonian Lecture: a farther investigation of the component parts of the blood. *Phil. Trans. R. Soc. Lond.* 110: 1-10.

Israel, O. (1902) Bemerkungen über die Anaplasie von Hansemann's und die Entstehung bösartiger Geschwülste, insbesondere des Carcinoms./Remarks on Hansemann's anaplasia and the origin of malignant tumours, in particular of carcinoma. *Virchow's Arch.* 167: 533-537. Translation in Bignold et al, 2007.

Koller, P. C. (1957) The genetic component of cancer. In: *Cancer,* vol 1, pp. 335-403. Ed RW Raven, Butterworth and Co, London.

Lea, D. E. (1962) *Actions of Radiations on Living Cells.* 2nd edn. Cambridge University Press, Cambridge UK.

Levine, M. D., Liotta, L. A. and Stracke, M. L. (1995) Stimulation and regulation of tumor cell motility in invasion and metastasis. Experientia Supplementum, *EXS* 74: 157-179.

Li, R., Sonik, A., Stindl, R., Rasnick, D. and Duesberg, P. (2000) Aneuploidy vs. gene mutation hypothesis of cancer: recent study claims mutation but is found to support aneuploidy. *Proc Natl Acad Sci U S A.* 97: 3236-3241.

Lima-de-Faria, A. (2003) *One Hundred Years of Chromosome Research, and What Remains To Be Learned.* Kluwer, Dordrecht.

Lobstein, J. F. (1829) *Traite d'anatomie pathologique.* F. G. Levrault, Paris.

Lockhart-Mummery, J. P. (1934) *The Origin of Cancer.* J. and A. Churchill, London.

Loeb, L. A. (1996) Many mutations in cancers. *Cancer Surv.* 28: 329-342.

Loeb, L. A. (2001) A mutator phenotype in cancer. *Cancer Res.* 61: 230-239.

Loeb, L. A. and Kunkel, T. A. (1982) Fidelity of DNA synthesis. *Annu. Rev. Biochem.* 51: 429-457.

Loeb, K. R. and Loeb, L. A. (2000) Significance of multiple mutations in cancer. *Carcinogenesis* 21: 379-385.

Loeb, L. A., Springgate, C. F. and Battula, N. (1974) Errors in DNA replication as a basis of malignant changes. *Cancer Res.* 34: 2311-2321.

Marsh, D. J. and Zori, R. T. (2002) Genetic insights into familial cancers – update and recent discoveries. *Cancer Lett.* 181: 125-164.

Martinez, N., Alonso, A., Moragues, M. D., Ponton, J. and Schneider, J. (1999) The nuclear pore complex protein Nu88 is overexpressed in tumor cells. *Cancer Res.* 59: 5408-5411.

Montell, D. J. (2005) Anchors away! Fos fosters anchor-cell invasion. *Cell.* 121: 816-817.

Mothersill, C. and Seymour, C. B. (2006) Actions of radiation on living cells in the "post-bystander" era. *EXS* 96:159-177.

Muller, H. J. (1922) Variation due to change in the original gene. *The American Naturalist* 56: 32-50.

Müller, J. (1838) *On the Nature and Structural Characteristics of Cancer and those Morbid Growths which may be Confounded with It.* G. Reimer, Berlin. Trans C. West, 1840, Sherwood, Gilbert and Piper, London.

Murray, J. A. (1904) *Spontaneous cancer in the mouse.* Sci. Rpts Imp. Cancer Res. Fund No 3, Taylor and Francis, London, 159-174.

Nelson, R.L. and Mason, H.S. (1972) An explicit hypothesis for chemical carcinogenesis. *J. Theor. Biol.* 37: 197-200.

Pathak, S. and Multani, A. S. (2006) Aneuploidy, stem cells and cancer. *EXS* 96: 49-64.

Pattison, D. A. and Davies, M. J. (2006) Actions of ultraviolet light on cellular structures. *EXS* 96: 131-158.

Philips, F. S. (1950) Recent contibutions to the pharmacology of bis 2-haloethyl amines and sulfides. *Pharmacol. Rev.* 2: 281-323.

Rather, L. J. (1978) *The Genesis of Cancer. A Study in the History of Ideas.* The Johns Hopkins University Press, Baltimore.

Rosai, J., Carcangui, M. L. and Delellis, R. A. (1992) *Tumours of the Thyroid Gland. Armed Forces Institute of Pathology,* Washington DC. p 61.

Ross, W. C. J. (1962) *Biological Alkylating Agents.* Butterworths, London.

Rous, P. (1959) Surmise and fact on the nature of cancer. *Nature* 183: 1357-1361.

Sell, S. (1993) Cellular origin of cancer: dedifferentiation or stem cell maturation arrest? Environ. *Health Perspect.* 101 (Suppl 5): 15-26.

Silletti, S. and Raz, A. (1996) Regulation of autocrine motility factor receptor expression in tumor cell locomotion and metastasis. *Curr. Top. Microbiol. Immunol.* 213: 137-169.

Speyer, J. F. (1965) Mutagenic DNA polymerase. *Biochem. Biophys. Res. Comm.* 21: 6-8.

Stoler, D. L., Chen, N., Basik, M., Kahlenberg, M. S., Rodriguez-Bigas, M. A., Petrelli, N. J. and Anderson, G. R. (1999) The onset and extent of genomic instability in sporadic colorectal tumor progression. *Proc. Natl Acad. Sci. U S A.* 96: 15121-15126.

Tyzzer, E. E. (1916) Tumour Immunity. *J. Cancer Res.* 1: 125-155.

Virchow, R. (1858) *Die Cellular Pathologie.* 2nd edn. Trans F. Chance and published as "Cellular Pathology", reprinted 1971, Dover Publications, New York.

Virchow, R. (1863) *Die Krankenhaften Geschwülste/The Disease-related Tumours.* Vol 1, A. Hirschwald, Berlin.

Virchow, R. (1871) Die Cellularpathologie in ihere Begründung auf physiologische und pathologishe Gewebelehre / *The Cellular Pathology Based on Physiological and Pathological Tissue Teaching.* 4th edn, A. Hirschwald, Berlin.

Virchow, R. (1884) Ueber Metaplasie/On Metaplasia. *Virchow's Arch.* 97: 410-430.

Vogelstein, B. and Kinzler, K. W. (1998) *The Genetic Basis of Human Cancer.* McGraw-Hill, New York.

Warthin, A. S. (1913) Heredity with reference to carcinoma. *Arch. Int. Med.* 12: 546-555.

Watson, J. D. and Crick, F. H. (1953) Genetical implications of the structure of deoxyribonucleic acid. *Nature* 171:964-967.

Wheeler, G. P. (1962) Studies relating to the mechanisms of action of cytotoxic alkylating agents: a review. *Cancer Res.* 22: 651-688.

Wheelock, M. J. and Johnson, K. R. (2003) Cadherins as modulators of cellular phenotype. *Annu. Rev. Cell. Dev. Biol.* 19: 207-235.

Whitman, R. C. (1919) Somatic mutation as a factor in the production of cancer; a critical review of v. Hansemann's theory of anaplasia in the light of modern knowledge of genetics. *J. Cancer Res.* 4: 181-202.

Willis, R. A. (1948) *Pathology of Tumours.* Butterworth and Co, London.

Wogan, G. N., Hecht, S. S., Felton, J. S., Conney, A. H. and Loeb, L. A. (2004) Environmental and chemical carcinogenesis. *Semin. Cancer Biol.* 14: 473-486.

Wolff J. (1907) Die Lehre von der Krebskrankheit/ *The Science of Cancerous Disease from Earliest Times to the Present. Translation by B Ayoub, and with an Introduction by P Sarco.* Science History Publishers, Sagamore Beach, MA, USA, 1990.

In: New Research on Genomic Instability
Editor: Eleanor J. Glascow, pp. 121-137

ISBN: 1-60021-320-0

Chapter V

Genomic Instability of the APC Gene Found in Glioblastoma

Nives Pecina-Slaus[1,2*], Vili Beros[3], Tamara Nikuseva Martic[1,2] and Floriana Bulic-Jakus[2]

[1] Laboratory of Neurooncology, Croatian Institute for Brain Research, School of Medicine University of Zagreb, Zagreb, Croatia

[2] Department of Biology, School of Medicine, University of Zagreb, Zagreb, Croatia

[3] Department of Neurosurgery, University Hospital "Sisters of Charity", Zagreb, Croatia

Abstract

The etiology and pathogenesis of tumors of the central nervous system are still inadequately explained. This study analyses tumor suppressor gene—adenomatous polyposis coli (APC) in 28 patients with glioblastoma, the most aggressive form of glial tumors. APC protein has a structural role in adherens junctions, but also plays a signaling role as a negative regulator of the wnt pathway. Our interest in the APC gene stemmed principally from the findings that the wild-type APC protein is highly expressed in the central nervous system, and upon the finding that it is critically involved in particular syndromes, among which brain tumors play a significant role. Glioblastoma samples were tested for gene instability by PCR/loss of heterozygosity using RFLP method. Two polymorphic markers were used: an Rsa I polymorphic site in exon 11, and an Msp I polymorphic site in exon 15. The results of our analysis for both markers showed allelic loss of the APC gene in 40% of our sample out of 25 heterozygous patients (informativeness 89%). Another 20% of samples demonstrated allelic imbalance of the

* Corresponding author: Nives Pecina-Slaus. Laboratory of Neurooncology, Croatian Institute for Brain Research, School of Medicine University of Zagreb, Salata 3, HR-10000 Zagreb, Croatia, e-mail: nina@mef.hr, tel. +385 1 46 21 140, fax:+385 1 45 50 744; +385 1 49 20 050; +385 1 45 96 942

APC allele in tumor tissue. Altogether, there were 15 samples (60%) demonstrating instability of this tumor suppressor gene. Despite increasing knowledge on glioma biology and genetics, the prognostic tools for glioblastoma still need improvement. Our findings on genomic instability of the APC gene may contribute to better understanding of glioblastoma genetic profile and could be used as a prognostic marker of disease evolution and progression.

Key words: glioblastoma, adenomatous polyposis coli gene (APC), loss of heterozygosity, wnt signaling pathway

Introduction

The genome of tumor cells is affected by two crucial types of changes: the accumulation of somatic mutations and the acquisition of genomic instability. The usual incidence rate of spontaneous somatic mutations is much lower to the rates of genetic changes observed in tumor cells and this increased frequency is the result of genomic instability that characterizes such cells. It is now believed that the increase in frequency of genetic changes is to be at least 100 times. Although it is usually assumed that the genetic instability is a later event in tumor progression, the question whether it may be an early causative event [1] in the formation of a specific tumor type is still speculative.

The well-known two-hit hypothesis for tumor suppressor genes in hereditary cancer, that has nowadays transcedented to the multi-hit hypothesis, describes that as many as seven distinct and sequential mutations are needed for the development of malignant tumor, whereas fewer could give rise to benign precursors. Kinzler and Vogelstein [2] investigating colo-rectal carcinoma provided evidence for this hypothesis and showed that the mutations could accumulate over a span of years to decades. On the other hand, the body of evidence indicates that the mutation rate of normal human cells is much too low to account for the accumulation of this number of genetic alterations over a single individual's life time [3]. We know that multiple genetic changes are required for carcinogenesis and the inadequate incidence rate observed during human life time seeks explanation. The answer to this discrepancy is given in the so-called mutator hypothesis. A mutator gene is a gene whose mutation increases the level of genetic changes of the individual's genome [4]. The mutator gene can be any gene, but usually comes from the group of genes whose products are involved in DNA repair mechanisms or genes whose products control the fidelity of DNA replication. Mutations of those genes are referred to as mutator mutations, and cancer cells are said to exhibit a mutator phenotype with an increased rate of genomic instability.

There are two categories of genomic instability observed in human cancers: microsatellite instability (MIN) and chromosomal instability (CIN). MIN involves the defects in the repair of short mismatches in the replication/repair machinery [5] resulting from the slippage of DNA polymerase as it synthesizes segments of short repeats. The consequential instability observed in tumor is sometimes termed as RER+ tumor phenotype from Replication Error Positive phenotype. RER-positive tumor samples can be visualized on high resolution gels after electrophoresis techniques as samples with bands on different positions in comparison to bands of autologous normal tissue. CIN occurs at the chromosome level and is caused by chromosomal mis-segregation due to improper alignment of chromosomes in mitosis. The resulting daughter cells comprise aneuploidy and allelic imbalance. Besides instability at the whole chromosomal level, there are other forms of CIN. Translocations, segmental duplications and deletions, as well as gene amplifications, can also be considered a form of CIN, although mechanisms that cause such alterations are distinct from partitioning at mitosis [1, 6].

Our study analyzed genomic instability in brain tumor glioblastoma, with regard to the role of the tumor suppressor gene—adenomatous polyposis coli (APC). Protein product of this gene is a component of the adherens junction. Its central part binds to E-cadherin, the most important molecule of cell-cell adhesion, via beta-catenin molecule [6, 7]. APC is illustrative of multiple roles that certain tumor suppressors play in a cell. Besides its structural role in cellular architecture, the protein product of the APC gene is also one of the key players of the wnt signaling pathway. The wnt/wingless pathway was first discovered in mouse and *Drosophila* and is one of the most interesting signal transductions, in which key components have multiple functions. In vertebrate cells, it is named after Wnt proteins, a family of highly conserved secreted signaling molecules that regulate cell-to-cell interactions during embryogenesis. Insights into the mechanisms of Wnt action have emerged from several systems: genetics in *Drosophila* and *Caenorhabditis elegans*; biochemistry in cell culture; and ectopic gene expression in *Xenopus* embryos. Many Wnt genes in the mouse have been mutated, leading to very specific developmental defects. As currently understood, Wnt proteins bind to receptors of the Frizzled family on the cell surface. Through several cytoplasmic relay components, the signal is transduced to beta-catenin, which then enters the nucleus to activate transcription of Wnt target genes [9, 10]. Although the main signaling molecule of the pathway is beta-catenin, APC is a critical component of the beta-catenin destruction machinery heading to the proteasome, and acts as a negative regulator of the wnt pathway. When wnt ligand is absent, beta-catenin binds to APC protein and is being destroyed. Another tantalizing compartment of the wnt signaling pathway lies in downstream transcriptional activation. In response to wnt signaling, or under the circumstances of mutated APC, beta-catenin is stabilized, accumulates in the cytoplasm and enters the nucleus, where it finds a partner, a member of the DNA binding protein family LEF/TCF (lymphoid enhancer factor-T cell factor). Together they activate new gene expression programs. Target genes for beta-catenin/TCF encode c-MYC and cyclin D1 proteins [11], explaining why constitutive activation of the wnt pathway can lead to cancer. The schematic illustration of key components of the wnt signaling is shown in figure 1.

It has been well documented that wnt genes, together with other components of the wnt signaling pathway, are implicated in cancer [10], especially in neoplasms of epithelial origin. Mutations in the adenomatous polyposis coli gene are responsible for familial adenomatous polyposis and the majority of sporadic colorectal cancers. The APC gene (chromosome 5q21) [12] is organized in 16 translated exons and encodes a 2843 amino acid protein that is expressed in specific (frequently post-mitotic) epithelial and mesenchymal cells of several fetal and adult human tissues. The final protein product of APC gene with a mass of 311.8 kDa shows little similarity to other known proteins. There are three known APC protein isoforms that arise through a complicated pattern of alternative splicing. The conventional

APC splice isoform is present in many cell types, while other splice forms are present only in differentiated cells [13]. These alternative splice forms exclude exon 1 but include alternate exons 5 'of the missing exon 1 [14]. Two of the splice variants, BS (brain-specific) and 0.3 (exon 0.3) contain in-frame translation initiation codons, and the protein isoforms lack exon 1 encoded homodimerization domain. It is interesting that these protein isoforms are expressed in neuronal cells and other tissues with nondividing cell populations. Carson et al. [15] demonstrated that all isoforms cause similar reduction in beta-catenin activity and show no difference in the subcellular localization, and concluded that different isoforms are not trancriptionally different but rather involve unidentified protein functions of the specific domains. The majority of identified mutations of the APC gene are frameshift mutations [16] that almost always lead to a truncated protein product, devoid of C-terminal peptide sequences. A number of tumors, including colorectal, gastric, lung, hepatocellular, renal and laringeal tumors demonstrate loss of heterozygosity (LOH) of this gene [2, 17-19]. It has been shown that tumor suppressors are differentially expressed during normal embryonic development [20] and that specific developmental pathways are often inappropriately regulated during tumorigenesis [21, 22].

Our interest in elucidating the role of tumor suppressor gene—adenomatous polyposis coli (APC) stemmed principally from the findings that wild-type APC protein is expressed in the central nervous system [23, 24] and that there are strong indications that wild-type APC protein is critically involved in initiation of neuronal differentiation [25]. Moreover, new knowledge on wnt signaling shows that wnt proteins regulate critical developmental processes of normal brain development [26-30]. Processes that include cellular adhesion and synaptic rearrangements require the expression of molecular components of the wnt pathway [31]. As we have already pointed out, the wnt pathway has repeatedly been implicated in tumorigenesis and lately in brain tumorigenesis as well [32]. These findings inspired us to focus our interest on the key player of the wnt signaling, APC tumor suppressor gene. The majority of brain tumors arise sporadically except those that are part of defined clinical syndromes. APC protein is critically involved in particular syndromes, like the Tourcot's syndrome, which includes the development of primary brain tumors such as medulloblastomas and gliomas [33, 34]. Cytoplasmically associated to E-cadherin is beta-catenin, yet another important molecule in brain tumorigenesis. Activating mutations of beta-catenin gene are found in a variety of tumors suggesting that beta-catenin gene acts as an oncogene. In 2003. beta-catenin was identified as a critical factor for dendritic morphogenesis [35]. Mutations of this gene have been reported in sporadic medulloblastoma [36] and recently this molecule was proposed prognostic factor for medulloblastoma [37].

The etiology and pathogenesis of the tumors of the central nervous system are still inadequately explained. Our knowledge on genetic background of specific histopathological type of brain tumor still needs to be elucidated although great progress has been achieved along with the increasing advances in molecular genetics. Glioblastoma is the most aggressive form of glial brain tumor. The majority of glioblastomas arise without clinical or histological evidence of less malignant lesion and this kind of glioblastoma is categorized as primary or *de novo* glioblastoma. They manifest in older patients (mean age 55 years) after a short clinical history. In contrast, secondary glioblastomas develop more slowly by malignant progression from diffuse or anaplastic astrocytoma and they manifest in younger patients (mean age 40 years) [38]. At present specific genetic alterations found to be associated with glioblastoma are emerging [39-41]. Both types of glioblastoma have disrupted p53 and Rb1 pathways, although in different ways. Primary glioblastomas have amplification of the 12q14 region encompassing the CDK4 and MDM2 genes which results in their over-expression and the dysfunction of both pathways. In addition to the genetic abnormalities involving p53 and Rb1 pathways, glioblastomas exhibit loss of alleles from 10q in over 90% of cases. This genetic region encompasses PTEN tumor suppressor gene. Moreover, amplification of the epidermal growth factor receptor (EGFR) gene has been reported in about 35% of glioblastoma [40, 42, 43]. Secondary glioblastomas generally have no wild type p53 due to loss of one allele and mutation of the other allele and also loss of functional Rb1 pathway in a similar manner. Nevertheless, the genetic basis of primary as well as reoccurring glioblastoma is unclear, and much work is still required to determine the final list of genes involved, as well as the point in time of their activation/inactivation. In this chapter we offered a new candidate to fill in the puzzle of genetic basis of glioblastoma. Novel genes and molecular candidates involved in mechanisms of brain tumor formation will offer improvement in comprehension, diagnosis and treatment of this disease.

Materials and Methods

Twenty-eight glioblastoma samples together with 28 autologous blood tissues were collected from the Department of Neurosurgery, University Hospital "Sisters of Charity", Zagreb, Croatia. Using the magnetic resonance imaging (MRI) tumor lesions were found in different cerebral regions (predominantly temporal and parietal region), with the surrounding zone of perifocal oedema (table 1). During the operative procedure the tumor was maximally reduced using a microneurosurgical technique. The patients had no family history of brain tumors and did not undergo chemotherapy or radiotherapy prior to surgery. Collected tumor tissues were frozen in liquid nitrogen and transported to the laboratory, where they were immediately transferred at -75°C. The peripheral blood samples were processed immediately. The histopathological classification was glioblastoma. The local Ethical Committee approved our study and patients gave their informed consent.

DNA Extraction

Tumor sample for DNA isolation was the part of obvious tumor mass evaluated by the neurosurgeon and based on macroscopic appearance and tissue color, density, and consistency on gross section. The sample was also evaluated for the percentage of tumor cells by pathologist and consisted of more than 85% of tumor cells.

Approximately 0.5 g of tumor tissue was homogenized with 1 ml extraction buffer (10 mM Tris HCl, pH 8.0; 0.1 M EDTA, pH 8.0; 0.5% sodium dodecyl sulfate) and incubated with proteinase K (100 μg/ml; Sigma, USA; overnight at 37°C). Phenol chloroform extraction and ethanol precipitation followed.

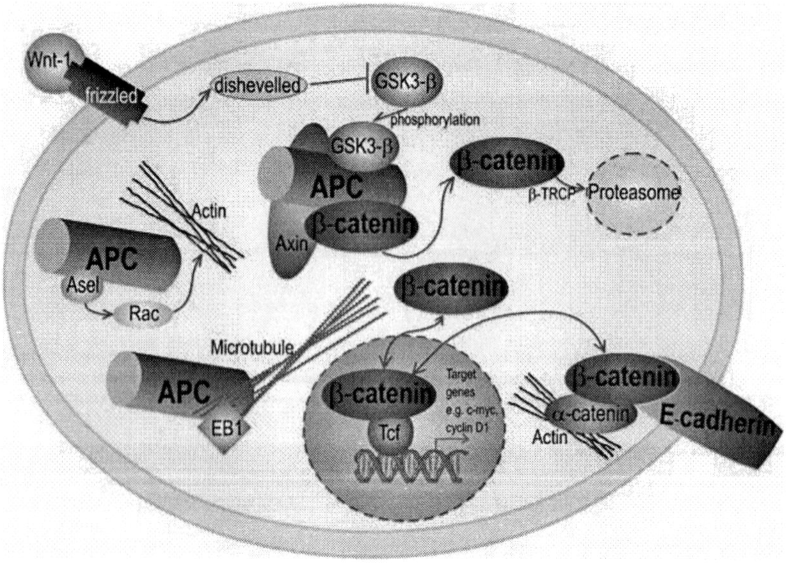

Figure 1. Wnt signaling pathway demonstrating its key components: frizzled (receptor of WNT proteins; GSK3-beta (glycogen-synthase-kinase 3-beta); Adenomatous polyposis coli (APC); beta-catenin. E-cadherin mediated adhesion is also shown since it modulates wnt signaling by sequestering cytoplasmic beta-catenin.

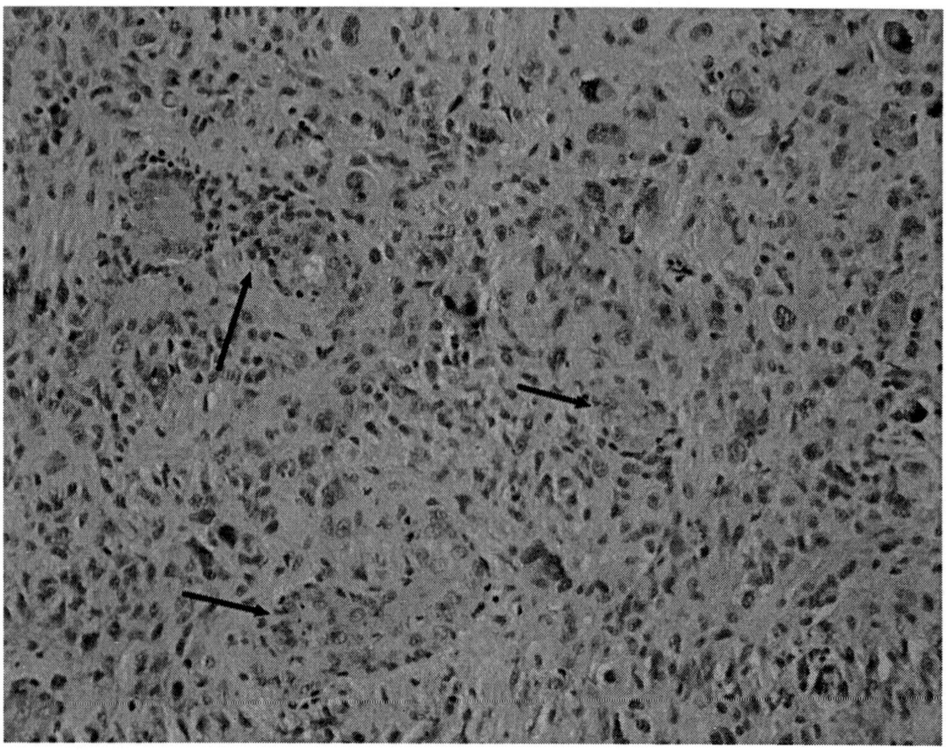

Figure 2. Glioblastoma multiforme (HE, x 200). Dense cellularity, striking pleomorphism, brisk mytotic activity and prominent microvascular proliferation (arrows).

Blood sample was used to extract lymphocyte DNA. Five ml of blood was lysed with 7 ml distilled water and centrifuged (15 min/5000 *g*). The pellet was then processed as for DNA extraction from the tissue samples.

Polymerase Chain Reaction

The optimal reaction mixture (25 μl) for APC's exon 11 amplification was: 20 pmol of each primer (5'-GGACTACAGGCCATTGCAGAA-3' and 5'-GGCTACATCTCCAAAA GTCAA-3'), 200 μM of each dNTP, 200-400 ng template DNA, 1 μl (0.5 U) of Taq polymerase, 1 mM $MgCl_2$, 5 μl 10 X reaction buffer II (500 mM KCl, 100 mM Tris-HCl, pH 8.3). Polymerase chain reaction (PCR) conditions were: initial denaturation, 4 min/95°C; denaturation, 1 min/94°C; annealing, 2 min/58°C; extension, 1.5 min/72°C; 35 cycles.

To amplify fragment of the exon 15 of the APC gene we used in a volume of 25 μl: 5 pmol of each primer (5-ATGATGTTGACCTTTCCAGGG-3 and 5-CTTTTTTGG CATTGCGGAGCT-3), 200 ng template DNA, 2.5 μl 10X buffer II, 1.5 mM $MgCl_2$, 25 μM of each dNTP, 0.2 μl (5U/μl) of Taq polymerase. PCR conditions were: initial denaturation, 5 min/95°C; denaturation, 30 s/95°C; annealing, 30 s/57°C; extension, 45 s/72°C; final extension, 72°C/7 min; 30 cycles. The PCR products for both markers were analyzed on 2% agarose gels.

Loss of Heterozygosity

Loss of heterozygosity (LOH) of the APC gene was detected on the basis of restriction fragment length polymorphism (RFLP) of the PCR products. Two different polymorphisms were investigated. One is an Rsa I polymorphic site in exon 11, and the other is an Msp I polymorphic site in exon 15. PCR amplification of exon 11 generated a 133- bp fragment that is cleaved to 85- and 48- bp fragments by Rsa I restriction if the polymorphic site is present, and remains uncleaved if the site is absent [44]. The amplified fragment of exon 15 is 550 bp long and is cleaved with the Msp I restriction endonuclease to two 250 bp fragments if the restriction site is present. LOH/Rsa I was demonstrated only in informative (heterozygous) persons when the tumor DNA showed loss of either the single uncut band (133 bp) or of the two cut bands (85+48 bp) compared to autologous blood tissue. For Msp I polymorphism heterozygous patients demonstrated two bands (550+250 bp), while LOH was shown when either band was missing in comparison to the autologous blood tissue. Samples that demonstrated quantitatively weaker allelic band in tumor tissue than in normal blood DNA were described as samples with allelic imbalance (A.I.).

PCR aliquots (10-15 μl) were digested with 6 U Rsa I (Gibco, USA; 12 h at 37°C) and with 6 U Msp I (AGS, Germany, overnight at 37°C) and were electrophoresed on Spreadex gels EL 300 in the SEA 2000 submarine electrophoresis apparatus (Elchrom scientific, Switzerland) at 120V. Temperature of the running buffer was kept constant at 55°C. The samples with LOHs were additionaly electrophoresed on 15% polyacrylamide gels stained with silver.

RESULTS

From 28 glioblastoma samples 79% appeared to have developed *de novo* , while six samples developed from previously diagnosed tumors of lower malignancy grade, two from diffuse astrocytoma (GII), one from anaplastic astrocytoma (GIII), two from oligoastrocytoma and one from oligodendroglioma. The histopathological diagnosis of glioblastoma (malignancy grade IV) was based on recognition of poorly differentiated astrocytic tumor cells with brisk mytotic activity and marked nuclear atypia. Differentiated elements were intermingled with bizzare multinucleated tumor giant cells. Prominent microvascular hyperplasia, in which proliferating blood vessels come to be lined by cells heaped up in disorderly fashion and transformed into glomeruloid or solid tufts (figure 2), and necrosis were also recognized as diagnostic features. The age of patients varied from 31 to 77 years (mean age = 59.1), and there were 15 females and 13 males. Tumor localization and the duration of symptoms are shown in table 1.

Genomic Instabilities

The Rsa I polymorphic site in APC's exon 11 was analyzed first. From 28 glioblastoma samples analyzed, 18 (64.3%) were informative for this polymorphism. For Msp I polymorphism in APC's exon 15, we found 20 heterozygous patients (71.4%). The results of our analysis showed 7 samples with LOH/exon 11 of the APC gene (39%) when tumor DNA was compared to autologous constitutive blood DNA. Another three samples demonstrated allelic imbalance (quantitatively weaker allelic band in tumor tissue than in normal blood DNA). Genetic changes of exon 15 demonstrated 3 samples with LOHs (15%) and another two samples with allelic imbalance.

When summing the informativeness provided by both markers, our sample showed (25/28) 89% of heterozygosity. The results of our analysis for both markers showed allelic loss of the APC gene in 10 glioblastoma samples out of 25 heterozygous patients (40%). Another 20% of samples demonstrated allelic imbalance of the APC allele in tumor tissue. Altogether, there were 15 samples (60%) demonstrating instability of this tumor suppressor gene. Three glioblastoma samples had both LOH at exon 11 and 15, while four demonstrated LOH exclusively on exon 11, and another three A.I. were detected in this genetic area. Two A.I. found on exon 15 did not show losses on exon 11. Genomic instabilities of tumor suppressor gene investigated and the polymorphic status for both markers are summarized in table 1. LOHs of the APC gene that both markers revealed are shown in Figure 3 A (exon 11) and B (exon 15).

When we correlated our molecular findings with the demographic variables we observed that genomic instability of the APC gene was confined to younger patients.

Table 1. Genetic changes of the APC tumor suppressor gene, anatomical location of glioblastoma and the duration of symptoms

Patient No.	APC gene exon11/Rsa I	APC gene exon15/MspI	Localization	Symptom duration/ months
1[1]	heterozygous	A.I.	FTP left	6
2	homozygous b	A.I.	FTP left	2
3	A.I.	heterozygous	P left	1
4	A.I.	heterozygous	TP right	7
5	A.I.	heterozygous	T right	0,5
6	heterozygous	heterozygous	FTP right	0,5
7	heterozygous	heterozygous	T left	1
8	heterozygous	heterozygous	FTP left	0,5
9	heterozygous	heterozygous	FT left	0,5
10	heterozygous	heterozygous	P right	1
11[2]	homozygous b	homozygous d	P right	2
12	homozygous a	heterozygous	bifrontal	1
13	homozygous a	heterozygous	FPO right	3
14	homozygous a	heterozygous	FTP left	2
15	homozygous a	heterozygous	T right	3
16	homozygous a	heterozygous	T left	2
17	homozygous b	heterozygous	TO right	2
18[1]	LOH	heterozygous	P left	12
19	heterozygous	homozygous c	T left	10
20[3]	heterozygous	homozygous c	T right	18
21	LOH	homozygous c	P left	1
22[4]	LOH	homozygous c	TO right	3
23	LOH	LOH	TP left	1
24	LOH	LOH	T left	3
25[3]	LOH	LOH	FTleft	2
26	LOH	ND	P left	2
27	homozygous b	ND	FP right	2
28	homozygous a	ND	FP left	0,5

Heterozygous = without LOH or A.I; a = both alleles 133 bp; b = both alleles have restriction sites (85+48 bp); c = both alleles 500 bp; d = both alleles have restriction sites (2 x 250); A.I. = allelic imbalance; [1]from diffuse astrocytoma (GII); [2]from anaplastic astrocytoma (GIII); [3]from oligoastrocytoma; [4]from oligodendroglioma; FTP= frontotemporoparietal region; P = parietal region ; TP = temporoparietal region ; T = temporal region; FPO = frontoparietooccipital region; TO = temporooccipital region; FT = frontotemporal region; FP= frontoparietal region; ND = not determined.

Discussion

The mechanisms of brain tumor initiation and progression have not yet been completely investigated and elucidated. As with other tumors, the formation of brain tumors is the result of multiple consecutive genetic changes that represent a critical factor in tumor evolution. No consistent genetic abnormalities have been detected that would indicate the genetic profile of glioblastoma [40]. Histopathologically, an unambiguous distinction of glioma subtypes has remained elusive, but they clearly evolve through different genetic pathways. It also remains to be shown whether these subtypes differ significantly with respect to the molecular genetic alterations underlying the oncogenesis and progression of glioblastoma. With this in mind, we investigated a new candidate gene, tumor suppressor gene APC, in a set of twenty eight glioblastomas. Although APC has been thought of primarily as a colon-specific tumor suppressor gene, its association with certain brain tumors and its expression in the CNS suggests that it performs important functions in these tissues also.

Genes involved in formation and acquisition of the full metastatic potential of specific tumor are not only those responsible for cell proliferation and survival, but also genes responsible for the control of cell adhesion and cell motility [45]. It is now apparent that tumor malignancy can, in certain aspects, be explained by alterations in the adhesive properties of neoplastic cells. One of the most important hallmarks of malignant gliomas is their invasive behavior. Despite modern diagnostics and treatment the median survival time does not exceed 15 months which is in accordance to extreme invasive characteristics of glioblastoma. Demuth and Berens in their review paper [46] state that glioblastoma recur predominantly within 1 cm of the resection cavity mainly due to the fact that at the time of surgery, cells from the bulk tumor have already invaded normal brain tissue. The problem of new rise of the malignant cells left behind the surgical resection of diffuse gliomas is one of the main disadvantages of therapeutic decision-making. The culprit of the highly invasive phenotype of human gliomas is thought to be associated to the cadherin group of adhesion molecules. Classical cadherins such as E-type and N-type are involved in forming both adherens and synaptic junctions in the nervous system. Shimamura and Takeichi [47] found that E-cadherin is transiently expressed in restricted regions of the mouse embryonic and adult brain. Although E-cadherin molecule is a well known and almost universal suppressor of invasion, little is known on the role of cell-cell adhesion in astrocytes and its alteration in migrating and invasive glioblastomas. In 2002. [48] Perego and co-workers demonstrated that disorganization of cadherin mediated junction, in which APC is included, is required to promote migration and invasiveness in glioblastoma cell lines.

The results of the analysis reported in this chapter suggested that APC plays an important role in the mechanisms of development and progression of this specific tumor of the central nervous system. Our previous investigations are also in favor of this presumption. We showed that neuroendocrine tumor pheochromocytoma exhibits microsatellite instabilities of the E-cadherin gene, which is the main adherens junction member molecule and an indirect member of the wnt signaling [49]. Preliminary analysis on oligoastocytoma showed gross deletion of the APC gene [50]. which is also indicative of involvement of this tumor suppressor gene in the formation and development of central nervous system tumors.

The analysis described in the present chapter showed that 60% of glioblastoma samples demonstrated LOH and allelic imbalance of the APC gene, which indicated that gross deletions of APC are part of genetic profile of these tumors. Allelic imbalances, quantitatively weaker allelic bands in tumor tissue than in normal blood, that we observed in 20% of our sample probably was not the result of normal tissue DNA interference. Because each glioblastoma sample was maximally reduced during the operative procedure using a fine microneurosurgical technique, our samples tested for genomic instabilities consisted of 90% of tumor cells. The explanation for allelic imbalances is that not all tumor cells of the sample underwent this sort of alteration. Behaviorally, gliomas can be viewed as consisting of two discreet subpopulations of cells, the proliferative cells at the tumor core, and cells invading the brain parenchyma. Thus, the gene expression profile of the tumor core may not necessarily depict the profile of genes active in the invading rim. Demuth and Berens [46] identified genes differentially expressed in invasive glioma cells and illustrated the differing biology of the invasive cells in contrast to the tumor cells at the core. Findings on different glioblastoma cellular components could explain for the portion of glioblastoma that did not show instabilities of APC gene. The relatively high number of allelic losses of the APC gene in our sample may be attributed to random variation in tumors, but in our opinion the frequency of APC's genomic instability in our sample was too high to be random. The observed high frequency led us to conclude that gross deletion of the APC gene that our glioblastoma samples comprised are an important event in the mechanisms of glioblastoma formation. Unfortunately, it was still not clear whether those changes represent an initiation event or come along the path of glioblastoma progression.

In our study genetic alterations were more frequently confined to exon 11 of the APC gene. Three glioblastoma samples had both LOH at exon 11 and 15, which could indicate that the deletion is extensive and goes all the way to exon 15. The alterations that were uniquely found on exon 15 could indicate that this genetic area is affected, and the same applies to the alterations found exclusively on exon 11. Exon 11 is a much smaller exon than exon 15, the biggest exon of the APC gene which encompasses 3/4 of its coding region (6577 bp). We could suspect that losses of exon 11 would not have such severe consequences at the protein level as would losses of exon 15. The alterations of the APC gene found in our sample were not correlated with the sex of the patients. Nevertheless, we observed that genomic instability of the APC gene was confined to younger patients. The glioblastomas we considered primary because the diagnosis of glioblastoma was made at the first biopsy, without clinical or histopathologic evidence of a less malignant precursor lesion. Patients with glioblastoma de novo are older at diagnosis, compared to patients with secondary glioblastoma, and the mean age of our patient group (=59.1) was somewhat higher than the mean age reported in literature of 55 years.

The clearest example of wnt signaling involvement in brain tumor formation is the critical role of the tumor suppressor gene APC in the Tourcot's syndrome. Association of polyposis of the colon with a CNS malignancy is represented with this syndrome. Genetically predisposed patients are younger at diagnosis of glioblastoma compared to patients with sporadic disease. The relative risk of brain tumor formation is 23 times greater for Tourcot's syndrome patients with APC-mutation when compared to the general population between the ages of 0 and 29. Hamilton et al, [33] reported 14 families with Turcot's syndrome in whom patients with medulloblastoma had mutations in the APC gene, whereas patients with glioblastoma multiforme had germline mutations in one of the mismatch-repair genes. Mutation reports by Mori et al. [51] indicated lack of mutations in neuroepithelial tumors they investigated (glioblastomas included), but presence of germ line mutations of APC gene in 3 Turcot's syndrome patients. Other authors found germline mutations of APC gene in Tourcot's syndrome patients, but failed to detect them in the primary astrocytoma [52], or glioblastoma multiforme [53]. These results suggest that APC gene is associated with pathogenesis of Turcot's syndrome, but that at least one other gene is responsible for the genesis of neuroepithelial tumors in the CNS.

Somatic mutations of the APC gene were reported predominantly in sporadic medulloblastoma [34], and are much less frequent in glioblastoma. Nevertheless, Steigerwald and co-workers [13] found base change mutations in APC gene in two of 23 sporadic glioblastomas examined and a heterozygous G to A transition at position 148 of exon 0.3 in cell line SW 1088 from a human astrocytoma. This base change affects the amino acid sequence of the APC gene. One of the sporadic glioblastoma they investigated had an additional G following a string of seven Gs located at position 34 of exon 0.1 , but the effect on amino acid sequence could not be predicted. Microarray technology has been applied to the genome of glioblastoma, because it allows a large number of hybridizations to be performed simultaneously and was used for array comparative genomic hybridization by Roversi et al. [54]. Twenty five primary glioma cell lines were investigated for gross and subtle genetic changes involved in glioma initiation and progression at the whole genome level. In this extensive study the authors have found among many other candidate chromosomal regions, recurrent losses of 5q22.2-q23.3 region in 10 glioblastoma cell lines (IV WHO grade) they analysed, which is in accordance to our results.

So far little is known about the events that are involved in tumor initiation and early development of glioblastoma. Different subtypes of gliomas differ significantly in their age distribution, growth potential, tendency for progression and clinical course.

The work that we described in this chapter elucidated the potential contribution of genomic instability of APC tumor suppressor gene, to malignant transformation of the specific tumors of the central nervous system. The functional consequences of the observed instabilities that we found at the genetic level, would need to be confirmed at the protein level by analysing changes of APC's protein expression. We can speculate that wild-type APC protein is going to be absent in glioblastoma cells by inactivation of both alleles of the APC gene. One allele could be deleted, and the remaining one might have suffered another type of mutation, probably point mutation or a small deletion. On the other hand, only one allele may have suffered gross deletion, leaving the other one do the job of wild type protein expression. It is therefore unclear whether the clinical evolution of glioblastoma is due to quantitative reduction or complete loss of APC protein product. Whether genetic losses we found have consequences at the protein level is left for future studies.

Nevertheless, our study offers the APC tumor suppressor gene as a potential candidate involved in mechanisms of glioblastoma development. We consider our finding a specific but relevant contribution to understanding the pathophysiological mechanisms of brain tumor formation.

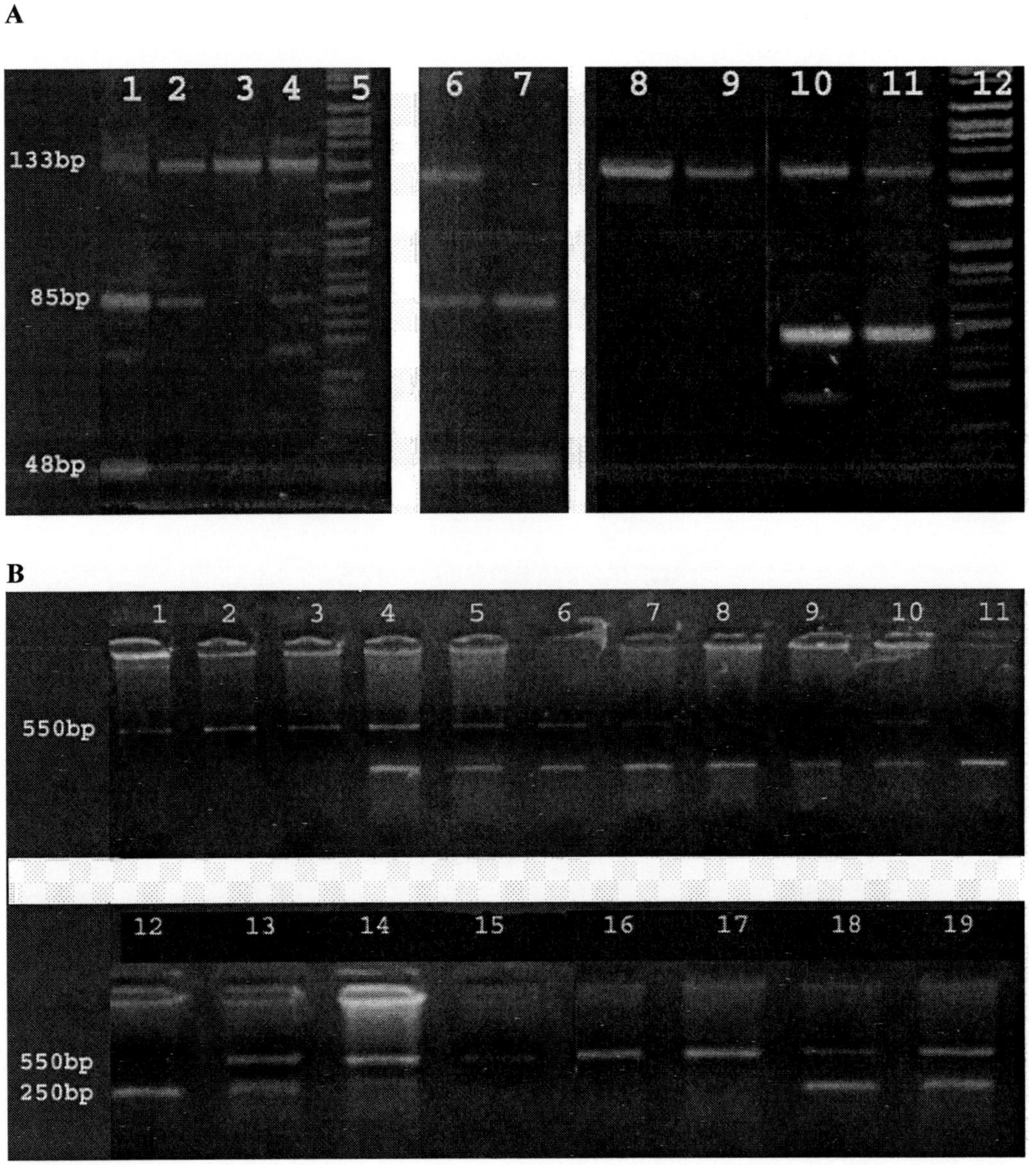

Figure 3. A. Loss of heterozygosity of APC gene in two patients with glioblastoma. Exon 11/RsaI/RFLP is demonstrated. Lanes 1, 2- heterozygous sample (tumor and blood); lane 3- LOH in glioblastoma patient (the digested/cut allele is missing); lane 4- informative blood sample of the same patient showing undigested/uncut allele 133 bp and cut allele (85+48 bp); lanes 5, 12 - standard M3 (Elchrom scientific); lane 6- informative blood sample; lane 7- LOH in the corresponding glioblastoma sample (uncut allele is missing); lanes 8, 9 – uninformative homozygous sample showing uncut allele (tumor and blood), lane 10- informative blood sample; lane 11- allelic imbalance in the corresponding glioblastoma (uncut allele is weaker, 48 bp fragments are not shown).
B. Loss of heterozygosity of APC gene in 3 patients with glioblastoma. Exon 15/MspI/RFLP is demonstrated. Lane 1 - standard DNA/50 bp ladder; lanes 2, 3, 16, 17 – uninformative homozygous samples showing uncut alleles of 550 bp (tumors and blood); lanes 4, 5, 6, 7 - heterozygous samples, both alleles, cut and uncut, are visible; lanes 8, 10, 13, 14, 19 - informative blood samples; lane 9- allelic imbalance in the corresponding glioblastoma (uncut allele is weaker); lane 11, 12- LOHs in the corresponding glioblastomas, uncut alleles are missing; lane 15 – LOH in the corresponding glioblastoma cut allele is missing; lane 18 – allelic imbalance in the glioblastoma corresponding to the blood in lane 19 (uncut allele is weaker).

Future Perspectives

It would be interesting to asses the consequence that the instabilities of the APC gene might have at the protein level. Future studies need to confirm the functional changes of the APC molecule by analyzing its protein expression or changes at APC's mRNA levels.

The question whether microsatellite genetic instabilities, that indicate involvement of mismatch repair, have a role in glioblastoma development and progression is also very tempting and should be considered for future work.

Further investigation regarding this gene on a bigger sample as well as employment of additional markers of the APC gene should be performed in the future to confirm our suggestions.

It would also be relevant to obtain data on wnt target genes, such as c-myc, cyclin D1 in future studies.

Acknowledgements

I would like to thank Prof. Dr. sc. Ivica Kostovic for critically reading the manuscript and for his constructive suggestions. This work was supported by grants 0108215 and 0108049 from the Ministry of Science and Technology, Republic of Croatia.

References

[1] Bignold LP. Embryonic reversions and lineage infidelities in tumour cells: genome-based models and role of genetic instability. *Int J Exp Pathol* 2005, *86*, 67-79.

[2] Kinzler KW, Vogelstein B. Lessons from hereditary colorectal cancer. *Cell* 1996, *87*, 159-170.

[3] Prochownik EV. Functional and physical communication between oncoproteins and tumor suppressors. *Cell Mol Life Sci* 2005, *62*, 2438-2459.

[4] Beckman RA, Loeb LA. Genetic instability in cancer: theory and experiment. *Semin Cancer Biol* 2005, *15*, 423-435.

[5] Ionov Y, Peinado MA, Malkhosyan S, Shibata D, Perucho M. Ubiquitous somatic mutations in simple repeated sequences reveal a new mechanism for colonic carcinogenesis. *Nature* 1993, *363*, 558-561.

[6] Gagos S, Irminger-Finger I. Chromosome instability in neoplasia: chaotic roots to continuous growth. *Int J Biochem Cell Biol* 2005, *37*, 1014-1033.

[7] Rubinfeld B, Souza B, Albert I, Muller O, Chamberlain SH, Masiarz FR *et al.* Association of the APC gene product with beta-catenin. *Science* 1993, *262*, 1743-1745.

[8] Pecina-Slaus N. Tumor suppresor gene E-cadherin and its role in normal and malignant cells. *Cancer Cell Int* 2003, *E3*, 17 (http://www.cancerci.com/content/3/1/17).

[9] Morin PJ, Sparks AB, Korinek V, Barker N, Clevers H, Vogelstein B, *et al.* Activation of beta-catenin-Tcf signaling in colon cancer by mutations in β-catenin or APC. *Science* 1997, *275*,1787-1790.

[10] Peifer M, Polakis P. Wnt signaling in oncogenesis and embriogenesis - a look outside the nucleus. *Science* 2000, *287*, 1606-1609.

[11] He TC, Sparks AB, Rago C, Hermeking H, Zawel L, da Costa LT, *et al.* Identification of c-MYC as a target of the APC pathway. *Science* 1998, *281*, 1509-1512.

[12] Groden J, Thliveris A, Samowity W, Carlson M, Gelbert L, Albertsen H, *et al.* Identification and characterization of the familial adenomatous polyposis coli gene. *Cell* 1991, *66*, 589-600.

[13] Steigerwald K, Santoro IM, Kordich JJ, Gismondi V, Trzepacz C, Badiali M, *et al.* A distinct splice form of APC is highly expressed in neurons but not commonly mutated in neuroepithelial tumours. *J Med Genet* 2001, *38*, 257-262.

[14] Bardos J, Sulekova Z, Ballhausen WG. Novel exon connections of the brain- specific (BS) exon of the adenomatous polyposis coli gene. *Int J Cancer* 1997, *73*,137-142.

[15] Carson DJ, Santoro IM, Groden J. Isoforms of the APC tumor suppressor and their ability to inhibit cell growth and tumorigenicity. *Oncogene* 2004, *23*, 7144-7148.

[16] Sieber OM, Tomlison IP, Lamlum H. The adenomatous polyposis coli (APC) tumour suppressor - genetics, function and disease. *Mol Med Today* 2000, *6*, 462-469.

[17] Pecina-Slaus N, Pavelic K, Pavelic J. Loss of heterozygosity and protein expression of APC gene in renal cell carcinomas. *J Mol Med* 1999, *77*, 446-453.

[18] Pecina-Slaus N, Gall-Troselj K, Slaus M, Radic K, Nikuseva-Martic T, Pavelic K. Genetic changes of the E-cadherin and APC tumour suppressor genes in clear cell renal cell carcinoma. *Pathology* 2004, *36*, 145-151.

[19] Pecina-Slaus N, Kljaic M, Nikuseva-Martic T. Loss of heterozygosity of APC and CDH1 genes in laryngeal squamous cell carcinoma. *Pathol Res Pract* 2005, *201*, 557-563.

[20] Maric S, Bulic-Jakus F, Jezek D, Juric-Lekic G, Kos M, Vlahovic M. Expression of the proliferating cell nuclear antigen and protein products of tumour suppressor genes in the human foetal testis. *Andrologia* 2004, *36*, 24-30.

[21] DiCiommo D, Gallie BL, Bremner R. Retinoblastoma: the disease, gene and protein provide critical leads to understand cancer. *Sem Canc Biol* 2000, *10*, 255-269.

[22] Kato N, Shibuya H, Fukase M, Tamura G, Motoyama T. Involvement of adenomatous polyposis coli (APC) gene in testicular yolk sac tumor of infants. *Hum Path* 2006, *37*, 48-53.

[23] Brakeman JS, Gu SH, Wang XB, Dolin G, Baraban JM. Neuronal localization of the Adenomatous polyposis coli tumor suppressor protein. *Neuroscience* 1999, *91*, 661-672.

[24] Wedgwood S, Lam WK, Pinchin KM, Markham AF, Cartwright EJ, Coletta PL. Characterization of a brain-selective transcript of the Adenomatous polyposis coli tumor suppressor gene. *Mamm Genome* 2000, *11*, 1150-1153.

[25] Dobashi Y, Katayama K, Kawai M, Akiyama T, Kameya T. APC protein is required for initiation of neuronal differentiation in rat pheochromocytoma PC12 cells. *Biochem Biophys Res Commun* 2000, *279*, 685-691.

[26] Joyner AL. Engrailed, Wnt and Pax genes regulate midbrain-hindbrain development. *Trends Genet* 1996, *12*, 15-20.

[27] Patapoutian A, Reichardt LF. Roles of wnt proteins in neural development maintenance. *Curr Opin Neurobiol* 2000,*10*, 392-399.

[28] Judas M, Milosevic JN, Rasin MR, Heffer-Lauc M, Kostovic I. Complex Patterns and Simple Architects: Molecular Guidance Cues for Developing Axonal Pathways in the Telencephalon. In: Kostovic I, editor. *Guidance Cues in the Developing brain.* Berlin Heidelberg: Springer-Verlag; 2003; p.1-32.

[29] Lie DC, Colamarino SA, Song HJ, Desire L, Mira H, Consiglio A, *et al.* Wnt signalling regulates adult hippocampal neurogenesis. *Nature* 2005, *437*, 1370-1375.

[30] Pozniak CD, Pleasure SJ. A tale of two signals: Wnt and Hedgehog in dentate neurogenesis. Sci. STKE 2006 (319), pe5. [www.stke.org/cgi/content/full/sigtrans; 2006/319/pe5]

[31] Hall AC, Lucas FR, Salinas PC. Axonal remodeling and synaptic differentiation in the cerebellum is regulated by WNT-7a signaling. *Cell* 2000, *100*, 525-535.

[32] Howng SL, Wu CH, Cheng TS, Sy WD, Lin PC, Wang C, *et al.* Differential expression of Wnt genes, beta-catenin and E-cadherin in human brain tumors. *Cancer Lett* 2002, *183*, 95-101.

[33] Hamilton SR, Liu B, Parsons RE, Papadopoulos N, Jen J, Powell SM, *et al.* The molecular basis of Turcot's syndrome. *N Engl J Med* 1995, *332*, 839-840.

[34] Koch A, Waha A, Tonn JC, Sorensen N, Berthold F, Wolter M, *et al.* Somatic mutations of WNT/wingless signaling pathway components in primitive neuroectodermal tumors. *Int J Cancer* 2001, *93*, 445-449.

[35] Yu X., Malenka R.C. Beta-catenin is critical for dendritic morphogenesis. *Nature Neurosci* 2003, *6*, 1169-1177.

[36] Yokota N, Nishizawa S, Ohta S, Date H, Sugimura H, Namba H, *et al.* Role of wnt pathway in medulloblastoma oncogenesis. *Int J Cancer* 2002, *101*,198-201.

[37] Ellison DW, Onilude OE, Lindsey JC, Lusher ME, Weston CL, Taylor RE, *et al.* United Kingdom Children's Cancer Study Group Brain Tumour Committee. Beta-catenin status predicts a favorable outcome in childhood medulloblastoma: the United Kingdom Children's Cancer Study Group Brain Tumour Committee. *J Clin Oncol* 2005, *23*, 7951-7957.

[38] Kleihues P, Louis DN, Scheithauer BW, Rorke LB, Reifenberger G, Burger PC, *et al.* The WHO classification of tumors of the nervous system. *J Neuropathol Exp Neurol* 2002, *61*, 215-225.

[39] Holland E. Gliomagenesis: Genetic alterations and mouse models. *Nat Rev Genet* 2001, *2*, 120-129.

[40] Collins VP. Brain tumours: classification and genes. *J Neurol Neurosurg Psychiatry* 2004, *75* (Suppl 2), ii2-11.

[41] Fukushima T, Favereaux A, Huang H, Shimizu T, Yonekawa Y, Nakazato Y, *et al.* Genetic alterations in primary glioblastomas in Japan. *J Neuropathol Exp Neurol.* 2006, *65*, 12-18.

[42] Ekstrand AJ, James CD, Cavenee WK, Selinger B, Pettersson RF, Collins VP. Genes for epidermal growth factor receptor, transforming growth factor alpha, and epidermal

growth factor and their expression in human gliomas in vivo. *Cancer Res* 1991, *51*, 2164-2172.

[43] Behin A, Hoang-Xuan K, Carpentier AF, Delattre JY. Primary brain tumours in adults. *Lancet* 2003, *361*, 323-331.

[44] Pecina-Slaus N, Slaus M. Genetic polymorphism in exon 11 of the APC tumor suppressor gene in a Croatian sample. *Homo* 2000, *51*, 151-155.

[45] Gumbiner BM. Cell adhesion: The molecular basis of tissue architecture and morphogenesis. *Cell* 1996, *84*, 345-357.

[46] Demuth T, Berens ME. Molecular mechanisms of glioma cell migration and invasion. *J Neurooncol* 2004,*70*, 217-228.

[47] Shimamura K, Takeichi M. Local and transient expression of E-cadherin involved in mouse embryonic brain morphogenesis. *Development* 1992, *116*, 1011-1019.

[48] Perego C, Vanoni C, Massari S, Raimondi A, Pola S, Cattaneo MG, *et al.* Invasive behaviour of glioblastoma cell lines is associated with altered organisation of the cadherin-catenin adhesion system. *J Cell Sci* 2002, *115*, 3331-3340.

[49] Pecina-Slaus N, Nikuseva-Martic T, Gall-Troselj K, Radic K, Hrascan R. Replication Error-positive Samples Found in Pheochromocytomas. *In vivo* 2005, *19*, 359-366.

[50] Pecina-Slaus N, Beros V, Houra K, Cupic H. Loss of heterozygosity of the APC gene found in a single case of oligoastrocytoma. *J Neurooncol* 2006,*78*, 213-215.

[51] Mori T, Nagase H, Horii A, Miyoshi Y, Shimano T, Nakatsuru S, *et al.* Germ-line and somatic mutations of the APC gene in patients with Turcot syndrome and analysis of APC mutations in brain tumors. *Genes Chromosomes Cancer.* 1994, *9*,168-172.

[52] Barel D, Cohen IJ, Mor C, Stern S, Shapiro R, Shomrat R, *et al.* Mutations of the adenomatous polyposis coli and p53 genes in a child with Turcot's syndrome. Cancer Lett 1998, *132*,119-125.

[53] Suzui M, Yoshimi N, Hara A, Morishita Y, Tanaka T, Mori H. Genetic alterations in a patient with Turcot's syndrome. *Pathol Int* 1998, *48*, 126-133.

[54] Roversi G, Pfundt R, Moroni RF, Magnani I, van Reijmersdal S, Pollo B *et al.* Identification of novel genomic markers related to progression to glioblastoma through genomic profiling of 25 primary glioma cell lines. *Oncogene* 2006, *25*, 1571-1583.

In: New Research on Genomic Instability
Editor: Eleanor J. Glascow, pp. 139-180

ISBN: 1-60021-320-0

Chapter VI

Radiation Hormesis and the Control of Genomic Instability

Bobby R. Scott*
Lovelace Respiratory Research Institute, Albuquerque, NM

Abstract

Life on earth evolved in a low-level ionizing radiation environment comprised of terrestrial, solar, and cosmic radiations. A beneficial evolutionary consequence of the interaction of environmental low-level, sparsely ionizing radiations (e.g., X-rays, gamma rays, and beta particles) with mammalian organisms on earth is the suppression of genomic instability via radiation-hormesis-associated protective biological processes. Radiation hormesis as used here involves low-dose-induced protection from biological harm (e.g., cancer) that would otherwise occur spontaneously. High doses inhibit protection. This chapter focuses on the control of genomic instability and associated stochastic biological effects (e.g., mutant and neoplastically transformed cells and cancer) in mammals via radiation-hormesis-associated protective processes (adapted protection). Cancer risk increases as the numbers of mutant and neoplastically transformed (precancerous) cells increase. Mutant and neoplastically transformed cells increase as the number of cells with genomic instability increases. Thus, protective biological processes that suppress genomic instability also suppress mutant and neoplastically transformed cells and, thereby, prevent some spontaneous and environmental stressor-associated cancers that otherwise would have occurred.

This chapter describes how a system of cooperative protective biological processes that are activated (or stimulated) by low doses of sparsely ionizing radiation likely suppress genomic-instability-associated diseases. The protective processes that make up this adapted protection include normal high-fidelity DNA repair/apoptosis (presumed to

* Address Correspondence to: Bobby Scott, Ph.D., Lovelace Respiratory Research Institute, 2425 Ridgecrest Dr. SE, Albuquerque, NM 87108, Phone: 505-348-9470, Fax: 505-348-8567, E-mail: bscott@lrri.org

be p53-dependent), an auxiliary apoptosis process (presumably p53-independent) that selectively removes cells with genomic instability including mutant and neoplastically transformed cells, and the immune system. The auxiliary apoptotic component relates to what is called the *p*rotective *a*poptosis-*m*ediated (PAM) process, which is considered to involve reactive oxygen and nitrogen species and specific cytokines (e.g., transforming growth factor β1). When activated by a low dose of sparsely ionizing radiation, the PAM process removes genomically unstable cells from our bodies and, along with activation of the immune system, can prevent some later occurring spontaneous cancers and other genomic-instability-associated diseases. Natural background sparsely ionizing radiation above a threshold dose activates the PAM process. The indicated hormesis-associated protective system and its activation by natural background radiation therefore guards against genomic instability and its adverse consequences to humans and other mammals. Reducing background radiation sources to the extent that radiation doses remain below the protective threshold will likely harm us rather than provide a benefit! Reduced life expectancy and significant increases in all genomic-instability-associated diseases would be the expected consequence of reducing background radiation to the degree indicated.

Keywords: Ionizing radiation, adaptive response, genomic instability, hormesis

INTRODUCTION

This chapter relates to forms of radiation that cause ionizations in matter and to a type of radiation-induced biological damage called genomic instability. The focus is on the role radiation hormesis (an adaptive response) has in the control of spontaneous and environmental stressor-associated genomic instability. The radiation types mainly of interest are all forms of sparsely ionizing radiation that include X-rays, gamma rays, beta particles, and protons. Combined exposures to sparsely ionizing radiation and highly ionizing radiations such as alpha particles are also discussed, as they are frequently encountered in the environment.

First, some background on radiation hormesis and basic concepts associated with ionizing radiation is presented in the subsections below. The background information is followed by a discussion of radiation biological effects and how they are mitigated. A mathematical model that characterizes different genomic instability states induced by radiation and their consequences at the cellular level is then described and used to evaluate the expected deleterious consequences of living in a reduced natural background radiation environment. Some public health implications of radiation hormesis for current and future generations are also discussed.

Radiation Hormesis

Hormesis has been formally defined as low-dose stimulation and high-dose inhibition (Calabrese and Baldwin, 1999, 2003a,b; Calabrese and Cook, 2005; Kant *et al* , 2003). Survival of all organisms on earth depends upon their ability to withstand environmental stresses, an ability that involves adaptive changes in cells and molecules. Results from animal studies and studies involving human populations suggest that hormesis is an effective means of protecting against different diseases including diabetes, cardiovascular disease, cancers, and neurodegenerative disorders (Mattson, 2005). The term radiation hormesis, as used here, involves low-dose-induced protection from biological harm (Luckey, 1991, 1994, 2003; Ducoff, 2002). High doses can inhibit protection, leading to a U- or J-shaped dose-response curve typical of hormetic effects (Figure 1). The lower shaded portion of the figure for which the dose-response curve has a U shape represents the hormetic zone (dose range shows protective effects). The curve can also be flat over this zone showing a dose-independent character (Scott 2006a,b).

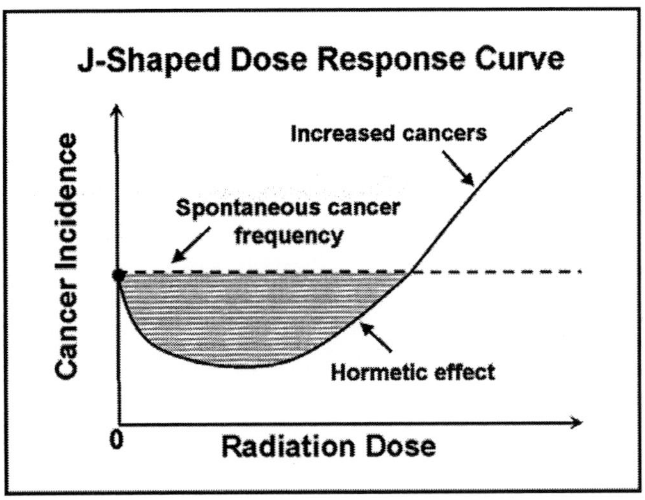

Figure 1. Typical J-shaped dose-response curve for a hormetic dose-response curve for cancer induction by irradiation.

Electromagnetic and Particulate Ionizing Radiation

Ionizing radiation spans the universe (Bonner, 2003). There are two basic forms of ionizing radiation (Scott and Guilmette, 2005): electromagnetic and particulate. Electromagnetic radiation is comprised of uncharged photons that interact with electrons in matter causing ionizations if photon energy is high enough. For photon energy (E) < 3.89 electron volts (eV), ionizations do not occur. For $3.89 < E < 24.6$ eV, the photon radiation may be either ionizing or non-ionizing, depending on the material irradiated. Photons with E

> 24.6 eV are ionizing electromagnetic radiation. Examples of ionizing electromagnetic radiations are X-rays, some forms of ultraviolet (UV) radiation, and gamma rays. Because UV-C (a form of ultraviolet radiation) exceeds the ionization energy of many elements, it is often considered ionizing radiation. Visible light involves photon energies from 1.77 to 3.10 eV and, therefore, is considered non-ionizing radiation. This form of electromagnetic radiation (a component of sunlight) is essential for the existence of the vast numbers of species of plants and animals (including humans) that inhabit the earth.

Examples of particulate ionizing radiation are alpha and beta particles emitted by radioisotopes and protons ejected from the sun. Neutrons do not directly cause ionizations but cause them indirectly through secondary charged particles such as protons (e.g., from water in biological tissue) that are dislodged by neutrons.

Natural background ionizing radiation on earth comes from the following three sources: the sun (solar radiation), outer space (cosmic rays), and terrestrial sources (e.g., radionuclides in our bodies and environment, and radon in the home). While most solar radiation is electromagnetic, the sun also produces particulate radiation (solar cosmic rays), including protons, which vary with the solar cycle.

All organisms on earth are constantly bombarded by cosmic radiation from outside our solar system. This radiation is comprised of charged particles ranging in atomic mass from protons to iron nuclei. These particles interact in the atmosphere creating secondary radiation that rains down and includes X-rays, electrons, protons, alpha particles, neutrons, pions, and muons. Our exposure to cosmic rays increases each time we take an airline flight. Persons living at high elevations such as in Denver, Colorado, and Salt Lake City, Utah, receive higher exposures to cosmic rays than do persons residing in Miami, Florida, or New Orleans, Louisiana.

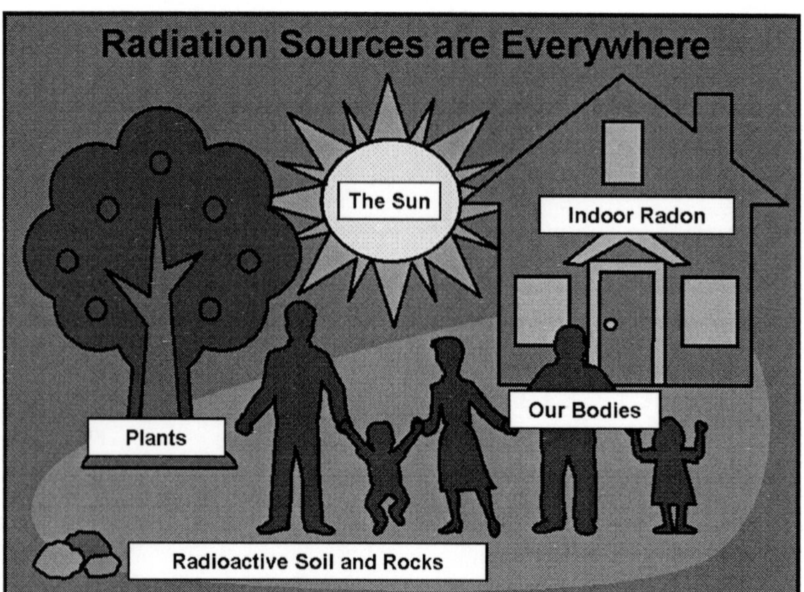

Figure 2. Sources of radioactivity on earth include soil and rocks, our bodies, plants, and radon in our homes.

All organisms on earth are continuously exposed to varying amounts of natural radiation. Radioactivity (the capacity to emit particulate or electromagnetic ionizing radiation forms) is everywhere on earth (Figure 2). This includes radioactivity in our bodies (e.g., associated with potassium-40), in plants (e.g., associated with carbon-14), in soil and rocks (e.g., from uranium and thorium isotopes), and in our homes and businesses (e.g., associated with radon). Thus, all humans are continuously exposed to radiation arising from naturally occurring terrestrial radioactivity and from the cosmos. This is true prior to birth and through one's entire life. Further, natural background ionizing radiation is likely playing a beneficial role (indirect) in the maintenance and preservation of life on earth through suppressing genomic instability; this topic is the focus of this chapter.

THE NUCLEAR AGE AND MEASURES OF RADIATION EXPOSURE AND DOSE

The nuclear age is considered by some (Draganić *et al.* 1990) to have started in 1896 with the discovery of radioactivity by Antoine-Henry Becquerel (1852-1908) and continues today. In 1897, X-rays were discovered by Konrad Wilhelm Roentgen (1845-1923). These research giants had such an impact on the field of radiation research that special units of radioactivity and electromagnetic radiation exposure bear their names. A presently used unit (International System [SI] of Units) of radioactivity is the Becquerel (abbreviated Bq) which represents 1 radioactive disintegration per second. A presently used unit for exposure to ionizing electromagnetic radiation (e.g., X- and gamma rays) is the Roentgen (abbreviated R).

While the R is a measure of exposure to electromagnetic radiations (X- and gamma rays), other measures are used to characterize what is called radiation dose (i.e., absorbed dose), which is a measure of radiation energy deposited in a given mass of matter divided by that mass. The absorbed dose applies not only to electromagnetic radiation but also to particulate radiation (e.g., beta and alpha radiations).

A widely used SI unit of dose is the gray (Gy), which represents 1 Joule/kg. Other related SI units include the mGy (1/1,000th of a Gy) and kGy (1000 Gy). An absorbed dose of gamma rays or X-rays that would be expected to be lethal to 50% of humans is called the median lethal dose (LD_{50}) and is estimated to be 3 Gy (3,000 mGy) for uniform exposure over the body at a high rate (Scott *et al.*, 1988; Scott and Dillehay, 1990).

A typical diagnostic chest X-ray delivers about 0.05 mGy to the chest. A computed tomography (CT) scan may deliver an absorbed X-ray dose of about 1 mGy to the body (NCRP, 1997).

LINEAR ENERGY TRANSFER

The ionization patterns produced by differently charged particles (e.g., electrons, protons, alpha particles) relate to their linear energy transfer (LET). LET is the average energy lost in traversing a small thickness of the target material of interest divided by that thickness. For typical X-rays (which produce secondary ionizing electrons), LET ranges from about 0.2 to 15 keV/μm; for fast neutrons (which produce secondary ionizing protons) from about 8 to 40 keV/μm; and for alpha particles (charged helium nuclei) it is greater than about 260 keV/μm. The efficiency with which a particular type of radiation produces biological damage depends strongly on the ionization pattern in biological tissue and, therefore, on LET.

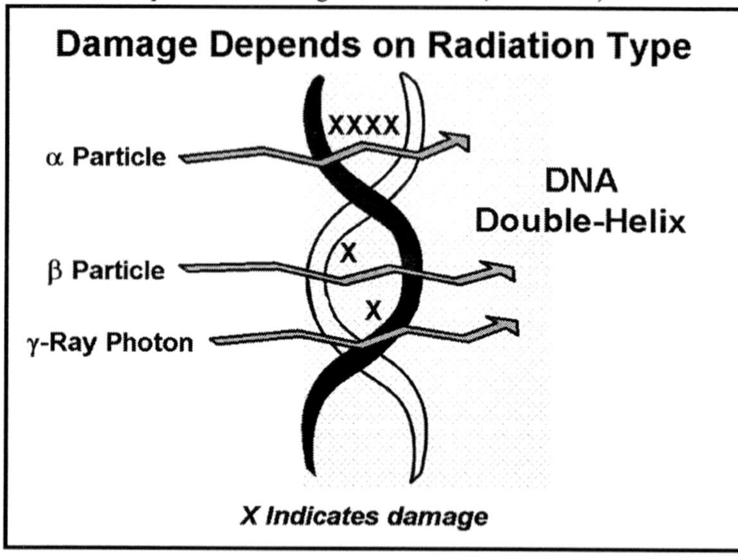

Figure 3. Comparison of the different patterns of damage to DNA caused by commonly encountered radiations in the environment. Highly ionizing (high-LET) alpha (α) radiation produces clustered, complex damage to DNA. Sparsely ionizing (low-LET) beta (β) and gamma (γ) radiations do not produce as much damage as α particles when compared at the same average radiation dose to target cells.

X-rays, gamma rays, beta particles (free energetic electrons), and protons are sparsely ionizing and considered low-LET radiation while the highly ionizing alpha particles are considered high-LET radiation. The pattern of damage to DNA (considered the critical target for genomic instability) depends on LET (Figure 3).

Low-dose Radiation-induced Stochastic Radiobiological Effects

Stochastic biological effects are governed by probabilistic relationships and include effects such as the number of mutant cells produced in a tissue by a low dose and the number of cancers induced in a population of irradiated persons. Low-dose radiation-induced biological effects in cellular communities (*in vivo* or *in vitro*) depend on the type of radiation, dose rate history, the spatial distribution of radiation energy deposition events, and the target cellular community characteristics (e.g., *in vivo* or *in vitro* targets). Low-dose radiation effects at the molecular level include alterations in expressed genes and proteins, oncogene activation, suppressor gene inactivation, chromosomal aberrations, mutations, alterations of metabolic functions, and changes in intercellular signaling characteristics. Low-dose irradiation can also produce a variety of stochastic effects at the cellular level that include apoptosis (programmed cell death), neoplastic transformation (an early step in cancer induction), and altered cell-cycle regulation. Effects at the systemic level include cancer and other mutation-associated diseases.

Largely due to the Office of Science (BER), U.S. Department of Energy-sponsored Low Dose Radiation Research Program, extensive knowledge has been gained about three important classes of low-dose radiation-induced effects: (1) genomic instability, (2) adaptive response, and (3) bystander-cell response (i.e., bystander effect).

Genomic Instability

Damage to DNA from internal or external sources can lead to genomic instability among mammalian cellular communities (Morgan *et al.,* 1996; Wright 1998; Mothersill and Seymour, 1998a; Mills *et al.,* 2003). Cancers in humans harbor numerous mutations considered to result from inherent genomic instability. Possibly, this instability was initially at a low level and then propagated over successive cell generations. Several pathways to genomic instability have been described: disruption of nucleotide-excision repair, base-excision repair, mismatch repair, recombinational repair (homologous recombination, non-homologous end joining) and chromosomal segregation, and changes in the status of CpG-island methylation (Sieber *et al.* 2003). That mutational events in these pathways can drive tumorigenesis is demonstrated by the existence of human cancer syndromes with germline mutations in genes that contribute to maintaining genomic integrity. These syndromes include xeroderma pigmentosum, ataxia telangiectasia, Nijmegen breakage syndrome, hereditary non-polyposis colorectal cancer, and Bloom syndrome (Sieber *et al.* 2003).

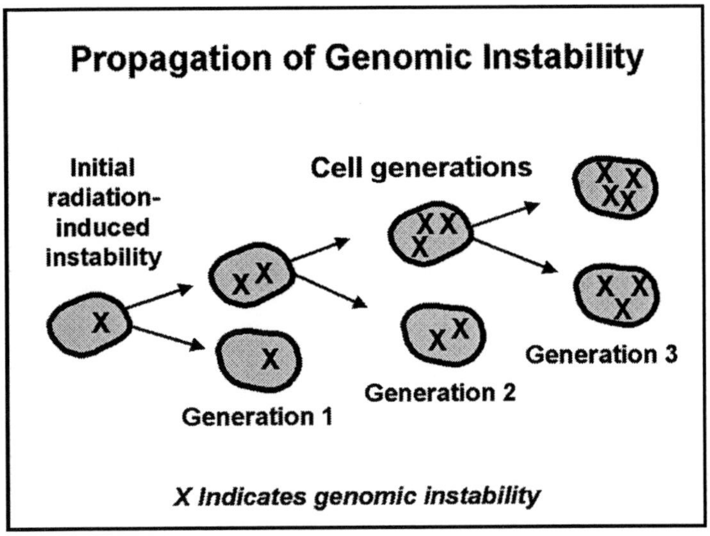

Figure 4. Propagation of PPI (persistent problematic instability) over subsequent cell generations (1, 2, and 3) after PPI is induced in the original parent cell. The instability increases as the number of X's increase. Daughter cells by generation are aligned vertically. Neoplastic transformation potential is assumed to increase as the level of instability increases.

Ionizing radiation and other environmental stresses can induce genomic instability. Three forms of genomic instability have been distinguished in the context of modeling low-dose radiation-induced stochastic effects (Scott 2004; Scott *et al.* 2004; Scott *et al.,* 2006a,b): normal minor instability (NMI), transient problematic instability (TPI), and persistent problematic instability (PPI) among hypersensitive cells in the target cell population. NMI is not considered harmful. TPI is considered potentially harmful but can be eliminated via cooperative protective processes (e.g., activated high-fidelity, presumed p53-dependent DNA repair/apoptosis). However, misrepair (or failure to undergo repair) of damage can lead to harmful PPI (mutant cells) which can be passed to cell progeny and propagate (Figure 4), increasing their chances for undergoing spontaneous neoplastic transformation (an early step in cancer occurrence). PPI corresponds to what is called genomic instability in much of the published literature on the topic.

The indicated three forms of instability are features of the previously published $NEOTRANS_3$ model (Scott, 2004) for low-dose radiation-induced stochastic biological effects. The model is discussed later.

Deleterious Bystander Effects

Recently, it was learned that bystander cells not directly hit by radiation can also develop genomic instability via other mechanisms in addition to reactive oxygen species (ROS) produced by irradiation. The instability is reflected in the presence of mutations, chromosomal abnormalities, and neoplastically transformed cells. Since the non-hit cells are bystanders, the effect has been given the special name "bystander effect" (Figure 5).

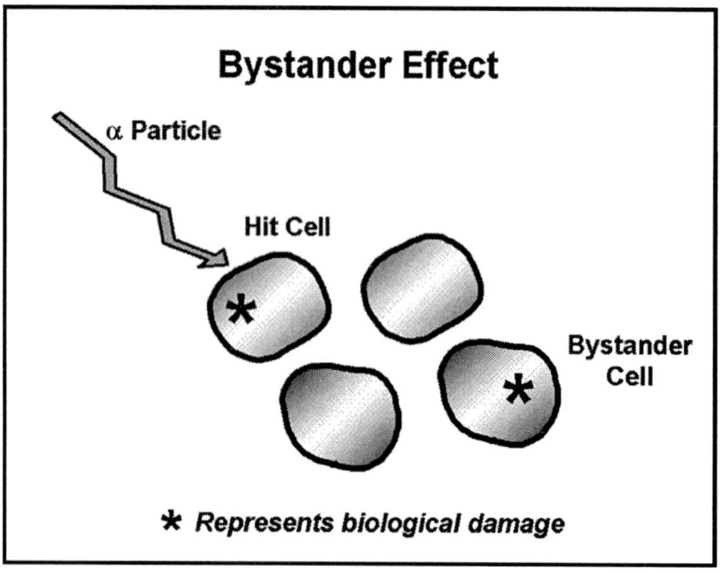

Figure 5. Bystander effect (deleterious in this case) whereby a cell not hit by radiation is somehow damaged (* indicates new damage resulting from irradiation) as a result of a radiation hit on another nearby cell. Some bystander effects are, however, protective (e.g. increased DNA repair capacity in a bystander cell [not shown in figure]).

For conducting high-LET alpha-particle research, new devices called microbeams were developed that allow a single cell to be hit *in vitro* by charged helium ions (the same as alpha particles) and for the nearby and distant non-hit cells to be examined for possible biological effects. Microbeam studies have shown that some non-hit cells can develop different manifestations of genomic instability. Biological substances released by the hit cells appear to play an important role in occurrence of bystander effects. Simply irradiating the medium containing the cells does not lead to bystander effects (Mothersill and Seymour, 1997, 1998b). Also, the indicated deleterious bystander effects are basically low-dose phenomena because as dose increases beyond a point, all cells in the target tissue are hit by radiation in normal circumstances (e.g., exposure to high levels of radon, which involves alpha and gamma irradiations of many cells).

Adaptive Response and Protective Bystander Effects

Researchers using cells from bacteria, yeast, algae, nematodes, fish, plants, insects, amphibians, birds and mammals (including wild deer), rodents, or humans show radio-adaptive processes in response to low doses of low-LET radiation. Low doses increased cellular DNA double-strand break repair capacity, reduced the risk of cell death, reduced radiation- or chemically-induced chromosomal aberrations and mutations, and reduced spontaneous or radiation-induced malignant transformation *in vitro* (Mitchel, 2006).

Mitchel (2006) points out each of the following:

In animals, a low, whole-body dose of low-LET radiation increased the cancer latency and restored a portion of the life that would have been lost either due to spontaneous or radiation-induced cancer in the absence of the low dose.

In genetically normal fetal mice, a prior low dose protected against radiation-induced birth defects.

In genetically normal adult male mice, a low dose prior to a high dose protected their offspring from heritable mutations produced by the large dose.

Mitchel (2006) further point out that the indicated results show that low doses of low-LET radiation induce protective effects and that these induced responses have been tightly conserved throughout evolution, suggesting that they are basic responses critical to life.

Classical two-dose, adaptive-response studies have involved administering a small adapting dose that after a time delay is followed by a larger test dose (Figure 6). Biological effects obtained are then compared to those observed after only administering the test dose. A reduced frequency of observed stochastic biological effects with the combined exposure is then interpreted to represent adaptation caused by the first dose. For such studies, it is clear that induced increased DNA repair capacity likely plays a major role in the indicated reduction of the frequency of stochastic effects (Wolff 1996; Wolff *et al.* 1988; Mitchel, 2004). However, this induced repair adaptation only occurs after a significant time delay. In addition, for brief, very low single doses of low-LET radiation, there is growing evidence that repair of the very small number of cells that are damaged does not occur (Rothkamm and Löbrich, 2003). However, cells with spontaneous and radiation-induced genomic instability (e.g., mutants, cells with activated oncogenes, neoplastically transformed cells) can be selectively eliminated via apoptosis activated by the small radiation dose (Scott, 2004, 2006a,b). This small dose can result in a reduction in the number of genomically unstable cells to well below the spontaneous level, which represents a single-dose adaptive response (Figure 7). However, this form of adaptation is considered to be associated with a novel p53-independent apoptosis pathway (Scott, 2006a,b). Because this form of protection can occur at low-LET radiation doses for which the vast majority of the target cell population receives no radiation hits, it is considered to be a protective bystander effect.

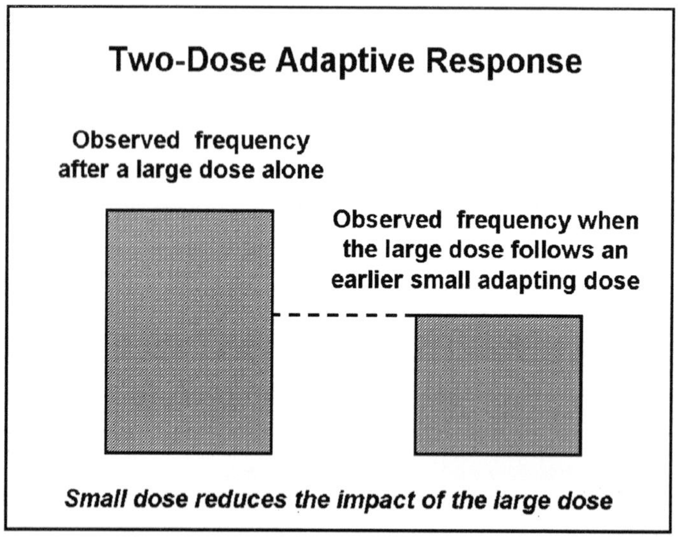

Figure 6. Features of a two-dose adaptive response study. Protection is likely associated with both activated high fidelity DNA/repair (presumably p53 dependent) and with the PAM process (presumably p53 independent). The small adapting dose needs to be low-LET radiation. The larger test dose can be any type of radiation or some other genotoxic agent.

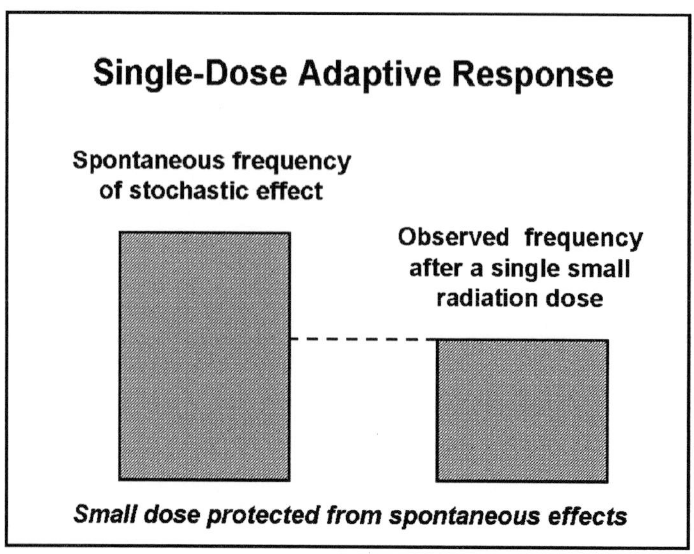

Figure 7. Features of a single-dose adaptive response study involving low-LET radiation. Protection is presumed to be associated mainly with the PAM process.

Azzam *et al.* (1996) and Redpath *et al.* (2001) introduced the indicated single-dose adaptive-response protocol. In their initial experiments, the yield of biological effect (neoplastic transformation *in vitro*) was reduced for irradiated cells when compared to the spontaneous frequency, which was a surprise to the investigators. Similar results were reported by Hooker *et al.* (2004) for inversion mutation induction in spleen of pKZ1 mice exposed *in vivo* to low doses of 250-kVp X-rays. However, their mutation data also showed an elevated mutation frequency at ultra low doses (< 0.01 mGy), suggesting activation of low-fidelity DNA repair (or no repair), rather than high-fidelity repair for this dose zone. Some novel theoretical results are presented later related to comparing expected consequences of single- and two-dose adaptive response studies when radiation-hormesis-associated protection (via apoptosis) is accounted for.

That apoptosis protects from stochastic effects such as cancer via eliminating genomically unstable cells has been proposed by others (Tubiana *et al.,* 2005). Hanahan and Weinberg (2000) in their classic paper, *The Hallmarks of Cancer,* point out the following: "Collectively, the data indicate that the cell's apoptotic program can be triggered by an overexpressed oncogene. Indeed, elimination of cells bearing activated oncogenes by apoptosis may represent the primary means by which such mutant cells are continually culled from the body's tissues."

Protective Processes and the Control of Genomic Instability

For mammals to have survived the hostile environment of primitive earth, mammalian cells had to acquire mechanisms of protection against solar, cosmic, and terrestrial radiation as well as against other environmental stresses (e.g., extreme changes in temperature, severe food shortages, and severe droughts). The fitness of mammalian organisms was therefore likely determined, in a large part, by their ability to compensate for mild and moderate levels of stress (Mattson, 2005). Numerous genes and their encoding proteins evolved over time to mediate adaptive responses to internal and external stresses (Mattson, 2005). Thus, mammalian cellular communities are not complacent when threatened by environmental stresses such as ionizing radiation. The community reacts through a system of protective processes (when they are activated by low doses of radiation) to limit deleterious biological consequences of irradiation (e.g., instability in the genome, mutations, neoplastic transformations, and cancer occurrence [*in vivo*]) (Feinendegen *et al.,* 2004; Scott, 2004).

The protective processes include (1) detecting (sensing) DNA damage; (2) correcting the damage after its detection via activated high-fidelity DNA repair; and (3) removing cells with residual DNA damage (e.g., damaged cells that fail to undergo repair or cells that misrepair their DNA damage) via apoptosis. At the systemic level, immune system activation provides additional protection from stochastic effects such as cancer (Sakai *et al,* 2003; Tubiana *et al.,* 2005; Tubiana, 2005). Immunosurveillance systems are able to eliminate clones of transformed cells as has been observed in tumor cell transplants. The importance of immunosurveillance is also demonstrated by the large increase in the incidence of several types of cancers among immunodepressed individuals (Tubiana *et al.,* 2005). Antioxidants and cell differentiation also contribute to mammalian defenses against stress-associated harm.

For mammalian cells to avoid mutations and neoplastic transformations, DNA repair enzymes must efficiently detect lesions in the genome and remove them. This is a monumental task since thousands of damages occur spontaneously in the genome of each cell each day (Ljungman, 2005).

There are many sources of DNA damage. In addition to external sources such as ionizing radiation and genotoxic chemicals, there are also intrinsic sources that include replication errors, spontaneous chemical changes in DNA, programmed double-strand breaks (in lymphocyte development), and DNA-damaging agents that are normally produced in cells. The latter category includes ROS (reactive oxygen species) such as hydrogen peroxide, nitric oxide, super oxide anion, and hydroxyl radical. ROS can cause lipid peroxidation, protein damage, DNA single- and double-strand breaks, and DNA adducts and cross-links (Lombard *et al.*, 2005).

A recent publication by Coates *et al.* (2005) indicates that cells are estimated to produce approximately 10^{10} ROS each day. This would be expected to result in approximately 20,000 oxidatively damaged DNA bases. In addition, Coates *et al.* (2005) indicated that spontaneous reactions of DNA with water causes depurination and steady-state levels of abasic sites ranging from 50,000 to 200,000 per cell. Both types of endogenous damage are found in varying amounts in different tissues, being highest in the brain and followed by the colon, heart, liver, lung, and kidney, respectively. Sophisticated DNA repair pathways limit the consequences of such damages in mammals.

Feinendegen *et al.* (2004) points out that at background radiation levels, electron tracks resulting from ionization produce ROS that stimulate DNA repair/apoptosis. This points to an important role of natural background radiation in adapted protection.

DNA double-strand breaks may be the most disruptive form of damage to nuclear DNA (Chu, 1997). Double-strand breaks are chromosomal lesions that can lead to genomically unstable cells (e.g., mutants, neoplastically transformed cells), or cell death. They may occur by extrinsic insult from environmental sources or may occur intrinsically as a result of cellular metabolism or a genetic program (Mills *et al.*, 2003). Double-strand breaks are repaired by homologous recombination and by non-homologous end-joining (NHEJ). NHEJ is the predominant mechanism in eukaryotes.

Cells therefore respond to DNA damage through the activation of repair and damage signaling pathways. While the repair proteins directly interact with DNA lesions, the DNA damage signaling pathways could be activated by the effect the DNA lesions have on DNA replication, transcription, or chromatin topology (Ljungman, 2005). Misrepair of DNA damage can lead to mutations and associated genomic instability (Tubiana *et al.*, 2005). The failure to undergo repair can lead to apoptosis (Rothkamm and Löbrich, 2003). Thus, DNA repair and apoptosis cooperate in controlling genomic instability in mammalian cell communities.

Biological-Based Mathematical Model of Genomic Instability

Our research group previously published several models that relate to low-dose ionizing-radiation-induced genomic instability. Here, the most recent model ($NEOTRANS_3$) is presented and used for new low-dose radiation applications. The model links bystander effects, genomic instability, and adaptive response (Scott, 2004). The word NEOTRANS is short for neoplastic transformation. In addition to allowing for quantification of radiation-induced genomic instability, three other radiobiological effects (mutations, neoplastic transformation, and apoptosis) can also be quantified using the model. The model incorporates different states of genomic instability and can be technically characterized as a state-vector model (Scott, 1977; Schöllnberger *et al.*, 2001; Crawford-Brown and Hofmann 2002), although state-vector characterization is not used here.

$NEOTRANS_3$ Model's Genomic Instability States

A key feature of the $NEOTRANS_3$ model is the inclusion of a *p*rotective *a*poptosis-*m*ediated (PAM) process (Scott, 2004). The PAM process is a bystander effect involving crosstalk between genomically unstable (e.g., mutants, transformants, micronucleated cells, etc.) and genomically stable cells, leading to intercellular induction of apoptosis among genomically unstable cells. The PAM process appears to require several hours to be fully activated after a brief low-LET dose and seems to last for at least a day after activation (Scott, 2004). The process appears not to be activated by high-LET radiation based on limited data.

Doses (D) above a small stochastic threshold (called StoThresh), D_{PAM}, activate the PAM process, which then selectively removes precancerous cells (spontaneous and newly emergent transformants) as well as some other genomically unstable cells from the target population. Above a dose-rate-dependent, low-LET StoThresh, D_{off}, PAM is inactivated. Thus, the PAM process is hormetic, taking place in the hormetic zone where the individual-specific, hormetic StoThresh D_{PAM} and D_{off} occur. Both D_{PAM} and D_{off} vary for different *in vitro* replicates and for different individuals (i.e., *in vivo*). Thus, they are properly characterized by distributions rather than point estimates. Bayesian analysis methods facilitate obtaining distributions for D_{PAM} and D_{off} as well as for other $NEOTRANS_3$ model parameters which can vary for different *in vitro* replicates and for different individuals (Scott, 2004, 2006a,b).

Minimum and maximum values for D_{PAM} are given by $D_{PAM}\{min\}$ and $D_{PAM}\{max\}$, respectively. Corresponding values for D_{off} are $D_{off}\{min\}$ and $D_{off}\{max\}$. Members of long-residing mammalian populations in regions of high natural background radiation may have higher values for $D_{off}\{min\}$, lower values for $D_{PAM}\{max\}$, and higher DNA repair capacity due to local-environment-specific adaptation.

The $NEOTRANS_3$ model (Figure 8A and 8B), links neoplastic transformation potential to the genomic instability status of hypersensitive cells. Figure 8A shows transitions when the PAM process is activated by low-LET radiation ($D_{PAM}\{max\} < D < D_{off}\{min\}$) and Figure 8B shows corresponding transitions when the PAM process is not activated ($D < D_{PAM}\{min\}$). Whether or not normal DNA repair/apoptosis (presumed p53 related) occurs after very low radiation doses in the absence of the PAM process is unknown. This is reflected by the question mark and starred parameters in Figure 8B. It is possible that at zero natural background radiation no DNA or apoptosis would be activated (Feinendegen *et al.*, 2004). The notations, $\mu_I(p53)$, $\phi_I(p53)$, and $\eta_I(p53)$, in Figure 8A indicate a presumed dependence of the indicated pathways on p53. The $NEOTRANS_3$ model also integrates genomic instability, bystander effects, and adaptive response (Scott, 2004).

In addition to a normally stable genome for resistant cells, $NEOTRANS_3$ involves three types of genomic instability considered to be important among hypersensitive cells that respond to low radiation doses: (1) NMI (normal minor instability) associated with normal cell function and normal genome status and not considered harmful; (2) TPI (transient problematic instability) associated with genomic damage that may sometimes be repaired without error but can be misrepaired leading to mutant cells; and (3) PPI (persistent problematic instability) that arises from misrepair and is carried by the mutant cells that arise. Such constructs were given the general name "genomic instability state" model (Scott, 1997; Shöllnberger *et al.* 2001). Note that PPI can be passed to progeny, increasing their potential for neoplastic transformation. As previously indicated, genomic instability can also propagate (increase) over subsequent generations of PPI cells (Figure 4).

For the dose ranges of interest, it was assumed that hypersensitive cells are much more likely to be impacted than are resistant cells in the target cell population. Thus, newly induced mutants and transformants are much more likely to arise from hypersensitive cells than from resistant cells. The fraction of hypersensitive cells is given by f_I (not presented in Figure 8). Low-LET radiation-induced genomic damage to cells with NMI leads to new cells with TPI. Target genes considered include all DNA repair, apoptosis, and cell cycle regulatory genes. Misrepair, error-free repair, and apoptotic pathways are available for cells with TPI. However, the apoptosis associated with TPI (called "normal apoptosis") is presumed to be associated with p53 (Scott, 2006a,b).

Bauer (2000) summarized what is known about the intercellular induction of apoptosis among fibroblasts based on numerous studies by his research group (Jürgensmeier *et al.* 1994; Bauer 1995, 1996, 2000; Langer *et al.* 1996; Hipp and Bauer 1997). The protective process involves a sophisticated system of interdependencies and interactions of ROS and nitrogen species. The release of transforming growth factor beta (TGF-β1) by transformed cells is a key early event (Figure 9). Non-transformed cells, when activated, release a novel peroxidase (P) and nitric oxide (•NO). Superoxide anions (O_2^-•) generated and released by transformed cells participate in the intercellular signaling and also make transformed cells the target for intercellular induction of apoptosis (i.e., transformed cells are selectively removed via apoptosis). Chloride ions (Cl^-) and hydrogen peroxide (H_2O_2) also participate in the intercellular signaling. The interactions of these molecules result in two currently known major signaling pathways to apoptosis that are based on hypochlorous acid (HOCl)/hydroxyl radicals (•OH) and •NO/peroxynitrite ($ONOO^-$). H_2O_2 plays a key role by fostering the HOCl/•OH pathway and by inhibiting the •NO/$ONOO^-$ pathway.

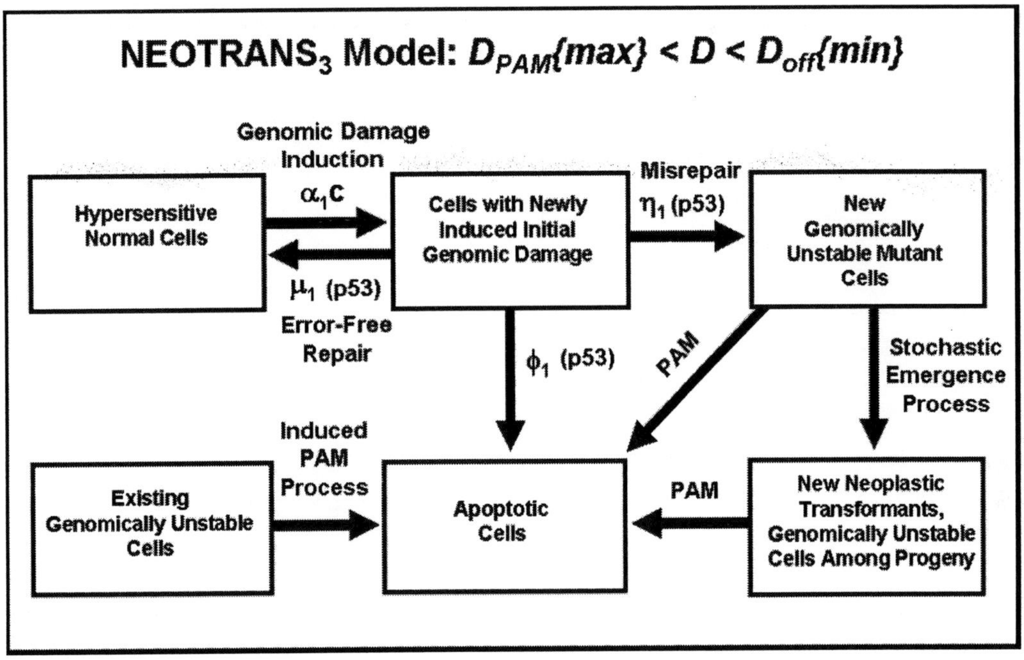

Figure 8A. NEOTRANS$_3$ model that focuses on the response of a hypersensitive subfraction of cells in a population containing mostly resistant cells when the auxiliary PAM process is activated (i.e., $D_{PAM}\{max\} < D < D_{off}\{min\}$). Pathways related to activated normal DNA repair/apoptosis are presumed to depend on p53. The parameters $\mu_1(p53)$, $\eta_1(p53)$, and $\phi_1(p53)$ reflect this dependence.

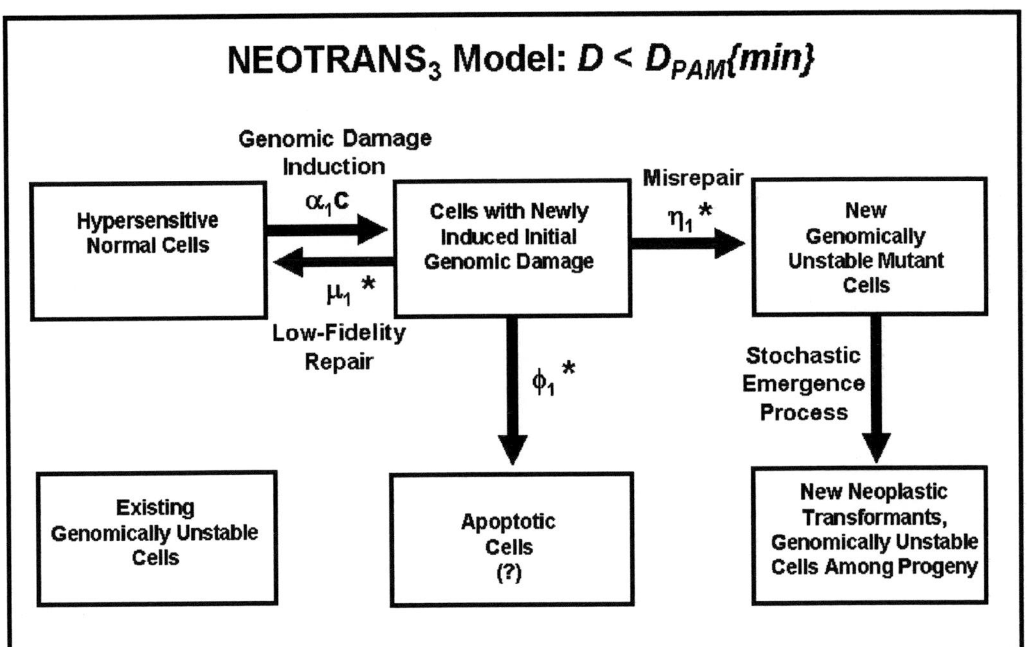

Figure 8B. Model transitions when the PAM process is not activated ($D < D_{PAM}\{min\}$). Starred parameters (μ_1^*, η_1^*, and ϕ_1^*) and the question mark reflect uncertainty about whether or not normal DNA repair/apoptosis occurs after ultra-low doses of radiation. Model parameters and abbreviations are explained in the text.

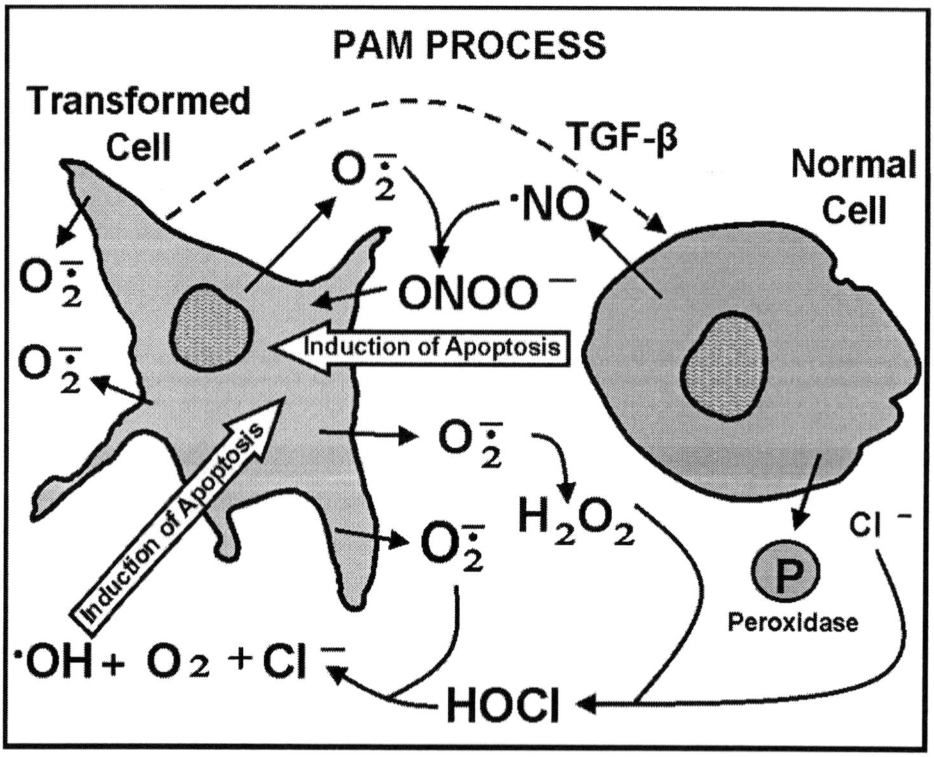

Figure 9. PAM process as it relates to transformed and untransformed fibroblast. The process is explained in the text.

Additional pathways to apoptosis may be associated with the PAM process. The path may depend on the cell type to be eliminated via apoptosis (mutants, neoplastically transformed cells, micronucleated cells, etc.), its local environment, and the nature of the damage to DNA (Scott, 2004). Misrepair leads to mutant cells with PPI that have a heightened potential to produce progeny that undergo neoplastic transformation.

The parameter $\alpha_I(p53)$ in Figure 8A is multiplied by the radiation dose rate c (Figure 8A) and the product accounts for low-dose-induced incremental genomic damage among the hypersensitive cells in the population. The parameter $\mu_I(p53)$ governs the rate of commitment of damaged hypersensitive cells to the activated error-free repair pathway. This repair pathway is associated with normal repair, which is presumed to be activated by radiation and is presumably related to p53. The corresponding parameter for the misrepair pathway (which competes with the error-free pathway in normal repair and is presumed to be related to p53) is $\eta_I(p53)$. Misrepair leads to a variety of viable mutants (cells with PPI).

The parameter $\phi_I(p53)$ (also presumably related to p53) governs the rate of commitment of newly damaged cells (including lethal mutations) to the normal apoptotic pathway. However, unlike the PAM process, *normal apoptosis is not a bystander effect* (Scott, 2004). In addition, the PAM process operates at low doses (but not very low doses) of low-LET radiation while normal apoptosis may occur both at low and high doses as well, as after high-LET irradiation. The PAM process appears not to occur after low doses of high-LET alpha irradiation based on limited information (Scott *et al.*, 2004).

Typical units for α_l are mGy^{-1}. Typical units for parameters μ_l, η_l, and ϕ_l are min^{-1}. For the remainder of the chapter, the p53 association for these parameters has been omitted. The parameter f_l is dimensionless. These parameters are stochastic (i.e., have distributions) but currently are not time dependent. However, parameters μ_l, ϕ_l, and η_l are allowed to vary over rather broad dose regions. This allows for thresholds for activation and inactivation of key biological processes (e.g., low-dose activation of high-fidelity DNA repair/apoptosis). The starred parameters μ_l^*, η_l^*, and ϕ_l^* in Figure 8B reflect uncertainty about whether the associated pathways operate after ultra low radiation doses (i.e., $D < D_{PAM}\{min\}$). Based on limited data, high-fidelity DNA repair appears to be absent after X-ray doses < 0.01 mGy (Scott, 2006b; Scott et al., 2006).

With the NEOTRANS$_3$ model the PAM process, normal apoptosis, and normal repair (error-free component) work together in guarding against genomic instability:

- Normal apoptosis removes damaged cells before mutations arise.
- The error-free component of normal repair eliminates DNA damage and thereby suppresses the occurrence of genomic instability.
- The PAM process removes genomically unstable cells (mutants, neoplastic transformants, micronucleated cells, etc.) and, therefore, prevents the occurrence of some cancers and other genomic-instability-associated diseases.

Thus, the indicated team of biological protectors when activated by low-dose radiation, provides powerful protection against spontaneously occurring genomic-instability associated diseases. Further, they provide a biological basis for radiation hormesis when considering effects such as genomic-instability-associated diseases including cancer.

Analytical Solutions for the NEOTRANS$_3$ Model for Low Doses

Previously published (Scott 2004, 2006a; Scott *et al.,* 2006) analytical solutions for the NEOTRANS$_3$ model are presented below for ultra-low or low absorbed doses, *D*, of low-LET radiation. The solution has also been used for low dose rates. The expected frequency *Y(D)* of biological effects (genomically unstable cells [e.g., mutants, neoplastically transformed cells]) per surviving cell is given by:

$$Y(D) = LNT(D) = Y_0 + [(1-Y_0)]kD, \text{ for } D < D_{PAM},$$
$$Y(D) = HR(D) = (1-f_0)LNT(D), \text{ for } D_{PAM} \le D < D_{off},$$
$$Y(D) = LNT(D), \text{ for } D \ge D_{off}, \quad (1)$$

where the slope parameter *k (*a pseudo parameter*)* is given by:

$$k = f_l \alpha_l \eta_l \Omega/(\mu_l + \eta_l + \phi_l). \quad (2)$$

Again, D_{PAM} and D_{off} are StoThresh, as they vary between experimental replicates and between different individuals. The pseudo parameter *k* incorporates other basic parameters of the NEOTRANS$_3$ model as indicated in Equation 2. Individual parameters f_l, α_l, η_l, μ_l, ϕ_l, and

Ω could not be adequately resolved from the currently available data so pseudo parameter k is estimated. Equation 2 shows that f_l (proportion of hypersensitive cells), α_l (radiosensitivity of the genome), and Ω (propensity of cells with PPI to undergo spontaneous neoplastic transformation) have similar influences on k (which reflects overall radiosensitivity for inducing PPI). As these parameters increase, k increases. With increasing DNA repair/apoptosis fidelity, $\mu_l + \phi_l$ increases, causing k to decrease (reflecting efficacy of protection against PPI).

The hormetic risk (HR) function $HR(D)$ accounts for hormetic effects. The function $LNT(D)$ corresponds to the risk function that is based on the linear, no-threshold (LNT) hypothesis. An LNT function was used to extrapolate from risk estimates for stochastic effects observed at high doses to projected (hypothetical) elevated risk at low doses (NCRP 2001; NRC 2005). However, this practice is quite controversial (Higson, 2004; Tubiana *et al.* 2005; Tubiana, 2005).

As previously indicated, parameters $D_{PAM}\{min\}$ and $D_{PAM}\{max\}$ are used to represent lower and upper bounds for the StoThresh D_{PAM}, respectively. Correspondingly, $D_{off}\{min\}$ and $D_{off}\{max\}$ are bounds for the StoThresh D_{off}. For $D_{PAM}\{max\}$ within the natural background dose zone and $D_{off}\{min\}$ above this zone, which is assumed to be the case here, Equation 1 becomes:

$$Y(D) = Y_0, \text{ for } D = b \text{ (i.e., the natural background radiation dose)},$$
$$Y(D) = HR(D), \text{ for } b < D < D_{off},$$
$$Y(D) = LNT(D), \text{ for } D \geq D_{off}, \qquad (3)$$

With the current version of the NEOTRANS$_3$ model, when necessary the pseudo parameter k is allowed to vary over different dose regions (zones) to account for different fidelities in DNA repair/apoptosis. For example, for a dose zone over which high-fidelity DNA repair/apoptosis is assumed, k will be smaller than for a dose zone over which low-fidelity DNA repair/apoptosis is assumed. However, for the applications of the NEOTRANS$_3$ model presented here, k is fixed.

Note that for $D \geq D_{off}$, Equation 1 reduces to the LNT dose-response function for a dose range over which k is constant. This appears to be the case for alpha radiation-induced neoplastic transformation ($D_{off}\{max\}$ appears to be very small or zero [Scott 2004, 2006a,b]) based on limited data. $Y(D)$ can represent either mutation or neoplastic transformation frequency with the pseudo slope parameter k differing for these two endpoints. Also, relative risk (RR) relations derived for neoplastic transformation can be adapted for application to cancer induction (Scott, 2006b) as is done later in this chapter.

The parameter f_l in Equation 1 is the fraction of the spontaneously occurring, genomically unstable cells (e.g., spontaneous transformants or mutants) removed via the radiation-induced PAM process and is given the special name "protection factor" (*PROFAC*) (Scott, 2003). A similar *PROFAC* applies to cancer induction and accounts for protection associated with immune system stimulation *in vivo* as well as protection associated with the PAM process. For cancer induction, the *PROFAC* gives the expected fraction of the cancer cases that are prevented by radiation hormesis (Scott, 2006a,b). For $f_l = 0$ (i.e., no PAM), Equation 1 reduces to the LNT dose response over a dose range for which k is constant.

For neoplastic transformation, Ω represents the probability that a cell with radiation-induced PPI (mutant cells) will produce transformed progeny during a specified follow-up time. For mutations, Ω should be replaced by Ω^*, where Ω^* represents the probability for the occurrence of mutation of interest among all the mutations induced by irradiation.

Equations 1 and 3 represent a stochastic hidden Markov model (Isaev, 2004; Waterman, 2000). Equation 3 can be re-expressed in terms of the PAM process activation probability *PAM(D),* where *PAM(D)* gives the proportion of times the function *HR(D)* should apply among a large irradiated population with each member receiving the same dose D for $D > b$. *LNT(D)* is otherwise applied. The criteria for *HR(D)* applying is that D has to be contained in the stochastic hormetic interval, $D_{PAM} \leq D < D_{off}$ for the replicate sample or individual of interest . Thus,

$$Y(D) = Y_0, \text{ for } D = b \text{ (i.e., natural background radiation only)},$$
$$Y(D) = PAM(D) \bullet HR(D) + [1 - PAM(D)] \bullet LNT(D) \text{ for } D > 0. \quad (4)$$

Here the cumulative probability function, *PAM(D),* has been introduced for the first time. In our previous applications of the NEOTRANS$_3$ model, model parameters were allowed to vary through use of Bayesian inference methods implemented with Markov Chain Monte Carlo (MCMC). Prior distributions were assigned and posterior distributions derived (Scott 2004; Scott 2006b; Scott *et al.* 2006). Thus, depending on whether or not the PAM process is activated, the value assigned to *Y(D)* switched between *HR(D)* and *LNT(D)* for $D > b$, where b was assigned a value of zero. This lead to a solution (posterior mean) for *Y(D)* that was a weighted average (over the posterior distribution for parameters) of *HR(D)* and *LNT(D)*. The proportion of times *HR(D)* was selected is an estimate of *PAM(D)*. For replicate *in vitro* assays of neoplastic transformation, and for a very large number of replicates and a fixed dose D^*, the observed proportion having the PAM process activated would be an estimate of $PAM(D^*)$.

For combined exposure to low- and high-LET radiations one should replace the term kD in Equation 1 with the corresponding combination $k_L D_L + k_H D_H$, where subscripts L and H indicate the low- and high-LET radiations, respectively. The PAM process is considered to be associated only with the low-LET component of the dose (Scott 2006a,b) so that *PAM(D)* does not depend on D_H. Even for combined exposure to high- and low-LET radiation, $PAM(D_L)$ applies. Functions $Y(D_L,D_H)$, $HR(D_L,D_H)$, $LNT(D_L,D_H)$ replace the corresponding functions *Y(D), HR(D),* and *LNT(D)*, for combined exposure to low- and high-LET radiations. A *RR* function, $RR(D_L,D_H)$, for the combined exposure can be constructed by dividing the expression for $Y(D_L,D_H)$ by Y_0. This function could then be applied to low-LET radiation only, high-LET radiation only, or to combinations of both, depending on which radiations were involved.

Application of NEOTRANS$_3$ to Neoplastic Transformation

Equation 3 was applied previously (Scott 2004; Scott *et al.* 2006) to data for gamma- and X-ray induced neoplastic transformation of HeLa x skin fibroblast, human hybrid cells

(Redpath *et al.* 2001). Equation 1 was applied to data for inversion mutations in the spleen of pKZ1 mice (Hooker *et al.,* 2004) after brief high-rate exposure to X-rays and shown to adequately explain the hormetic dose-response curves observed (Scott 2004, 2006a; Scott *et al.*, 2006). Here, some of the neoplastic transformation data of Redpath *et al.* (2001) are revisited and some new results are presented.

Table 1. NEOTRANS$_3$ model parameter estimates for neoplastic transformation of HeLa x skin fibroblast, human hybrid cells by brief exposure to gamma rays (Scott *et al.*, 2006)

Y_0	$2.25 \times 10^{-5} \pm 6.33 \times 10^{-7}$
k	$4.04 \times 10^{-8} \pm 5.8 \times 10^{-9}$ mGy^{-1}
PROFAC	0.38 ± 0.05
D_{off}	Uniformly distributed over the interval [150mGy, 250 mGy)*

* The left bracket indicates a partially-closed interval.

Bayesian inference methods were used to obtain model parameter estimates (Scott 2004; Scott, 2006; Scott *et al.* 2006). Using Bayesian methods facilitated addressing StoThresh D_{PAM} and D_{off}. The NEOTRANS$_3$ model parameter estimates (Bayesian analysis posterior means) forY_0, *k*, *PROFAC*, and D_{off} previously obtained using Bayesian inference methods are presented in Table 1 for the neoplastic transformation data of Redpath *et al.* (2001). Associated Bayesian analysis posterior distributions are presented in Figure 10. Y_0, *k*, and *PROFAC* were approximately normal. The posterior distribution for D_{off} was essentially the same as the uniform prior distribution assigned. To improve on the distribution, a new study is needed with more doses in the critical region where StoThresh D_{off} appear to occur.

The StoThresh distribution for D_{PAM} could not be estimated directly from the data of Redpath *et al.* (2001) because doses were not low enough. The smallest dose used was 1 mGy while the dose range needed to adequately estimate the distribution of D_{PAM} is 0 to about 0.1 mGy. Thus, the distribution for D_{PAM} has to be estimated using data at very low doses. Only one data set is currently available for such doses of low-LET radiation. The data are for X-ray-induced mutations in pKZ1 mice (Hooker *et al.,* 2004). Our previous application of the NEOTRANS$_3$ model (Scott *et al.*, 2006) to the indicated data indicates a value for $D_{PAM}\{max\}$ around 0.01 mGy. For neoplastic transformation it is assumed that $D_{PAM}\{min\}$ is slightly above or within the natural background radiation zone and for practical consideration can be treated as zero. The indicated distribution can be simulated using a uniform distribution from 0 to 0.01 mGy for D_{PAM}. Experiments specifically designed to estimate the distribution of D_{PAM} are needed to improve on this distribution. Table 1 gives the estimates $D_{off}\{min\} = 150$ mGy and $D_{off}\{max\} = 250$ mGy for gamma irradiation of HeLa x skin fibroblast, human hybrid cells.

We have used Bayesian inference to obtain posterior distributions for *Y(D)* and associated NEOTRANS$_3$ model parameters. MCMC facilitates implementing Bayesian inference when having to address StoThresh (Scott 2004; Scott *et al.,* 2006). While MCMC also can be used for making model predictions, given a set of model parameter posterior distributions, using MCMC requires high-level computing for models as complex as NEOTRANS$_3$.

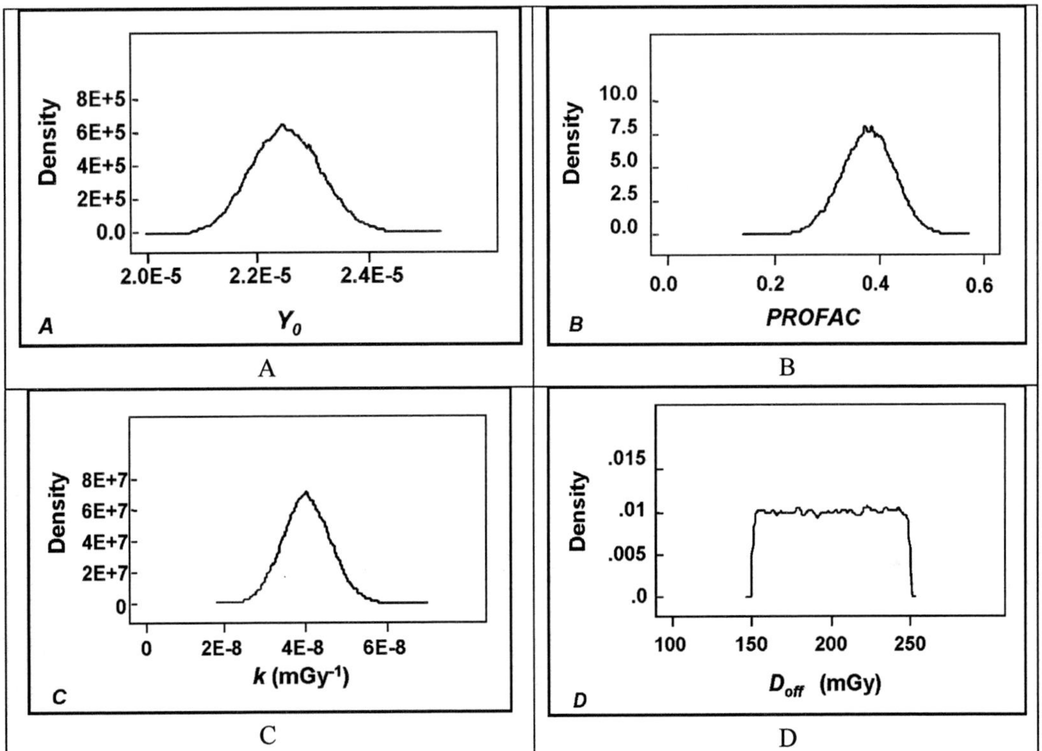

Figure 10 (A-D). Posterior distributions for NEOTRANS$_3$ model parameters Y_0, k, D_{off}, and *PROFAC* for gamma-ray exposure of HeLa x skin fibroblast, human hybrid cells.

Predictions of an average characteristic of a dose-response curve can, however, be made using previously obtained model parameter posterior distribution averages. Dose-response functions obtained in this way can be used to predict the outcomes of single- and two-dose adaptive response studies as well as the influence of reducing natural background radiation to near zero. Predictions can be made with low-level computations using these posterior distribution averages. However, one has to account for the influences of StoThresh D_{PAM} and D_{off}. This can be done using Equation 4, as illustrated below using the posterior averages presented in Table 1 and the indicated uniform distribution for D_{off}. Here, D_{PAM} is assigned the subjective uniform distribution over the interval [0 to 0.01 mGy), and here zero dose means no background radiation.

The PAM activation probability, *PAM(D),* has to be evaluated in order to apply Equation 4 and can be achieved using the piecewise approach indicated in Table 2. For the first partially-closed interval $[0, D_{PAM}\{max\})$, *PAM(D)* can be calculated as $D/D_{PAM}\{max\}$. This is just the cumulative distribution function for uniformly distributed D_{PAM} over the indicated interval. This first dose zone is called Transition Zone A, since it contains StoThresh D_{PAM} (Scott, 2006b; Scott *et al.*, 2006). For the next dose zone $[D_{PAM}\{max\}, D_{off}\{min\})$, the PAM process is modeled as being activated with probability 1 (Zone of Maximal Protection). In the next zone $[D_{off}\{min\}, D_{off}\{max\})$ StoThresh D_{off} occur, and so the PAM process is progressively suppressed (Transition Zone B) over this dose zone. For the next dose zone (LNT Zone) where $D \geq D_{off}\{max\}$,the PAM process is modeled as being completely inhibited (or not activated). The inhibition may relate to damage (direct or indirect) to mitochondria, which are presumed to participate in the PAM process. For the LNT Zone, $PROFAC = 0$ so that the LNT function applies. This is the dose zone where many of the epidemiological studies of radiation-induced cancer have focused. The LNT observations in such studies (for the LNT Zone) were presumed to apply at much lower doses. This presumption has led to the projection of phantom excess cancer risks after low doses of low-LET radiation (Scott 2005). Such phantom risk projection has driven radiation phobia to the extent that more than 100,000 abortions were reported in Eastern Europe shortly after the Chernobyl accident (Ketchum, 1987). Similar radiation-phobia driven casualties are likely after detonation of a dirty bomb, thanks largely to the previous misuse of the LNT hypothesis by others.

Table 2. Piecewise approach to estimating *PAM(D)*

Dose interval	*PAM(D)*	Dose Zone
$[0, D_{PAM}\{max\})$*	$D/D_{PAM}\{max\}$	Transition Zone A where StoThresh D_{PAM} occur
$[D_{PAM}\{max\}, D_{off}\{min\})$	1	Zone of Maximal Protection
$[D_{off}\{min\}, D_{off}\{max\})$	$1-([D-D_{off}\{min\}]/[D_{off}\{max\}-D_{off}\{min\}])$	Transition Zone B where StoThresh D_{off} occur
$[D_{off}\{max\}$, *higher dose)*	0	LNT Zone

* Left brackets indicate partially-closed intervals.

Equation 4, parameter estimates in Table 1, and a uniform distribution for *PAM(D)* over the interval [0 mGy,0.01 mGy) were used to obtain the smooth curve drawn in Figure 11 related to applying the $NEOTRANS_3$ model to the neoplastic transformation data of Redpath *et al.* (2001) for gamma-ray induced neoplastic transformation (*in vitro*) of HeLa x skin fibroblast, human hybrid cells. Calculations were carried out using an Excel spreadsheet. The curve is almost identical to the one we previously calculated by averaging *Y(D)* over posterior parameter distributions obtained in our Bayesian analysis using MCMC. Also shown is the linear extrapolation from high to low doses (based on *LNT(D)*). Note that the dose-response curve in Figure 11 has an approximate J-shape. The curve was modeled as rising rather steeply over Transition Zone B where StoThresh D_{off} occur. Only experimental data for a 24-h delay in platting were used. The 24-h delay allowed the PAM process sufficient time to eliminate genomically unstable cells. The lower portion of Figure 11 is enlarged in Figure 12. Both functions, $10^{-5}PAM(D)$ and *Y(D)*, are shown in Figure 13. *PAM(D)* was scaled down by 10^5 so that its shape could be viewed in the same figure with *Y(D)*.

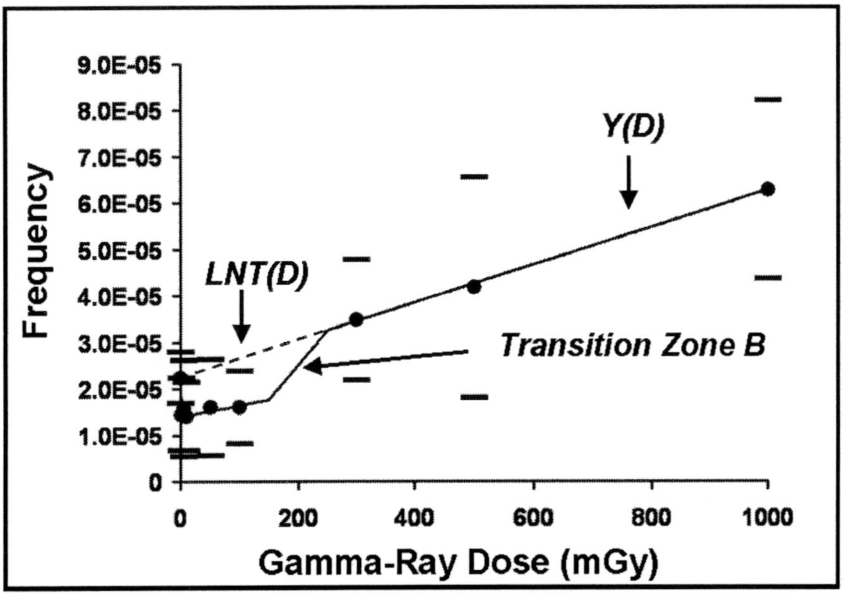

Figure 11. Application of Equation 4 to the neoplastic transformation data (solid points) of Redpath *et al.* (2001). The dashed curve represents a linear extrapolation from the high-dose data. Error bars are 95% confidence limits.

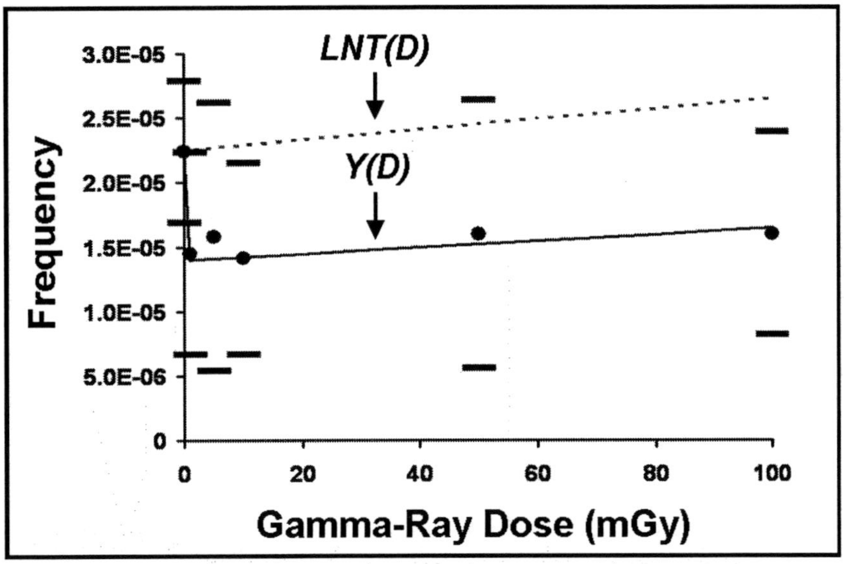

Figure 12. Low-dose portion of Figure 11. Data points in the range 1 to 100 mGy are in the Zone of Maximal Protection.

The PAM process represents a protective bystander effect. A general characteristic of bystander effects is the absence of dependence on radiation dose for a relatively wide range of low doses. Figure 14 shows results of having used individual experimental neoplastic transformation frequencies from replicate studies reported by Redpath *et al.* (2001) (the data yield the averages presented in Figure 11). For some replicates, no transformants were observed (*PROFAC* = *1*). This may relate to D_{PAM} being a StoThresh or to experimental limitations in detecting very low transformation frequencies. Only non-zero data in the dose range 1 to 100 mGy are presented. The indicated range falls within the hormetic zone where protective effects predominate. Controls (zero dose group) are excluded, but the mean and 95% confidence region for the control data are represented by horizontal lines. A correlation analysis of the data revealed that the transformation frequency was not correlated (R^2 = 0.02) with dose for the indicated dose interval [1,100 mGy].

The *PROFAC* also can be estimated using the individual low-dose (1 to 100 mGy) data points in Figure 11. Separate estimates of *PROFAC* were obtained from the non-zero transformation frequency data in this potion of the hormetic zone using 1 – ([average transformation frequency for irradiated sample]/[average transformation frequency for controls]). Results are plotted in Figure 15 along with 95% confidence limits.

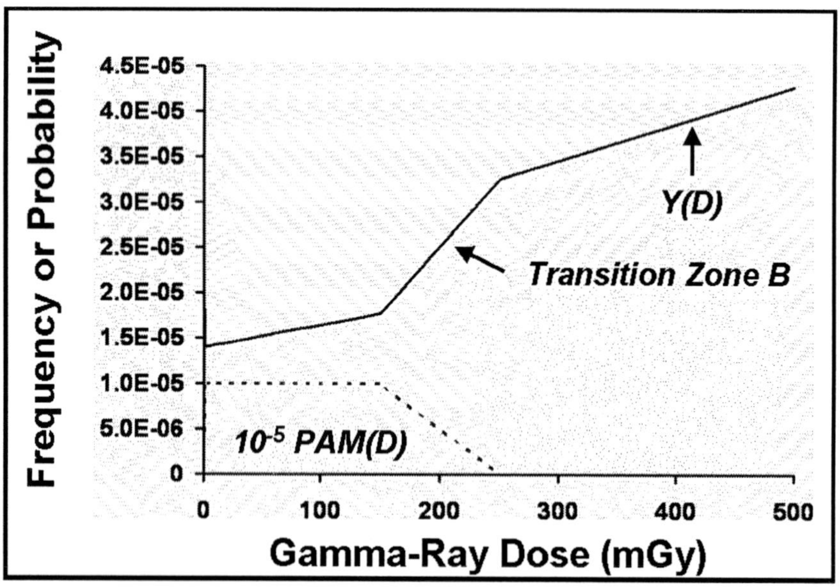

Figure 13. Comparison of curve shapes for *Y(D)* and *PAM(D)*. *PAM(D)* has been multiplied by 10^{-5} for presentation along with *Y(D)*.

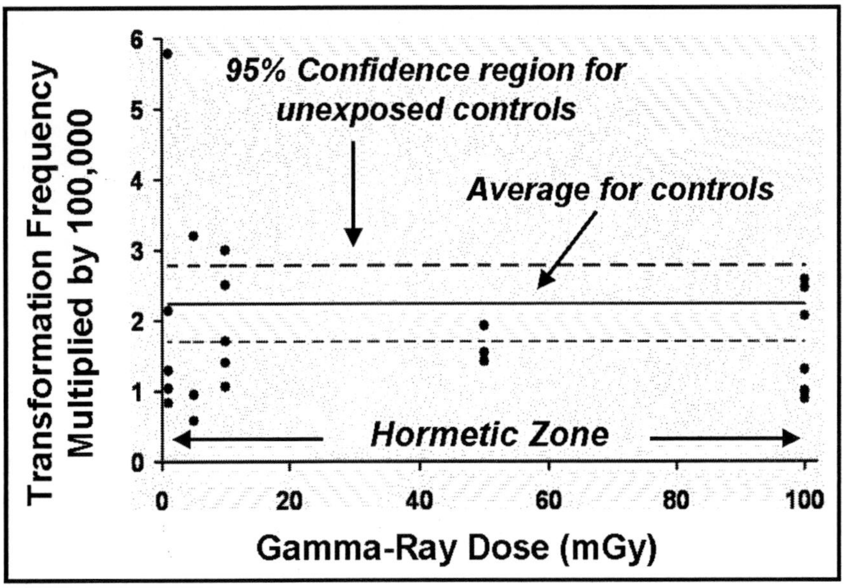

Figure 14. Transformation frequency in the hormetic zone is independent of dose (R^2 for correlation analysis was 0.02). These data points are the replicates that make up the averages in Figures 11 and 12.

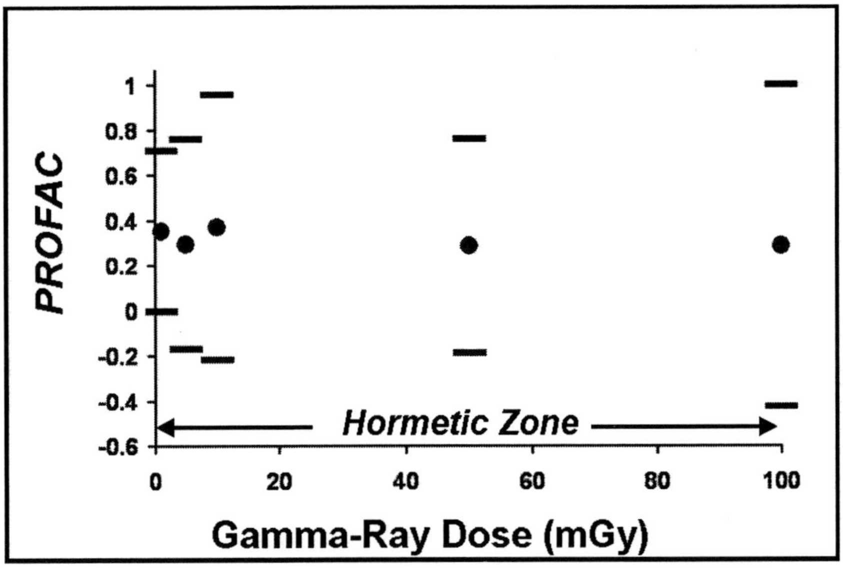

Figure 15. Individual estimates of *PROFAC* based on neoplastic transformation data from Redpath et al. (2001) in Figure 11. Doses used ranged from 1 to 100 mGy and are presumed to be in the hormetic zone.

NEOTRANS$_3$ Model Predictions for Adaptive Response Studies

Equation 4 can be used to predict the expected results of single- and two-dose adaptive response studies. Here, calculations were carried out for neoplastic transformation of Hela x skin fibroblast, human hybrid cells exposed to a single, adapting gamma-ray dose or to two doses of equal sizes separated by a 24-h period. It is assumed that the PAM process lasts for at least 24 h and after a 24-h delay can be reactivated to last for at least another 24-h period. It is also assumed that cells are plated 24 h after radiation exposure is completed.

Model parameters presented in Table 1 were used in the calculations. *PAM(D)* was evaluated as previously described, based on the piecewise scheme in Table 2. Both the single dose and first of two equal-size doses are designated here as *D1*. *Y(D1)* was calculated first. Then, the value obtained was substituted for Y_0 and a new *Y(D1)* was calculated with the new value for Y_0, yielding *Y(2D1)*. Thus, for the two-dose study, the total dose was *2D1*. Results obtained for a wide range of total doses are presented in Figure 16. Two important predictions arise: (1) Two doses of equal size in the hormetic zone separated by a 24-h interval between the doses causes greater protection than a single dose of size *D1*. (2) Transition Zone B shifts to larger doses in the two-dose adaptive response implicating an extended range of protective doses.

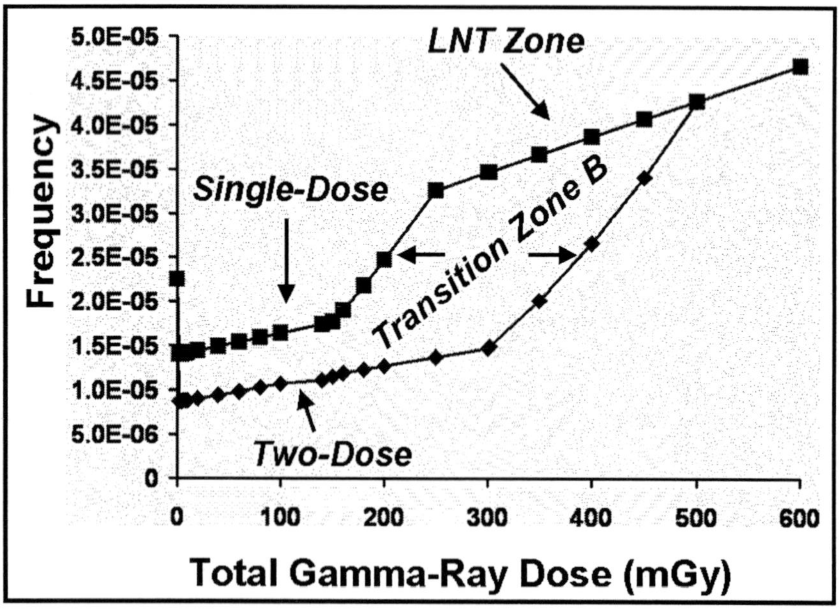

Figure 16. Predicted single- and two-dose adaptive response study results for gamma ray-induced neoplastic transformation *in vitro* of HeLa x skin fibroblast, human hybrid cells based on the NEOTRANS$_3$ model.

Figure 17 shows the curves in Figure 16 plotted as a function of the adapting dose *D1*. Note that the transition zone is predicted to be determined by the dose fraction size of the adapting dose *D1* even for the two-dose, adaptive-response study.

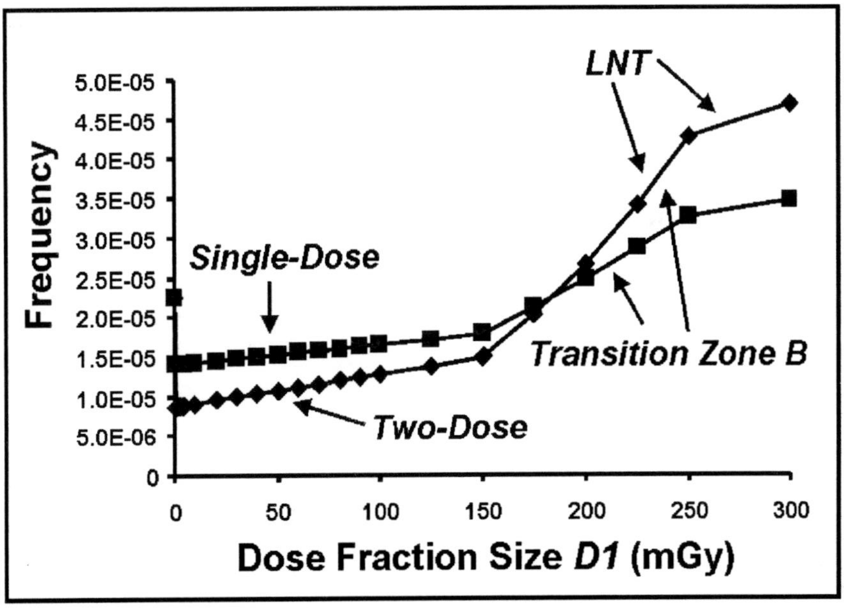

Figure 17. Data from Figure 16 plotted as a function of the dose fraction size *D1*. For the single-dose curve, the total *dose* = *D1*. For the two-dose curve, the *total dose* = *2D1*. The LNT portion of the two curves if extrapolated to a dose of zero would intersect with the spontaneous frequency.

These modeling results support the view that protracted exposure to low-LET radiation at very low rates (continuing small increments ΔD) over a prolonged period may over and over activate the protective PAM process and provide significant protection against stochastic effects such as cancer. Protracted exposure at a low rate would be expected to shift Transition Zone B, where the StoThresh D_{off} occur, to much higher doses. Repeated stimulation of the immune system would be expected to contribute to protection. Indeed, chronic exposure at a low rate to natural background low-LET radiation may be providing such protection against diseases that relate to genomic instability by eliminating the unstable cells via the PAM process and immune system stimulation before disease occurrence. In addition, the epidemiological and ecological data presented in Table 3 and discussed later are in agreement with this view.

Table 3. Central estimates of the percent of cancer cases (mortality or incidence for spontaneous cancer) avoided due to radiation-hormesis in irradiated human populations based on the estimated PROFAC*

Group	Effect	Percent cases avoided (100•PROFAC)
Chernobyl accident emergency workers (Ivanov et al., 2001)	Malignant cancers	13‡
Chernobyl accident emergency workers (Ivanov et al., 2001)	All causes of death	15‡
USA residents of high background states (Frigèrio and Stowe, 1976)	Cancers	15‡
British medical radiologist§ after 1955-1979 (Berrington et al., 2001)	All causes of death	32‡
USA residents having high residential radon (Jaworowski, 2001)	Lung cancers	35‡
USA nuclear shipyard workers (Jaworowski, 2001)	Leukemia	58‡
Canadian nuclear industry workers (Jaworowski, 2001)	Leukemia	68‡
USA DOE facilities workers (Jaworowski, 2001)	Leukemia	78‡
Russian Mayak plutonium facility workers (Scott, 2006b)	Lung cancers	86‖
Taiwanese in Cobalt-60 contaminated apartments (Chen et al., 2004)	Cancers	97‡

* *PROFAC* = protection factor

‡ *PROFAC* significantly > 0 ($p < 0.05$).

§ Evaluated relative to all men in England and Wales.

‖ The standard deviation for the Bayesian analysis posterior mean PROFAC of 0.86 was 0.07 (Scott 2006a).

Risk for Stochastic Effects at Reduced Natural Background Exposure

Here, Equation 4 is modified in order to evaluate *RR* for neoplastic transformation at below natural background exposures. Posterior-distribution-based parameter averages obtained (Table 1) for application of the $NEOTRANS_3$ model to data from Redpath *et al.* (2001) again are used. In carrying out these calculations, natural background radiation is presumed to protect (via adapted protection) from stochastic effects that would otherwise have occurred in the absence of background radiation. The adapted protection is considered to be associated with the low-LET component of the dose, since currently there is no evidence for high-LET radiation such as alpha particles by themselves activating the PAM process (Scott 2006a,b). There is evidence for protons (low-LET radiation) activating the PAM process (Elmore *et al.*, 2005). Neutrons (high-LET radiation) in their interaction with matter produce gamma rays (low-LET radiation), and these gamma rays could activate the

PAM process. For combined exposure to low doses and dose rates of alpha plus gamma radiations, the gamma-ray component of the dose appears to activate the PAM process (Scott 2006a). Thus, for combined exposure to low doses and dose rates of high- and low-LET radiations (including beta radiation), the low-LET component to the dose is expected to activate the PAM process.

The calculations carried out here illustrate the potentially harmful impact of reducing natural background radiation exposure. To facilitate computations and plotting of results, the stochastic effects normalized dose, *S*, is introduced here and used. This normalized dose is just the low-LET radiation-specific ratio, $D_{L,j}/D_{PAM}\{max\}_j$, where the subscript "*j*" is used to distinguish between different low-LET radiation types and energies (X-rays, gamma rays, protons, beta particles, etc.). Under the assumption of D_{PAM} being uniformly distributed over the closed interval $[0, D_{PAM}\{max\}_j]$, normalized doses, S_j, *for* $j = 1,2, \ldots n$, for the *n* different low-LET radiation sources can be added to obtain the total normalized dose *X* and this dose can be used to evaluate *RR* and to plot results for doses *S* that occur in Transition Zone A.

Functions $Y(S,D_H)$, $HR(S,D_H)$, $LNT(S,D_H)$, $RR(S,D_H)$, and $PAM(S)$ now correspond to the functions $Y(D_L,D_H)$, $HR(D_L,D_H)$, and $LNT(D_L, D_H)$, $RR(D_L,D_H)$, and $PAM(D_L)$, respectively. Based on Equation 4, one then obtains the following Zone-A specific equation:

$$Y(S,D_H) = PAM(S) \bullet HR(S,D_H) + [1\text{-}PAM(S)] \bullet LNT(S,D_H). \quad (5)$$

Here, no distinction is made between different high-LET radiations. The total high-LET absorbed radiation dose is given by D_H. However, for Transition Zone A, dose-response functions $HR(S,D_H)$ and $LNT(S,D_H)$ can be assumed to be essentially constant at values of $(1\text{-}PROFAC) \bullet Y_0^*$ and Y_0^*, respectively, based on our previous exploratory analyses. Y_0^* is the spontaneous transformation frequency at zero background radiation and here is presumed to be given by $Y_0/(1\text{-}PROFAC)$ so that at natural background radiation levels, Y_0 equals $(1\text{-}PROFAC) \bullet Y_0^*$. To facilitate calculations, it is assumed that there is no loss of DNA repair capacity at zero background radiation. However, with such a loss, Y_0^* could be many orders of magnitude greater than indicated here (Feinendegen *et al.,* 2004). In this case loss of background radiation could have a devastating effect on health.

For Transition Zone A, $RR(S,D_H) = RR(S)$, and is given after some algebra by

$$RR(S) = S - [(1\text{-}S)/(1\text{-}PROFAC)]\,, for\ S \leq 1. \quad (6)$$

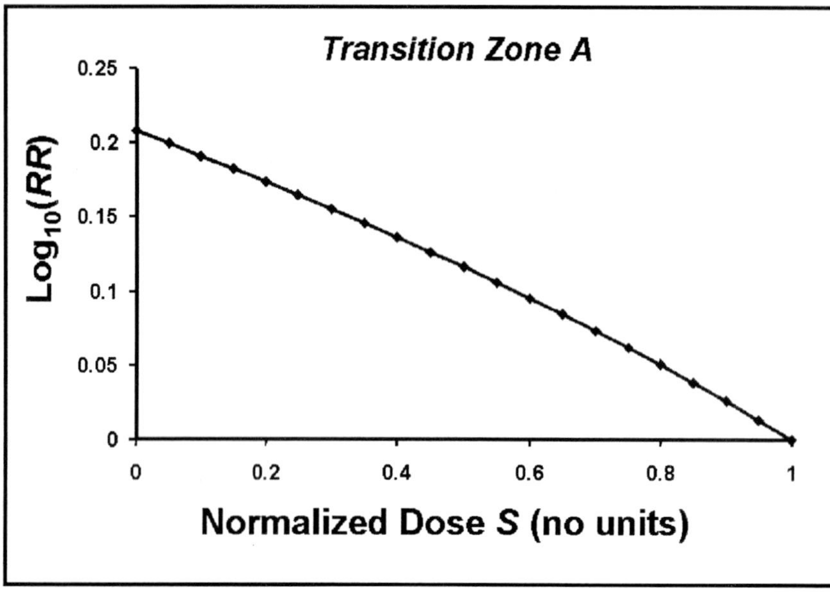

Figure 18. Simulated impact of reducing natural background ionizing radiation to zero on the *RR* for neoplastic transformation (HeLa x skin fibroblast, human hybrid cells *in vitro*) based on the $NEOTRANS_3$ model. Risk is evaluated relative to the spontaneous frequency reported by Redpath *et al.* (2001) at natural background. Relative risk is plotted as a function of the normalized dose, $S = D/D_{PAM}\{max\}$. *RR* increases, rather than decreases, as the radiation dose goes down from $S = 1$ (which corresponds to 0.01 mGy) towards zero. Similar results are expected for the *RR* for cancer induction and for other genomic-instability-associated diseases. The thresholds D_{PAM} for activating the PAM process occur in Transition Zone A where $0 < S < 1$.

Note that $RR(0) = 1/(1\text{-}PROFAC)$ and $RR(1) = 1$. Figure 18 show a plot of $ln_{10}[RR(S)]$ vs. *S* for neoplastic transformation of HeLa x skin fibroblast, human hybrid cells. The *log10* scale is used here and later to emphasize the limiting value of *RR* at *S* approaches zero.

Public Health Implications of Radiation Hormesis

Under the plausible assumption that cancer arises from cells with persistent problematic instability through a series of stochastic changes, independent of how the cells with persistent problematic instability originate, but dependent on the number of cells with this instability in an organ (a measure of total genomic instability), then the relative risk, $RR_T(S)$, for neoplastic transformation (*in vivo* or *in vitro*) would be expected to be proportional to the relative risk, $RR_C(S)$, for cancer occurrence, conditional on exposure to a dose *S*. This means that Equation 6 can be adapted for cancer *RR* evaluation simply by replacing Y_0 with the baseline cancer frequency *B* and Y_0^* with $B/(1\text{-}PROFAC)$, with the *PROFAC* evaluated based on cancer incidence or mortality data for the type of cancer of interest.

Redpath *et al.* (2001) have validated the proportionality relationship between RR_T and RR_C. For doses in the hormetic zone and for cancer risk assessment, the following relationships apply (Scott, 2006a):

$RR_C = 1$, at background radiation exposure. (7)

$RR_C = 1 - PROFAC$, for doses > background. (8)

Note that Figure 18 shows projected *RR* for gamma-ray induced neoplastic transformation of HeLa x skin fibroblast cells as a function of the normalized dose *S* as the gamma ray dose is reduced from $D_{PAM}\{max\}$ to zero (i.e. a background radiation dose of zero). Note that results point to reducing natural background as causing more neoplastic transformations rather than reducing their numbers. This relates to a loss of an opportunity for radiation-induced adapted protection. Increases could be orders of magnitude higher than indicated if DNA repair capacity is lost at zero natural background (Feinendegen *et al.*, 2004). Similar results are presented later for leukemia and lung cancer in humans.

Figure 19 shows *RR* of neoplastic transformation of HeLa skin x skin fibroblast human hybrid cells exposed to gamma rays (Redpath *et al.*, 2001), 28-kVp X-rays (Ko *et al.*, 2004), or 60-kVp X-rays (Redpath *et al.*, 2003) in vitro, for inversion mutations in pKZ1 mice exposed to 250-kVp X-rays (Hooker *et al.*, 2004), and lung cancer in gamma-ray exposed mice (Ullrich *et al.,* 1976) as a function of the normalized dose *S*. Only doses in the Zone of Maximal Protection and Transition Zone B are presented. For 1 < *S* < 10,000, all of the values for *RR* are less than one. This is also expected to be the case for cancer induction in humans that already contain significant numbers of spontaneously or other genomically unstable cells (e.g. long-time heavy smokers).

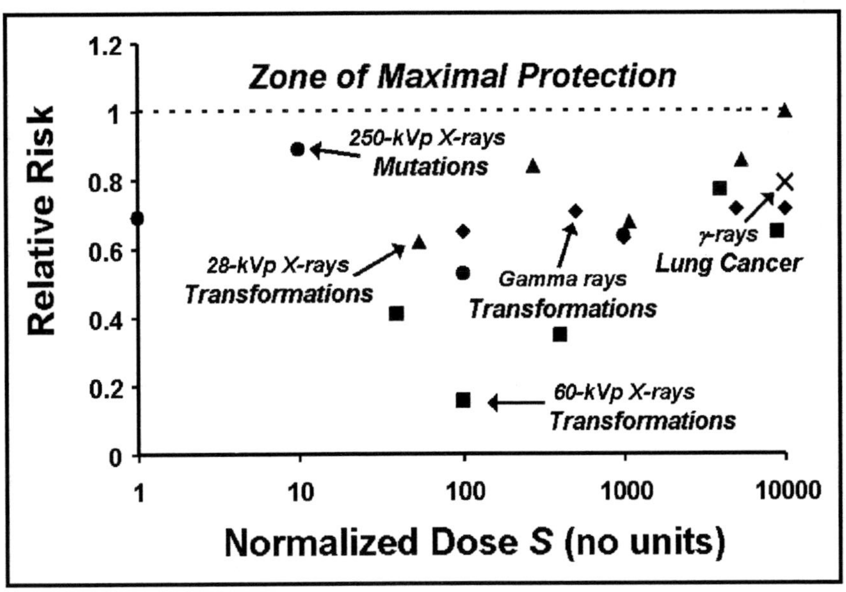

Figure 19. Relative risk for neoplastic transformation of HeLa skin x skin fibroblast human hybrid cells exposed to gamma rays (Redpath *et al.*, 2001), 28-kVp X-rays (Ko *et al.*, 2004), or 60-kVp X-rays (Redpath *et al.*, 2003) *in vitro*, for inversion mutations in pKZ1 mice exposed to 250-kVp X-rays (Hooker *et al.*, 2004), and lung cancer in gamma-ray exposed rats (Ullrich *et al*, 1976). The Zone of Maximal Protection involves a wide range of normalized doses *S* from 1 to about 1000. Higher doses presented relate to Transition Zone B. Here *S* = *1* corresponds to 0.01 mGy.

Background cosmic, solar radiations are higher at high altitudes such as in Denver, Colorado and in Salt Lake City, Utah than at sea level. Persons residing in high natural background radiation regions should be expected to have fewer cancers and other genomic-instability-associated diseases than persons residing in regions of reduced natural background radiation. Ecological research supporting this view has been published (Dissanayake, 2005; Luckey 1991; Frigèrio and Stowe 1976; Kant *et al.*, 2003).

Results presented in Figure 19 support the view presented here and elsewhere (Feinendegen *et al.*, 2004) that natural background radiation and radiation doses just above natural background radiation levels protect us from spontaneously-occurring, genomic-instability-associated diseases.

Nuclear workers chronically exposed to low levels of low- or low- plus high-LET radiation over prolonged periods (e.g., years) may have reduced cancers as compared to persons exposed at the lower end of the natural background radiation range. Long-time smoking adults and heavy alcohol consumers residing in regions mildly contaminated by beta- and/or gamma-emitting radionuclides (e.g., cesium-136, iodine-131) released during the Chernobyl accident may have reduced lung and liver cancers relative to matched populations residing in natural background (terrestrial radiation) regions. Longtime airline pilots and flight attendants may have reduced cancers and genomic-instability associated diseases when compared to matched populations that work at the surface of the Earth (low background region) where cosmic and solar radiation doses are lower.

Heavy long-time smoking and alcohol consuming Mayak Plutonium Facility workers chronically exposed at low rates to low doses of gamma plus alpha radiations over years appear to have reduced lung and liver cancers (Scott, 2006a; Tokarskaya *et al.,* 2002, 2006). In fact, the chronic exposure at low rates to gamma rays appears to have protected against both alpha-radiation and cigarette smoke-induced lung cancer (Scott, 2006a).

Activation of the PAM process requires crosstalk between genomically unstable and stable cells (Scott *et al.* 2003; Scott 2004) (see Figure 9). Thus, only persons harboring significant numbers of genomically unstable cells may benefit from low-dose activation of the PAM process. Low doses can, however, also stimulate the immune system, in which case cancer cells at an early stage of cancer development might be eliminated via the immune response (Sakai *et al.*, 2003). The *PROFAC* can also account for protection associated with immune system stimulation when modeling cancer induction *in vivo*. With its present use, the *PROFAC* gives the expected proportion of spontaneous cancer cases (e.g., cancer mortality, cancer incidence) avoided due to radiation hormesis-associated protective processes (i.e., PAM activation and immune system stimulation). The product, *100•PROFAC*, gives the expected percentage of cases avoided. The product also measures the expected percentage of spontaneous cancer-related deaths that are avoided. This relates to the cancer cases that would have otherwise occurred in the absence of adapted protection (i.e., hormesis).

Table 3 gives estimates of cancer cases avoided (as a percentage) for different human populations exposed to low- or low- plus high-LET radiation. The percentage is evaluated relative to the number of cases that would normally be expected. For example, if after radiation exposure *RR* is found to be 0.14, this would mean that 86% of cancers that normally would have been expected did not occur. The *PROFAC* in this case is 0.86. Values presented

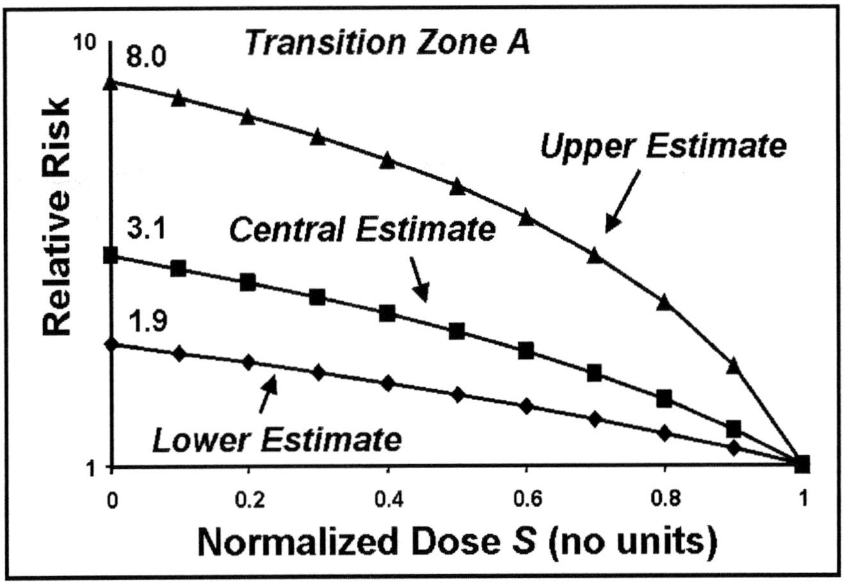

Figure 20. Simulated impact of reducing natural background radiation to zero on the *RR* for leukemia in adult humans. Central, lower, and upper estimates are presented. The lower and upper estimates define the 95% confidence region. No allowance has been made for a possible loss of DNA repair capacity at below natural background radiation levels.

in Table 3 for the *PROFAC* range from 13% to 97%. However, the 97% is somewhat controversial. For groups in Table 3 that represent nuclear workers, a small component of the assigned protection may be associated with what has been called the healthy worker effect (Sponsler *et al.,* 2005). It appears that in many poorly-designed, epidemiological studies of workers exposed to low doses of genotoxic agents, induced adapted protection was missed by simply assuming a healthy worker effect.

The radiation hormesis-associated protection indicated in Table 3 supports the view that low doses and dose rates of low- or low- plus high-LET radiation do not pose the kinds of health risk that others (NRC, 2005; NCRP, 2001), including the news media, have claimed based on the LNT assumption. Radiation hormesis-associated protective processes (adapted protection) that include activated high-fidelity DNA repair apoptosis, the auxiliary PAM process, and immune system stimulation are likely dominant factors in the response to low-dose radiation in mammals. Extrapolating from high-dose cancer risk data based on the LNT assumption appears not to be warranted from low-LET radiation or for combined exposure to low- and high-LET radiations (Scott, 2005; Tubiana, 2005; Tubiana *et al.,* 2005; Luckey 1991).

Figure 20 shows the expected impact of reducing natural background radiation to zero on the *RR* for leukemia mortality in adults when evaluated relative to leukemia mortality from spontaneous leukemia at the current natural background level. Three curves are shown (lower, central, and upper estimates) based on the *PROFAC* values presented in Table 3. For the three indicated values of 0.58, 0.68, and 0.78, the average and standard deviation are 0.68

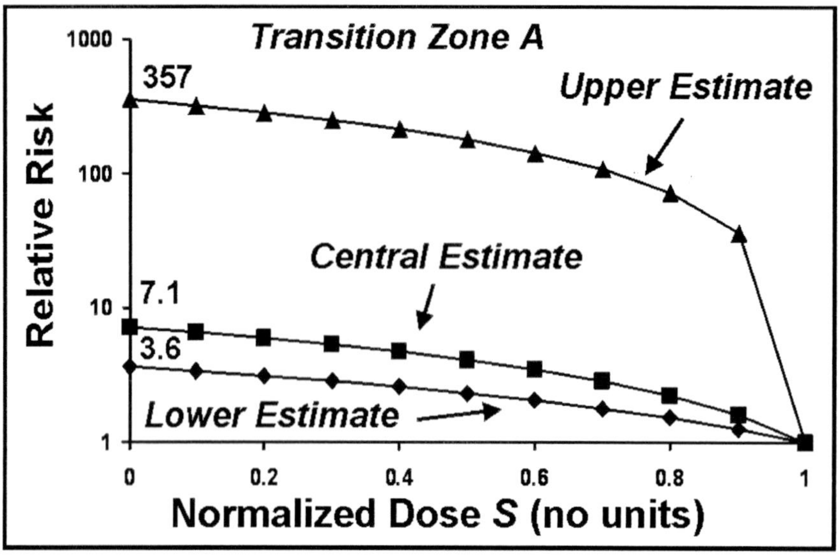

Figure 21. Simulated impact of reducing natural background radiation to zero on the *RR* for lung cancer in adults. Central, lower, and upper estimates are presented. The lower and upper estimates define the 95% confidence region. No allowance has been made for a possible loss of DNA repair capacity at below natural background radiation levels.

± 0.1. The average and associated 95% confidence values were used along with Equation 6 to obtain the three curves in Figure 20. Results are plotted as a function of the stochastic effects, normalized dose *S*. Corresponding curves for lung cancer in adults are presented in Figure 21, based on the *PROFAC* having a normal distribution with a mean and standard deviation of 0.86 ± 0.07 as indicated in Table 3 based on lung cancer data for Mayak workers. Between a 3-fold and > 350-fold increase in lung cancer death is projected for near zero background radiation. However, this risk projection does not account for the possible loss of DNA repair capacity at zero natural background radiation. With such a loss, *RR* could increase by many additional orders of magnitude as discussed below.

With our previous analysis (Scott *et al.*, 2006) of data for X-ray induced inversions mutations in spleen of pKZ1 mice exposed to doses in the range 0.001 to 0.01 mGy using the NEOTRANS$_3$ model, the pseudo parameter *k* was found to be approximately 5 orders of magnitude greater than for doses in the range 10 to 100 mGy. The background low-LET radiation dose during the radiation exposure can be assumed to be < 0.003 mGy. The 0.003 mGy value is just the average 1-day dose from background low-LET radiation. The indicated results can be interpreted to imply that the risk coefficient for cancer induction after ultra low lifetime doses < 0.01 mGy of low-LET radiation may be as high as 5 orders of magnitude larger than for exposure at above natural background levels in the LNT Zone. For example, for a risk coefficient of 10^{-5}/mGy for cancer induction in the LNT zone by low-LET radiation, the corresponding risk estimate at near zero background low-LET radiation would be ≈ $10^5 \bullet 10^{-5}$/mGy or a risk of about 1/mGy at near zero natural background ionizing radiation levels, due to the absence of adapted protection associated with current natural background ionizing radiation levels. Thus, an environment comprised of zero natural background ionizing radiation is likely to be a very unhealthy environment!

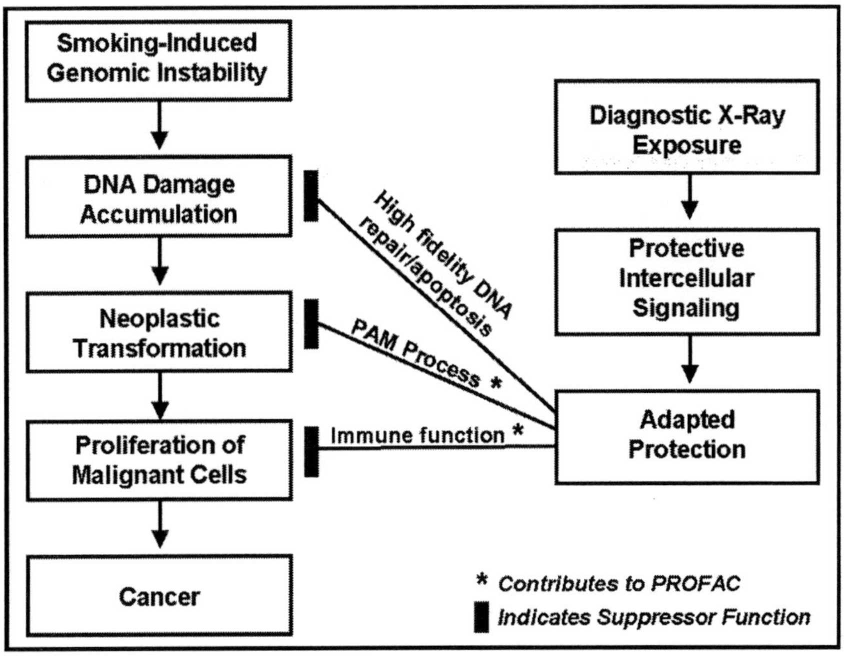

Figure 22. Explanation of how diagnostic X-rays are likely suppressing cancer in long-time, heavy smokers. Both activation of the PAM process and immune system stimulation contribute to the *PROFAC*.

Radiation doses from diagnostic X-rays and CT scans fall in the dose range ($D_{PAM}\{max\}$ to $D_{off}\{min\}$), i.e., the Zone of Maximal Protection. Thus, rather than causing harm to humans as some others have claimed (NRC 2005; NCRP, 2001), these doses may provide a benefit via activating the PAM process and stimulation of the immune system. Because the PAM process is transient and may not persist for much longer than a day (Scott *et al.*, 2003; Scott, 2004), repeated applications of X-rays would be expected to over and over activate the process, and thereby provide greater protection from stochastic effects such as cancer and other mutation-associated diseases (provided the total dose is not too high). Figure 22 summarizes how diagnostic chest X-rays may be suppressing cancer in long-time, heavy smokers via inducing adapted protection. Figure 23 shows how low doses and dose rates of low-LET radiation arising from natural radiation sources (cosmic, solar, and terrestrial radiation) may be suppressing spontaneous, and other, cancer occurrences in humans worldwide via inducing adapted protection. Similar suppression may also be occurring for other genomic-instability-associated diseases.

Cancer cells are resistant to undergoing the PAM process. However, researchers are seeking drugs that will reverse this resistance. Should they be successful, such drugs would make it possible to use low radiation doses in curing cancer (Cuttler, 2003). Multiple dose fractions at appropriate intervals (to be determined by new research) in combination with a drug that targets and sensitizes the cancer cells to undergo apoptosis (via the PAM process) might prove to be a highly desirable way (from the perspective of the patient) of treating cancer. Currently, very large, potentially lethal, radiation doses are used in treating cancer

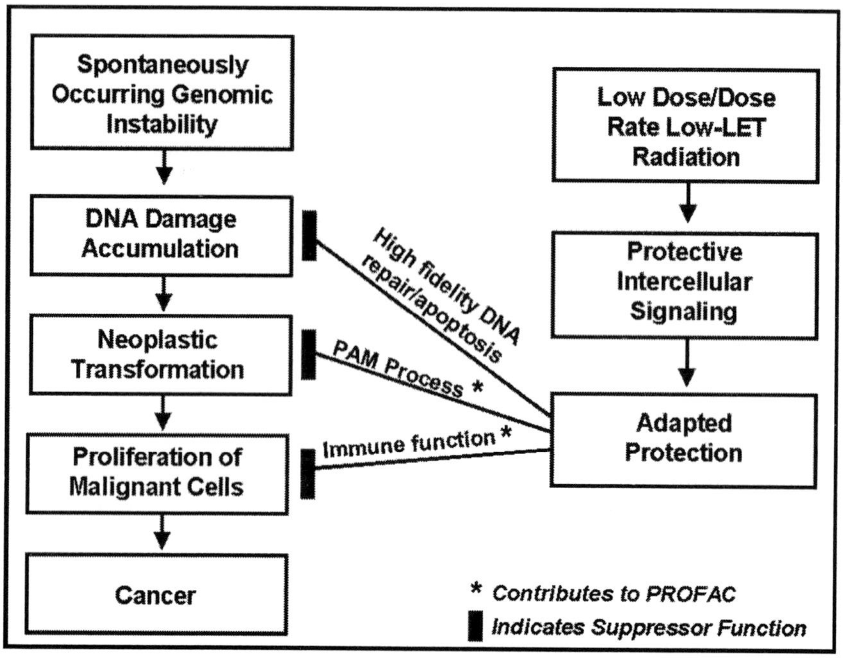

Figure 23. Explanation of how natural background radiation is likely suppressing cancer in humans around the world. Both activation of the PAM process and immune system stimulation contribute to the *PROFAC*.

(e.g., fractionated exposures to large doses). These very large doses have severe side effects. In the near future, cancer therapy using low-fractionated doses of low-LET radiation in combination with apoptosis-sensitizing agents may become more the rule than the exception.

Radiation hormesis (i.e., adapted protection) benefits society in a number of ways. However, regulatory agencies as well as many scientists have discounted hormesis as a real phenomenon. Some ways in which hormesis is discounted follow:

All radiation is assumed harmful without proof.

- Individuals receiving a low dose are included in the supposedly unexposed group.
- Low-dose data are excluded or ignored or unjustifiably assigned little statistical weight.
- Nonlinearity is ignored.
- Ecological data showing hormesis are discounted based on poor dosimetry.
- Protective processes such as activated DNA repair and apoptosis, and stimulation of the immune system are ignored.
- Evidence for lifespan prolongation is ignored.

Discounting or totally disregarding hormesis is having a high cost on society. As Edward Calabrese (Calabrese and Cook, 2005) has stated: "By dismissing hormesis, regulatory agencies such as EPA deny the public the opportunity for optimal health and avoidance of disease."

Conclusion

All organisms on earth evolved in a higher than current, but still relatively low-level, ionizing radiation environment comprised of terrestrial, solar, and cosmic radiations. A beneficial evolutionary consequence of the interaction of environmental low-level, low-LET radiation with mammalian organisms on earth is the suppression of genomic instability via radiation hormesis-associated protective biological processes. These processes include activated high-fidelity DNA repair/apoptosis, an auxiliary PAM process, and immune system stimulation. Reducing natural background radiation levels would be expected to cause harm to all mammalian organisms, including humans via loss of the opportunity for adapted protection! The harm would be expressed in increased cancers, other mutation-associated diseases, and reduced life expectancy.

Residing in regions of high natural background radiation should be associated with increased longevity, reduced cancers and other genomic-instability-associated diseases. For cancer, this was confirmed through ecological studies which were criticized based on poor radiation dose estimates. However, for the hormetic zone, dose is not needed to assess radiation-induced stochastic effects such as cancer because the *PROFAC* by itself should be sufficient for approximating risk under circumstances of adapted protection. Many of the ecological studies relate to doses in the Zone of Maximal Protection. Below this zone (i.e., in Transition Zone A), only the low-LET radiation dose is predicted to be important on Earth.

Low-level exposure to diagnostic X-rays are likely activating the PAM process and stimulating the immune system in individuals that harbor significant numbers of genomically unstable cells. The indicated adapted protection activation reduces deleterious stochastic effects such as cancer and other genomic-instability-associated diseases via ridding the body of the genomically unstable cells which may have arisen for example as a result of high-level cigarette or alcohol usage.

The PAM process and immune system stimulation could, in theory, be exploited to develop novel cancer treatment protocols that would involve fractionated low doses or low-LET radiation in combination with drugs that target cancer cells and render them susceptible to undergoing apoptosis. Use of fractionated low doses would likely be preferred by cancer patients, as this would eliminate the serious side effects associated with the high-dose radiation and chemotherapy currently used.

Continued use of the LNT risk function for assessing low-dose radiation health risk to humans is unjustifiable for low-LET radiation and for combinations of low- and high-LET radiation because the low-LET radiation activates the hormesis-associated system of protective processes (adapted protection) that protect all of us from deleterious stochastic radiobiological effects.

Acknowledgments

This chapter is dedicated to my new granddaughter, Kaylah Walker, her proud parents Renee and Paul Walker, her aunt Robin Scott, and her great grandmother Ethel Hill Scott. This research was supported by the Office of Science (BER), U.S. Department of Energy

(DOE) Grants DE-FG02-03ER63671 and DE-FG02-03ER63657. I am grateful to Ms. Jennifer Di Palma for editorial, graphical, computational, literature-searching, and other support and to Ms. Vicki Fisher for additional editorial assistance. I am especially grateful to Dr. Howard S. Ducoff for introducing me to the field of radiation hormesis research and to Dr. Edward Calabrese for providing a forum (annual BELLE Conference) for presenting my research results that relate to radiation hormesis. Our research group benefited from discussions with researchers around the world, some of which without hesitation provided adapted protection data to me for use in developing models presented here and elsewhere. These researchers include Drs. Leslie Redpath, Pamela Sykes, and Tanya Day. Strong support for this research was provided by the Directorate of our Institute and supporting staff, for which I am also very grateful. The views and conclusions contained herein are those of the author and should not be interpreted as necessarily representing the official policies or endorsement, either express or implied, of the DOE or of Lovelace Respiratory Research Institute. Hopefully, during Kaylah's lifetime, the benefits of low-dose radiation discussed in this chapter will be recognized and, where appropriate, utilized worldwide.

References

Azzam EI, de Toledo SM, Raaphorst P, Mitchel RE. Low-dose ionizing radiation decreases the frequency of neoplastic transformation to a level below the spontaneous rate in C3H 10T1/2 cells. *Radiat Res.* 1996;146:369-373.

Bauer G. Resistance to TGF-β-induced elimination of transformed cells is required during tumor progression. *Int J Oncol.* 1995;6:1227-1229.

Bauer G. Elimination of transformed cells by normal cells: Novel concept for the control of carcinogenesis. *Histol Histopathol.* 1996;11:237-255.

Bauer G. Reactive oxygen and nitrogen species: efficient, selective, and interactive signals during intercellular induction of apoptosis. *Anticancer Res.* 2000;20(6B):4115-4139.

Berrington A, Darby SC, Weiss JA, Doll R. 100 years of observation on British radiologists: mortality from cancer and other causes 1987-1997. *Br J Radiol.* 74;2001:507-519.

Bonner WM. Low-dose radiation: Thresholds, bystander effects, and adaptive response. *Proc Natl Acad Sci USA.* 2003;100(9):4973-4975.

Calabrese EJ, Baldwin LA. Reevaluation of the fundamental dose-response relationship. A new database suggests that the u-shaped, rather than the sigmoidal, curve predominates. *Biol Sci.* 1999;49(9):725-732.

Calabrese EJ, Baldwin LA. The hormetic dose-response model is more common than the threshold model in toxicology. *Toxicol Sci.,* 2003a;71:246-250.

Calabrese EJ, Baldwin LA. Hormesis: the dose-response revolution. *Annu Rev Pharmacol Toxicol.* 2003b;43:175-197.

Calabrese EJ, Cook RR. Hormesis: how it could affect the risk assessment process. *Hum Exp Toxicol.* 2004;24:265-270.

Calabrese EJ, Cook RR. Hormesis: how it could affect the risk assessment process. *Hum Exp Toxicol.* 2005;24:265-270.

Chen WL, Luan YC, Shieh MC, Chen ST, Kung HT, Soong KL, Yeh YC, Chou TS, Mong SH, Wu JT, Sun CP, Deng WP, Wu MF, Shen ML. Is chronic radiation an effective prophylaxis against cancer? *J Am Phys Surg.* 2004;9(1):6-10.

Chu G. Minireview. Double strand break repair. *J Biol Chem.* 1997;272:24097-24100.

Coates PJ, Lorimore SA, Wright EG. Cell and tissue response to genotoxic stress. *J Pathol.* 2005;205(2):221-235.

Crawford-Brown DJ, Hofmann W. Analysis of radon-induced lung cancer risk by a stochastic state-vector model of radiation carcinogenesis. *J Radiol Prot.* 2002;22:A61-A65.

Cuttler JM. Can cancer be treated with low doses of radiation? *J Am Phys Surg.* 2003;8(4):108-111.

Dissanayake C. Of stones and health: medical geology in Sri Lanka. *Science*, 2005;309:883-885.

Ducoff HS. Radiation Hormesis: incredible or inevitable? *Korean J Biol Sci.* 2002;6:187-193.

Draganić IG, Draganić ZD, Adloff J-P. *Radiation and radioactivity on earth and beyond.* Boca Raton, FL, CRC Press, Inc.;1990.

Elmore E, Lao X-Y, Ko M, Rogjtnar S, Nelson G., Redpath JL. Neoplastic transformati0on in vitro induced by low doses of 232 MeV protons. *Int J Radiat Biol.* 2005;81(4):291-297.

Feinendegen LE, Pollycove M, Sondhaus CA. Response to low doses of ionizing radiation in biological systems. *Nonlinearity.* 2004;2(3):143-171.

Frigèrio NA, Stowe RS. Carcinogenic and genetic hazard from background radiation. In: biological and environmental effects of low-level radiation. Proceedings of a Symposium on Biological Effects of Low-level Radiation Pertinent to Protection of Man and His Environment, *Volume II. Vienna: International Atomic Agency;*1976. p. 385–393.

Hanahan D, Weinberg RA. The hallmarks of cancer. Cell. 2000;100:57-70.

Higson DJ. The bell tolls for LNT. *Operational Radiation Safety.* November 2004;S47-S50.

Hooker AM, Bhat M, Day TK, Lane JM, Swinburne SJ, Morley AA, Sykes PJ. The linear no-threshold model does not hold for low-dose ionizing radiation. *Radiat Res.* 2004;162:447-452.

Hipp ML, Bauer G. Intercellular induction of apoptosis in transformed cells does not depend on p53. *Oncogene.* 1997;17(7):791-797.

Isaev A. *Introduction to mathematical methods in bioinformatics.* Berlin, Springer; 2004.

Ivanov VK, Gorski AI, Maksioutov MA, Tsyb AF, Souchkevitch GN. Mortality among the Chernobyl emergency workers: estimation of radiation risks (preliminary analysis). *Health Phy.* 2001;81(5):514-521.

Jaworowski Z. Ionizing radiation in the 20th century and beyond. Symposium "Entwicklungen im Strahleschutz." Munich; November 29, 2001. Available at: http://www.cns-snc.ca/branches/Toronto/radiation/. Accessed November 26, 2005.

Jürgensmeier JM, Höfler P, Bauer G. TGF-β induced elimination of transformed fibroblast by normal cells: independence of cell-to-cell contact and dependence on ROS. *Int J Oncol.* 1994;5:525-531.

Kant K, Chauhan RP, Sharma GS, Chakarvarti SK. Hormesis in humans exposed to low-level ionizing radiation. *Int J Low Rad.* 2003;1(1):76-87.

Ketchum L. Lessons of Chernobyl: SNM members try to decontaminate world threatened by fallout. *J Nuc Med.* 1987;28(6):933-942.

Ko SJ, Liao X-Y, Molloi S, Elmore E, Redpath JL (2004). Neoplastic transformation in vitro after exposure to low doses of mammographic-energy X rays: quantitative and mechanistic aspects. *Radiat. Res.*162:646-654.

Langer C, Jürgensmeier JM, Bauer G. Reactive oxygen species act at both TGF-beta-dependent and independent steps during induction of apoptosis of transformed cells by normal cells. *Exp Cell Res.* 1996;222(1):17-124.

Ljungman M. Activation of DNA damage signaling. *Mutat Res.* 2005;577:203-216.

Lombard DB, Chua KF, Mostoslavsky R, Franco S, Gostissa M, Alt FW. DNA repair, genome stability, and aging. Cell. 2005;120:497-512.

Luckey TD. Radiation hormesis, Boca Raton, Florida: CRC Press; 1991.

Luckey TD. A Rosetta stone for ionizing radiation. *Radiat Prot Man.* 1994;11(1):73-79.

Luckey TD. Radiation hormesis overview. *RSO Magazine;*2003:8(4), 22-40.

Mattson MP. Hormesis and disease resistance: Activation of cellular stress response pathways. *BELLE Newsletter.*2005;13(3):6-14.

Mills KD, Ferguson DO, Alt FW. The role of DNA breaks in genomic instability and tumorigenesis. *Immun Rev. 2003;*94:77-95.

Mitchel REJ. The bystander effect: recent developments and implications for understanding the dose response. *Nonlinearity.* 2004;2(3):173-183.

Mitchel REJ. Low doses of radiation are protective *in vitro* and *in vivo*: evolutionary origins. *Dose Response.* In press 2006.

Morgan WF, Day JP, Kaplan ML, McGhee EM, Limoli CL. Genomic instability induced by ionizing radiation. *Radiat Res.*1996;146:247-258.

Mothersill C, Seymour CB. Medium from irradiated human epithelial cells but not human fibroblast reduce the clonogenic survival of unirradiated cells. *Int J Radiat Biol.* 1997;71:421-427.

Mothersill C, Seymour CB. Review. mechanisms and implications of genomic instability and other delayed effects of ionizing radiation exposure. *Mutagenesis.* 1998a;13(5):421-426.

Mothersill C, Seymour CB. Cell-cell contact during gamma irradiation is not required to induce a bystander effecting normal human keratinocytes: evidence for release during irradiation of a signal controlling survival into the medium. *Radiat Res.* 1998b;149:256-262.

NCRP (National Council on Radiation Protection and Measurements). Ionizing radiation exposure of the population of the United States. NCRP Report 93: Bethesda, MD;1997.

NCRP (National Council on Radiation and Protection and Measurements). Evaluating the linear no-threshold dose-response model for ionizing radiation. NCRP Report 136: Bethesda, MD;2001.

NRC (National Research Council). Health risks from exposure to low levels of ionizing radiation. BEIR-VII Phase 2. The National Academies Press; 2005. Prepublication copy available from *www.nap.edu.*

Redpath JL, Liang D, Taylor TH, James C, Christie E, Elmore E. The shape of the dose-response curve for radiation-induced neoplastic transformation *in vitro*: evidence for an adaptive response against neoplastic transformation at low doses of low-LET radiation. *Radiat Res.* 2001;156:700-707.

Redpath JL, Lu Q, Lao X, Molloi S, Elmore E (2003). Low doses of diagnostic X-rays protect against neoplastic transformation in vitro. *Int. J. Radiat. Biol.* 79(4):235-240.

Rothkamm K, Löbrich M. Evidence for a lack of DNA double-strand break repair in human cells exposed to very low x-ray doses. *Proc Natl Acad Sci U S A.* 2003;100(9):5057-5062.

Sakai K, Hoshi Y, Nomura T, Oda T, Iwasaki T, Fujita K, Yamada T, Tanooka H. Suppression of carcinogenic process in mice by chronic low dose rate gamma-irradiation. *Int J Low Radiat.* 2003; 1(1):142-146.

Schöllnberger H, Scott BR, Hanson TE. Application of Bayesian inference to characterize risks associated with low doses of low-let radiation. *Bull Math Biol.* 2001;63:865-883.

Scott BR. Mechanistic state vector model for cell killing by ionizing radiation. *Radiat Environ Biophys.* 1977;14:195-211.

Scott, BR. A mechanistic model for neoplastic transformation of cells by high LET radiation and its implication for low dose, low dose rate risk assessment. *Radiat Prot Dosim.* 1997; 72:105-117.

Scott BR. A biological-based model that links genomic instability, bystander effects, and adaptive response. *Mutat Res.* 2004;568:129-143.

Scott BR. Low-dose radiation risk extrapolation fallacy associated with the linear-no-threshold model. *BELLE Newsletter.* 13(2), Part 2; December 2005:22-27.

Scott BR. Low-dose radiation-induced protective process and implications for risk assessment, cancer prevention, and cancer therapy. *Dose Response.* In press 2006a.

Scott BR. Stochastic thresholds: A novel explanation of nonlinear dose-response relationships. *Dose Response.* In press 2006b.

Scott BR, Dillehay LE. A model for hematopoietic death in man from irradiation of bone marrow during radioimmunotherapy. *Br J Radiol.* 1990;63:862-870.

Scott BR, Guilmette RA. Radiation toxicology, ionizing and nonionizing. In: Wexler P, editor. *Encyclopedia of toxicology,* Vol. 3. Oxford: Elsevier Limited; 2005. p. 601-614.

Scott BR, Hahn FF, McClellan RO, Seiler FA. Risk estimators for radiation induced bone marrow syndrome lethality in humans. *Risk Anal.* 1988;8:393-402.

Scott BR, Walker DM, Tesfaigzi Y, Schöllnberger H. Mechanistic basis for nonlinear dose-response relationships for low-dose radiation-induced stochastic effects. *Nonlinearity.* 2003;1(1):93-122.

Scott BR, Walker DM, Walker V. Low dose radiation and genotoxic chemicals can protect against stochastic biological effects. *Nonlinearity.* 2004;2(3):185-211.

Scott BR, Haque M., Di Palma J. Biological basis for radiation hormesis in mammalian cellular communities. *Int J Low Radiat.* Accepted 2006.

Sieber OM, Heinimann K, Tomlinson IPM. Genomic instability—the engine of tumorigenesis? *Nat Rev Cancer.* 2003;3:701-708.

Sponsler R, Cameron JR. Nuclear shipyard worker study (1980-1988): a large cohort exposed to low-dose-rate gamma radiation. *Int J Low Radiat.* 2005;1:463-478.

Tokarskaya ZB, Scott BR, Zhuntova GV, Okladnikova ND, Belyaeva ZD, Khokhryakov VF, Schöllnberger H, and Vasilenko EK. Interaction of radiation and smoking in lung cancer induction among workers at the Mayak Nuclear Enterprise. *Health Phys.* 2002;83(6); 833-846.

Tokarskaya ZB, Zhuntova GV, Scott BR, Khokhryakov VF, Belyaeva ZD, and Vasilenko EK (in press, 2006). Influence of alpha and gamma radiations and non-radiation factors on the incidence of malignant liver tumors among Mayak PA workers. *Health Phys.*

Tubiana M. Dose-effect relationship and estimation of the carcinogenic effects of low doses of ionizing radiation: The joint report of the Académie des Sciences (Paris) and of The Académie Nationale de Médicine. *Int Radiat Oncol Biol Phys.* 2005;63(2):317-319.

Tubiana M, Aurengo A, Averbeck D, Bonin A, Le Guen B. Masse R, Monier R, Valleron A-J, de Vathaire F. Dose-effect relationships and estimation of the carcinogenic effects of low doses of ionizing radiation. Académie des Sciences – *Académie Nationale de Médicine Report.* Paris, France; March 30, 2005.

Ullrich RL, Jernigan MC, Cosgrove GE, Satterfield LC, Bowles ND, Storer JB. The influence of dose and dose rate on the incidence of neoplastic disease in RFM mice after neutron irradiation. *Radiat Res* 1976;68(1):115–131.

Waterman MS. Introduction to computational biology. *Maps, sequences and genomes.*Boca Raton, FL, CRC Press, Inc.; 2000.

Wolff S, Afzal V, Wiencke JK, Olivieri G, Michaeli A. Human lymphocytes exposed to low doses of ionizing radiation become refractory to high doses of radiation as well as to chemical mutagens that induced double-strand breaks in DNA. *Int J Radiat Biol.* 1988;53(1):39-48.

Wolff S. Aspects of the adaptive response to very low doses of radiation and other agents. *Mutat Res.* 1996;358:135-142.

Wright EG. Radiation-induced genomic instability in hemopoietic cells. *Int J Radiat Biol.* 1998;74:681-688.

In: New Research on Genomic Instability
Editor: Eleanor J. Glascow, pp. 181-206

ISBN: 1-60021-320-0

Chapter VII

"Delayed Division Delay" Revealed by Time-lapse Photographic Study: A New Endpoint of Radiation-Induced Genomic Instability

Hiroshi Sasaki*
Kyushu University, Fukuoka, Japan

ABSTRACT

LET-dependent features of cell abnormality leading to death, lethal sectoring, genomic instability and cell cycle change were compared between α and X irradiated HeLa cells by means of a time-lapse photography. Cell abnormality induced by α irradiation tended to be manifested in the form of death without division (mitotic or interphase death), rather than incomplete division accompanied by anaphase bridge derived from dicentric chromosome (two-hit type damage) that was predominant for X irradiation. Frequency of incomplete division at irradiated generation was decreased remarkably (25.8% → 6.8%) by splitting X ray dose (4 Gy → 2 × 2 Gy), correlating well with the increase in cell survival (6.8% → 14.0%). However, in the case of α irradiation especially at low dose, two split doses (0.6 Gy → 2 × 0.3 Gy) significantly decreased cell survival (12.3% → 8.8%) with the accompanying increase in the frequency of incomplete division (7.2% → 13.5%) (reverse split dose effect). This suggests that an interaction between the damage by the 1st dose and the one by the 2nd dose gives rise to 2-hit type damage owing to irreparability of the damage and enhances cell killing at low dose where interphase cell death is not predominant. Lethal sectoring is the process for survival in which lethal damage remaining in irradiated cells is eliminated as lethal sector (offspring without reproductive integrity). This process occurs through the

* Corresponding author: Phone/Fax: +81-92-822-8354, E-mail; h.sasaki@hiz.bbiq.jp

postirradiation 1st to 4th divisions, resulting in the appearance of a clonogenic progenitor harboring no lethal damage. However, nonlethal damage remaining in the clonogenic progenitors led to an elevated incidence of delayed cell death in their progeny. From an analysis of the pedigrees of cells surviving α (0.45 Gy) or X irradiation (3 Gy) (20% survival dose) which were constructed by means of a long-term (~150 h) time-lapse observation, it became clear that the mean incidence was higher for α-particle (16.3%) than X-ray survivors (8.3%), indicating the greater potentiality for genomic instability by α particles. Evidence is available to suggest the intrinsic difference in the mechanisms for the induction of genomic instability by these two radiations; the association of misrepaired clustered DNA damage (CD) with α particles and the fixation of unrepaired potentially lethal damage (PLD) with X rays. A novel phenomenon "delayed division delay (DDD)" was noticed, though occasionally (~10% per cell), with the progeny during the postirradiation 1st–3rd generations. DDD was much longer in α- (mean: ~11 h) than X-irradiated cells (~4 h). DDD might be triggered by delayed DNA double-strand breaks (DSB) at the fragile break-misrejoining sites, and the latter was supposed to be induced by a higher-order conformational change of chromatin fibers during mitosis. This demonstrates the usefulness of DDD as a new endpoint of radiation-induced genomic instability. Contrary to DDD, a significant shortening (~5 h) of cell cycle time at the postirradiation 1st generation was found with the progeny of X-ray survivors harboring no lethal nor nonlethal damage.

Abbreviation

CD,	clustered DNA damage;
DDD,	delayed division delay;
GI,	genomic instability;
LET,	linear energy transfer;
LS,	lethal sectoring;
PLD,	potentially lethal damage.

Introduction

Radiobiological studies of cultured mammalian cells are resulting in abundant knowledge of radiation damage at the cellular level. In most of these studies, random or synchronized cell population was used as a material. For an example in the classical studies, the loss of colony-forming ability has been usually employed as a criterion of radiation-induced cell death. However, such studies are concerned only with the initial and final effects of radiation on proliferating cell populations; they do not clarify how irradiated cells divide and die, and how some progeny continue to divide and form viable colony. The better method of studying such diverse phenomena is a time-lapse observation of individual cells.

Important findings on cell abnormality leading to death have been obtained by this method. Incomplete division followed by cytoplasmic cell fusion mostly between sister cells born after X irradiation, first reported by Marin and Bender [1], resulted in binucleate cells without colony-forming ability, therefore playing an important part in X-ray-induced reproductive cell death. Later, we detected a nuclear thread bridging the sister cells prior to cell fusion by means of autoradiography with ^{3}H-thymidine and scanning electron microscopy [2]. Hurwitz and Tolmach [3] found cell death at mitosis (mitotic death or mitotic catastrophe) in X-irradiated HeLa cells; cells were arrested at mitosis for 8–10 h before disintegration.

Based on long-term time-lapse observations of X-irradiated mouse L cells, Thompson and Suit [4] constructed cell pedigrees and examined the colony-forming process of surviving cells and recognized the phenomenon "lethal sectoring" , which had been reported by James et *al.* [5, 6] who constructed the pedigrees of yeast surviving UV or X irradiation by isolating individual offspring, using a micromanipulator and found that the offspring without the colony-forming ability appeared during several postirradiation generations.

I have tried to elucidate the characteristic cellular responses after irradiation with α particles or accelerated high-LET charged particles compared to low-LET X rays. "Reverse split dose effect" was found with HeLa cells by splitting low dose of α particles (0.6 Gy → 2 × 0.3 Gy) ; cell survival decreased with the accompanying increase in the frequency of incomplete division (2 hit-type cell damage) [7]. "Dependence of induction of interphase cell death on LET" with maximum RBE (~3) around LET 200 keV/μm was recognized with CHO cells exposed to accelerated charged particles of various LETs [8]. I also constructed the pedigrees of HeLa cells surviving α or X irradiation but they remained unexplored for more than 20 years. Taking into consideration of "genomic instability" and "clustered DNA damage" both of which are apt to be induced by α particles, these pedigrees were analyzed in detail. Fortunately, I revealed a novel phenomenon "delayed division delay" [9].

MATERIALS AND METHODS

Cell and Culture

Human cervical carcinoma (HeLa S3-9 IV: a clone established by T. T. Puck) cells, kindly provided by Dr. T. Terasima, were used in most of the experiments. The cells were grown in F10 medium supplemented with 0.05% heart infusion broth (Difco) and 20% fetal calf serum. Chinese hamster ovary (CHO-K1) cells, provided by the RIKEN Cell Bank, were used in the experiments shown in Figs. 6, 7 and 14. The cells were grown in Ham's F12 medium supplemented with 10% fetal calf serum.

A temperature-sensitive mutant (tsBN2) derived from wild-type BHK21 (baby hamster kidney) cell line, kindly provided by Dr. T. Nishimoto, was used in the experiment shown in Figure 13. The cells were cultured in Dulbecco's modified Eagle medium (DME) supplemented with 10% fetal calf serum at 33.5℃ (permissive temperature). Premature chromosome condensation (PCC) is induced irrespective of the cell cycle by the incubation at nonpermissive temperature (40℃) in this ts mutant which has a single ts defect in the regulatory mechanism for chromosome condensation.

Irradiation

The radiation sources used were 200 kVp X rays (HVL, 0.92 mm Cu; dose rate, 0.7 Gy/ min), α particles from ^{241}Am point source (fluence rate, ~1.5 × 10^6/ cm^2/ min; dose rate, ~0.3 Gy/ min) and charges particles of various LETs accelerated by RIKEN Ring Cyclotron (RRC). At the time of α irradiation, a culture dish was put on an α source upside down after removal of the culture medium. The air path length of 12 mm from the α source to the cell surface and the thickness of the cells reduced the energy of α particles to 3.97 MeV (LET ~125 keV/μm in water), from 5.49 MeV.

Time-lapse Photography

For time-lapse observations of α or X irradiated cells, a plastic dish on which a few cells (~50) were grown in the center (diameter; ~1 cm) was placed in a glass container with ventilation ports and mounted on the stage of a phase-contrast microscope. In the case of irradiation with accelerated charged particles, cells were grown in a culture flask with ventilation ports at its screw cap. A small amount of 5% CO_2-95% air was supplied through these ports (Figure 1). Photographs were taken every 15 min (HeLa cells) or 30 min (CHO cells) with a 35-mm motor-driven camera. Since the exposures were of the same microscopic field during preirradiation and postirradiation period, the cell stage of asynchronously growing cells at irradiation could be estimated by elapsed time after cell division. The pedigrees of irradiated individual cells were constructed by reviewing the films. They were divided into survivor and nonsurvivior groups.

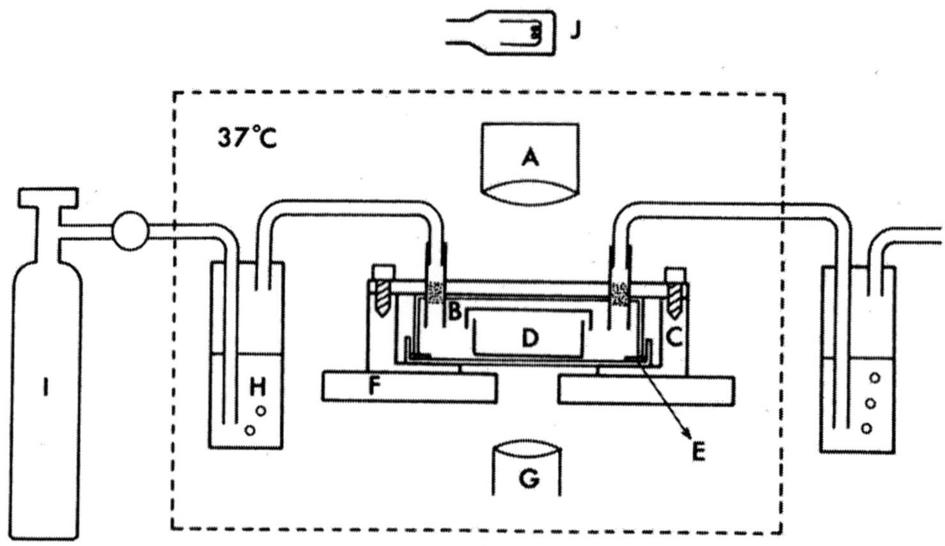

Figure 1. Cell culture system for time-lapse photography. A: Condenser. B: Glass container with ventilation ports. C: Lucite holder with screws. D: Culture dish. E: Packing. F: Stage of microscope. G: Phase-contrast objective lens (×5 or ×10). H: Water. I: Bomb filled with 5% CO_2-95% air. J: Halogen lamp.

Scanning Electron Microscopy

In the experiment shown in Figure 5-a and 5-b, HeLa cells were grown on cover slips. After fixation in 3% glutaraldehyde (sodium phosphate buffer, pH 7.3), followed by 1% osmium tetroxide, the samples were dehydrated in graded concentrations of ethanol and dried using a critical point drying device (HCP-2) and carbon dioxide as the transitional liquid. The specimens were coated with carbon and gold under rotation *in vacuo* (lower than 5×10^{-5} Torr) and were examined with a "JSM-2" scanning electron microscope at 25 kV accelerating voltage.

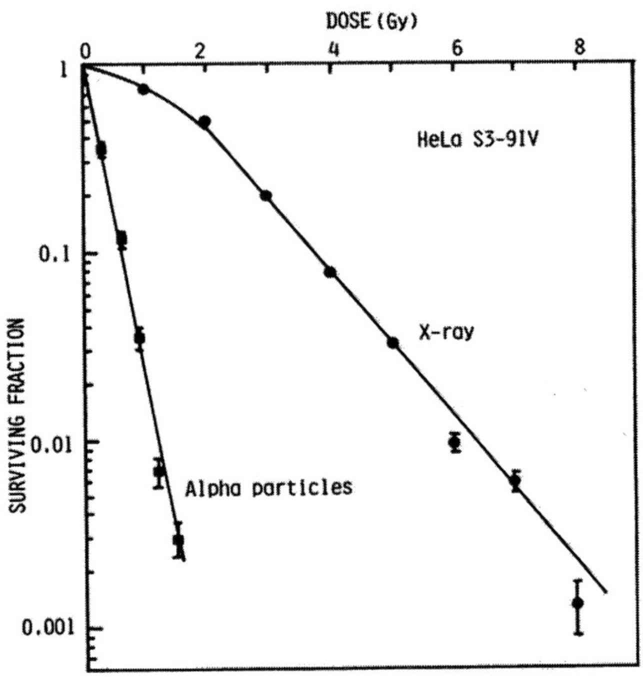

Figure 2. Survival curve of asynchronous HeLa cells irradiated with X rays or α particles. Vertical bars indicate standard deviations (Reproduced from Figure 2 of the reference [7]).

Autoradiography

In the experiment shown in Figure 5-c, HeLa cells were grown on cover slips, labeled with ^{3}H-thymidine (0.02 μCi/ ml, 15 Ci/ m mole) for 24 h and rinsed with fresh medium before X irradiation. Cells were fixed with Carnoy solution, stained with acetoorcein and coated with NR-M_2 dipping emulsion (Konishiroku Photo Industries Co. Ltd.). The specimens were exposed for about 6 months to β rays from ^{3}H-thymidine incorporated into cellular DNA.

RESULTS

Lethal Hit Number

In the investigations using the loss of colony-forming ability as an indicator of cell death, the effects of high- but not low-LET radiation are characterized by the reduction or absence of an initial shoulder in the cell survival curve (Deering and Rice [10]), diminished or no recovery from sublethal damage (Ngo *et al.* [11]), a lowered value of OER (Oxygen enhancement ratio) (Barendsen *et al.* [12]), and a smaller variation of radiosensitivity during the cell cycle (Hall *et al.*[13]). These radiobiological differences between high- and low-LET radiations have been explained reasonably on the basis of a linear-quadratic model which entails one- and two-hit lethal events and the predominance of the former for high-LET radiation (Kellere and Rossi [14]).

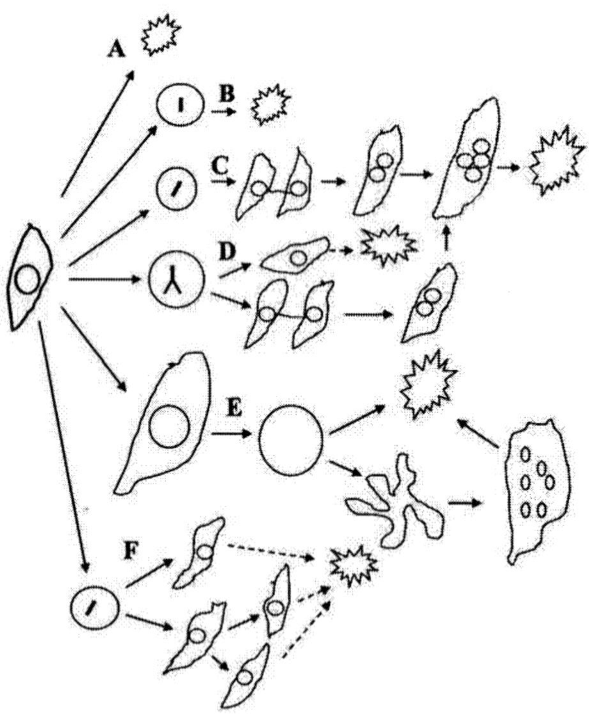

Figure 3. Cell abnormality leading to death in irradiated cultured mammalian cells. A: Cell-cycle arrest-independent interphase death. B: Death at mitosis (mitotic death or catastrophe). C: Incomplete division accompanied by sister cell fusion, which might be guided by an intercellular chromatin bridge. D: Multipolar division, presumably because of an abnormal replication of centrosome and usually accompanied by sister cell fusion. E: Cell-cycle arrest-dependent interphase death, characterized by giant cell formation with an accompanying fragmentation of the nucleus and of DNA. F: Death after normal (bipolar) division at least once (generally called reproductive death when the initial cell depicted by a bold line is located at an irradiated generation). A solid and broken arrow indicates either a direct or indirect (including other pathways) changes, respectively. E was predominantly induced by high-LET ionizing radiation (α particles, heavy-ions) at irradiated generation, whereas C was more frequent during postirradiation several generations after exposure to low-LET (X- or γ-rays) than high-LET ionizing radiation when compared at the same survival. (Reproduced from Figure 1 of the reference [9]).

Survival curves of HeLa cells irradiated with α particles and X rays are shown in Figure 1 [7]. The X-ray curve has an initial shoulder (D_q= 1.25 Gy) followed by an exponential part with a D_0 of 1.1 Gy, whereas the α-particle curve is completely exponential with a D_0 of 0.25 Gy. The fluence required to reduce the survival to e^{-1} (F_0) is 1.17 × 10^{-2}/ μm^2; therefore the inactivation cross-section (σ) is calculated to be 85 μm^2 according to the formula $\sigma = 1/ F_0$. Since an average geometrical cross-section of cell nucleus is about 220 μm^2 for the HeLa cells (Konishi *et al.* [15]), it is postulated that 2.6 hits (220/ 85) to the cell nucleus give rise to one lethal event. The reason why more than 1 hit of α particles are required in order to inactivate the reproductive integrity of the cell will be explained by the following facts; 1) some of the lethal damage is removed during several postirradiation cell divisions (lethal sectoring) and only irremovable lethal damage inactivates the cell, 2) some of the damage is nonlethal and responsible for genomic instability or completely repaired during division delay at irradiated generation.

Cell Abnormality and Death

Illustration of cell abnormality leading to death of irradiated cells (Figure 3) is based on the time-lapse observations mainly with HeLa cells including other cell lines such as CHO cells. According to the definition of interphase (premitotic) and reproductive (postmitotic) death, (A)-(B)-(E) and (C)-(D)-(F) correspond to the former and the latter, respectively. However, in HeLa but not CHO cells, death at mitosis (C: mitotic death or mitotic catastrophe) is characterized by a prolonged arrest at mitosis (8–10 hours) before disintegration and therefore is different from (A) and (E). Hopwood and Tolmach [16] considered deficient DNA synthesis as a cause of mitotic death. Death at interphase is divided into cell-cycle arrest-independent (A) and dependent (E) interphase death. The latter was predominantly induced by high-LET radiation [8]. Cells destined to undergo this type of interphase death were arrested before mitosis for more than 50 h for HeLa cells and 24 h for CHO cells and were transformed into giant cells (Figure 4-a and b). The morphological change in the cells that was associated with the initiation of interphase death was "rounding up" from the spread configuration (Figure 4-c and d). followed by fragmentation into multiple parts (Figure 4-e) and fusion, resulting in the formation of giant cells containing multiple micro nuclei, in regard to the CHO cells. These giant cells showed apoptotic features including DNA fragmentation. They then gradually disintegrated, but disintegration occurred from rounded configuration in the HeLa cells.

Dose response curve for the induction of interphase death in the CHO cells exposed to Ne ions (228 keV/μm) is different from the one exposed to X rays (Figure 5); the former increases linearly with dose, whereas the latter begins to increase when the dose exceeds ~10 Gy (threshold). This indicates the requirement, for the induction of interphase death, of multiple damage on the chromosome which may be accomplished by a single hit of heavy ion but demand higher dose in the case of X irradiation.

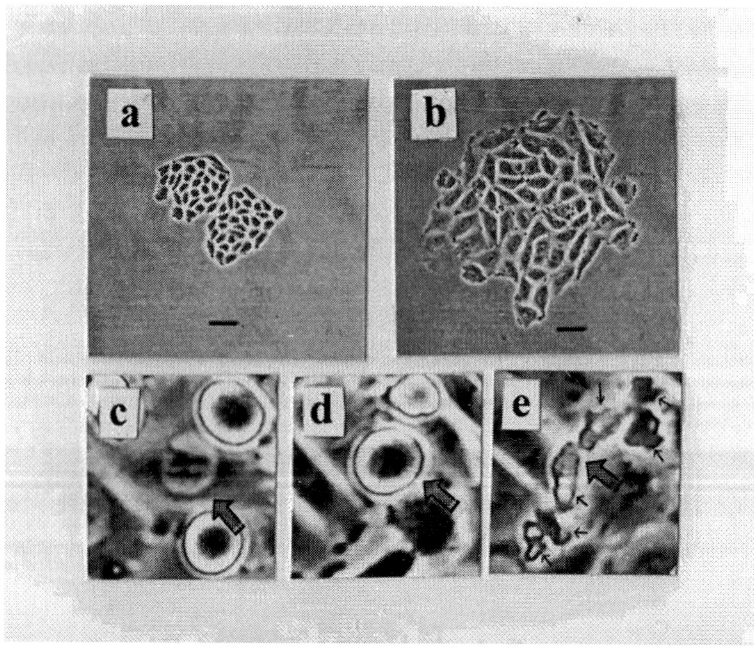

Figure 4. Time-lapse photomicrographs of the morphological changes leading to interphase death of CHO cells exposed to argon ions (LET: 320 keV/ μm) (a, b) and neon ions (LET: 230 keV/ μm) (c, d, e). a: Unirradiated cells. b: Giant cells derived from the same cells as in (a) 24 hours after exposure to 10 Gy of argon ions. c–e (The process of interphase death): A giant cell at higher magnification (arrow) 30 hours after exposure to 5 Gy of neon ions (c) rounded up (d) followed by fragmentation into six parts (e) (Reproduced in the edited form from Figure 1 of the reference [8]).

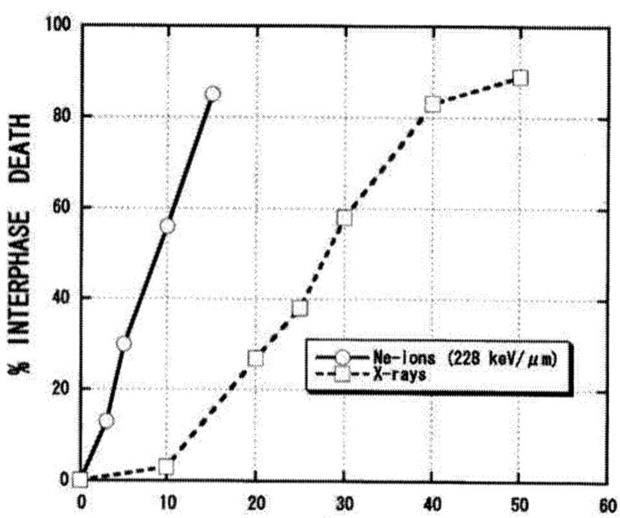

Figure 5. Dose-response curves for induction of interphase death in CHO cells, determined by means of time-lapse observation of 100 ~ 150 cells for each data point (Reproduced in the edited form from Figure 4 of the reference [8]).

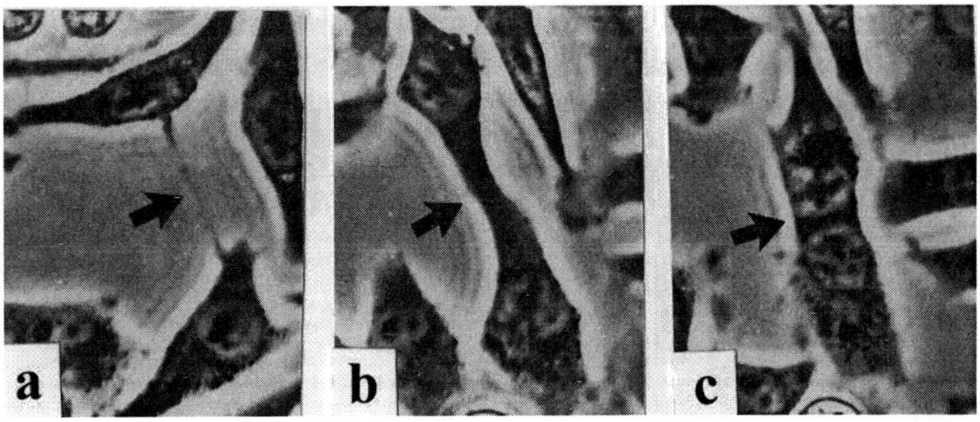

Figure 6. Time-lapse photomicrographs showing the process of sister-cell fusion in HeLa cells irradiated with 5 Gy of X rays. a: A protoplasmic bridge between sister cells 14 hours after division (23 hours after irradiation). b: A binucleate cell immediately after fusion (17 hours after division). c: The binucleate cell 1 hour after fusion (Reproduced in the edited form from Figure 1 of the reference [2]).

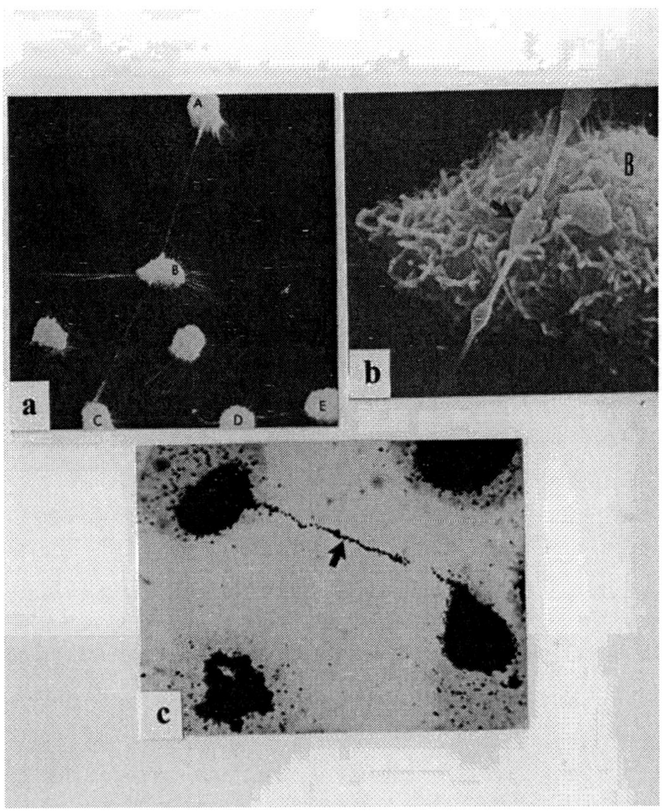

Figure 7. Photomicrographs of an intercellular protoplasmic bridge shown by scanning electron microscopy (a, b) and autoradiography (c) with HeLa cells exposed to 5 Gy of X rays. a: Two cells (A-C and D-E) are connected with a protoplasmic bridge (Magnification ×1,000). b: Cell B in (a) and a fine structure of a protoplasmic bridge on the cell (×10,000). c: Cells prelabeled with ^{3}H-thymidine (0.02 μCi/ ml, 15 Ci / m mole) for 24 h were fixed 24 hours after irradiation (×1,100) (Reproduced in the edited form from Figure 4 and 5 of the reference [2]).

An incomplete division (C in Figure 3) accompanied by protoplasmic cell fusion resulted in binucleate cell formation. This type of abnormal division took place frequently during postirradiation several generations in the progeny of the cells that could not survive, and predominantly after exposure to low-LET ionizing radiation such as X or γ rays (Hurwitz and Tolmach [17]). An intercellular thread bridging the sisters of HeLa cells prior to fusion was always noticed with time-lapse photographic observations under a phase-contrast microscope (Figure 6). A chromatin thread-like structure (~100 μm; ~0.7 μm in diameter) was confirmed by scanning electron microscopy as well as by autoradiography with ^{3}H-thymidine (Figure 7). From the examination of 248 cases, fusion took place most frequently between sister cells (94%), occasionally between cousin or niece cells (5%) and infrequently between the cells derived from a common ancestor cell three or four generations past (1%). The repetition of incomplete division and fusion led to the formation of a multinucleate cell that died also apoptosis. Small giant cells resulting from a fairly long delay in division often underwent multipolar division (D) (Dodson *et al.* [18]), most of which was tripolar for HeLa cells, although its frequency was much lower than that of incomplete division. Fusion tended to occur between two of the three sisters.

Reverse Split Dose Effect

The effects of two split doses on cell survival were compared between α particles and X rays (Figure 8). When two split doses of X rays (2 × 2 Gy) were given with various time interval (0–4 h), cell survival increased about two fold within 2–4 h because of the recovery from sublethal damage (Elkind and Sutton [19]). In the case of α particles (2 × 0.3 Gy), the survival decreased significantly (12.3% → 8.8%) during time interval between doses. However, in a larger dose of α particles (1.2 Gy), there was no significant difference in survival between single and two split exposures (single, 1.05±0.08%; two split with 3-h interval, 0.9±0.1%).

The fate of cells at irradiated generation with respect to their ability or inability to divide after exposure to single or split doses of α particles or X rays was examined with time-lapse photography (Table 1). After a single exposure to X rays, an incomplete division is predominant but the frequency was reduced remarkably by splitting the dose (25.8%[4 Gy] → 6. 8%[2×2 Gy]; 28.7%[6 Gy] → 10.9%[2×3 Gy]) with a concomitant increase in the frequency of normal (bipolar) division (68.0%[4 Gy] → 84. 3%[2×2 Gy]; 61.7%[6 Gy] → 79.7%[2×3 Gy]). This correlated with an increase in cell survival by split dose (Figure 8). On the other hand, the frequency of incomplete division and mitotic death was significantly increased by two split doses of 0.6 Gy α particles with the accompanying decrease in bipolar division which seemed to correlate with a slight but significant decrease in cell survival (reverse split dose effect). In the case of 1.2 Gy α particles, interphase death was predominantly induced by single dose but dramatically reduced by split dose with a concomitant increase in the frequency of mitotic death as well as bipolar division. However, the survival was not increased nor decreased by two split doses of 1.2 Gy α particles. This suggested the change of type from interphase to reproductive death; about 20% cells corresponding to the fraction of bipolar division increased by split dose exposure (67.3 – 48.8 = 18.5%) should divide but might die afterwards.

Table 1. Effects of two split doses of X rays and α particles on the fate of HeLa cells compared at similar survival (Surviving fraction [SF]: ~10%, ~1%)

		Cell Fate at Irradiated Generation (%)					Number of Cells Observed
		Division		Death			
		Bipolar	Multipolar	Incomplete	Mitotic	Interphase	
X rays (SF: ~10%)	4 Gy	68.0**	3.3	25.8**	1.4	1.4	147
	2×2 Gy	84.3	2.5	6.8	4.4	2.0	204
α particles (SF: ~10%)	0.6 Gy	82.5**	1.1	7.2*	6.0*	3.2	582
	2×0.3 Gy	71.5	0	13.5	12.0	3.0	133
X rays (SF: ~1%)	6 Gy	61.7**	3.2	28.7**	3.5	2.9	342
	2×3 Gy	79.7	2.3	10.9	5.2	1.9	477
α particles (SF: ~1%)	1.2 Gy	48.8**	1.5	6.3	6.8**	36.6**	205
	2×0.6 Gy	67.3	1.6	9.5	16.3	5.3	300

In split dose experiments, the second dose was given 3 h after the first dose. *Significant difference ($p < 0.05$ by χ^2 –test), **highly significant difference ($p < 0.01$) (Reproduced in the edited form from Table 1of the reference [7]).

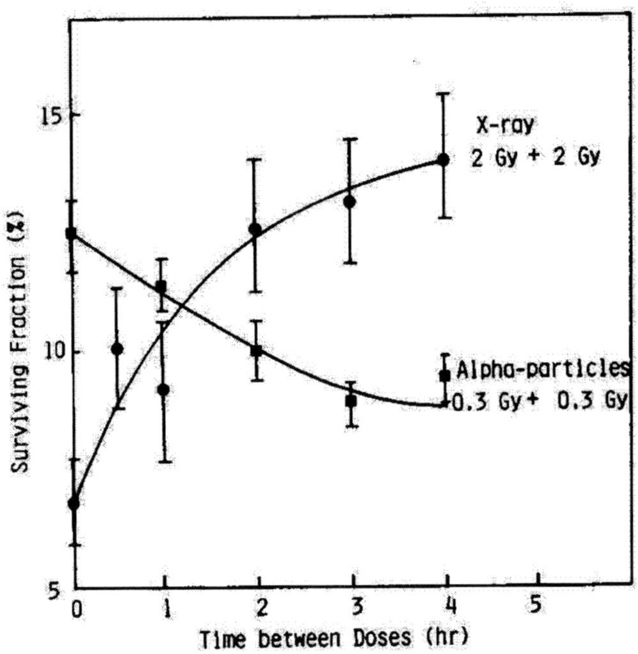

Figure 8. Effects of two split doses of X rays and α particles on the survival of HeLa cells. Vertical bars indicate standard deviations (Reproduced in the edited form from Figure 2 of the reference [7]).

A chromatin thread bridging the sister cells prior to cell fusion is considered to originate in dicentric chromosome that is a typical 2-hit type chromosome damage and formed by an illegitimate rejoining between two broken ends of centric chromosomes. Therefore, the decrease in incomplete division and the increase in bipolar division with a concomitant increase in survival after exposure to slit doses of X rays can be explained by the repair of damage induced by the 1st dose, which leads to chromosome break if not repaired, during the time interval between doses. On the other hand, the reverse split dose effect noticed with low dose (0.6 Gy) α particles will be interpreted reasonably by taking irreparability of the damage into account; if the damage leading to chromosome break is not repaired and a broken end of centric chromosome is produced during the time interval, its interaction with the damage by the second dose will result in the increase of 2-hit type damage.

Reverse phenomenon with high- relative to low-LET radiation has been identified also with respect to nonlethal radiation effect. Reverse low dose rate or split dose effect on the transformation of C3H 10T1/2 cells was reported by Hill *et al.* [20] who found that small doses (<2 Gy) of fission neutrons greatly increased transformation efficiency when applied at low dose rates or fractionated over several hours. However, Hieber *et al.* [21] could not observe such effect with α particles using the same cell line.

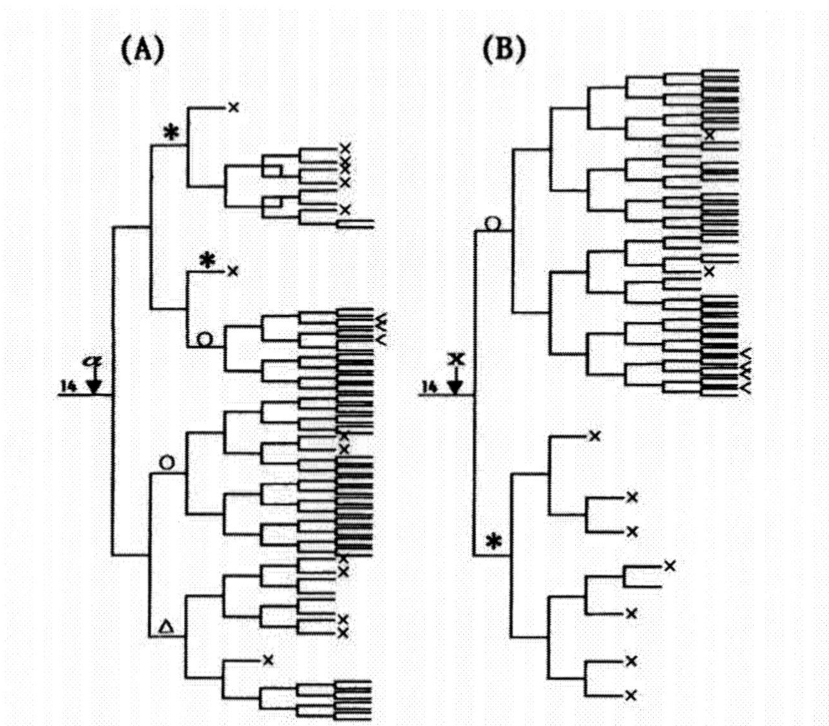

Figure 9. Pedigrees of HeLa cells surviving α- (0.45 Gy) or X-irradiation (3 Gy). (A) α-Irradiation; (B) X-irradiation. The cells were irradiated at late S/G_2 (14 h after division). Asterisks *, lethal sector; open circles,

stable clonogen; open triangles, unstable clonogen; ×, cell death; <, the cells divided more after the 7th division (Reproduced from Figure 2 of the reference [9]).

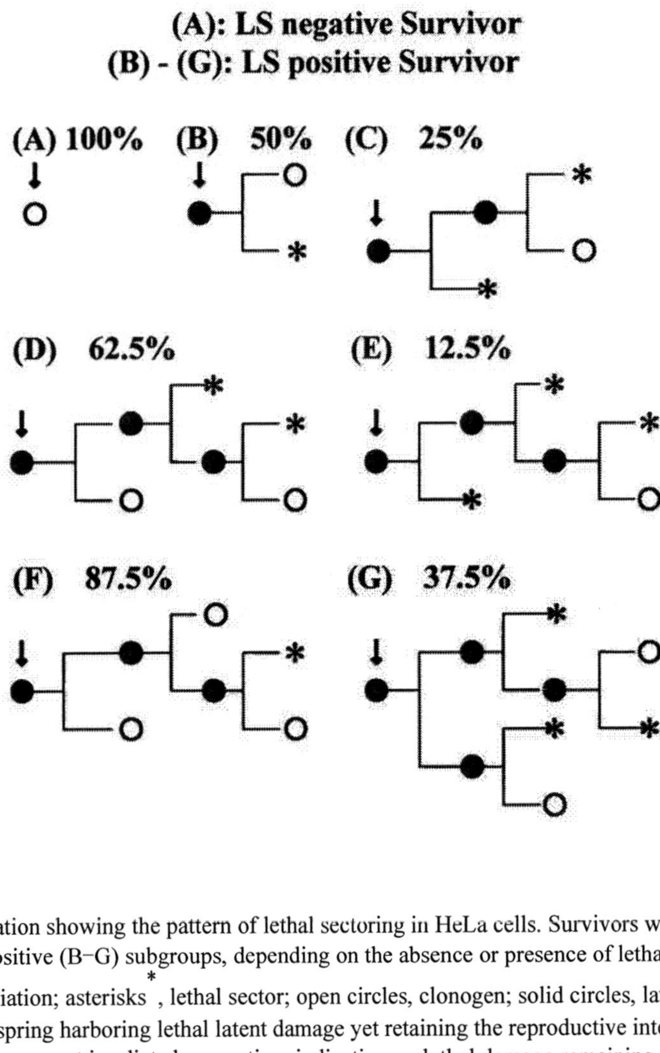

Figure 10. Illustration showing the pattern of lethal sectoring in HeLa cells. Survivors were divided into LS negative (A) or positive (B–G) subgroups, depending on the absence or presence of lethal sectoring in the pedigrees. ↓, irradiation; asterisks *, lethal sector; open circles, clonogen; solid circles, latent cell (the survivor or its offspring harboring lethal latent damage yet retaining the reproductive integrity). (A) Appearance of clonogen at irradiated generation, indicating no lethal damage remaining in the irradiated cell. (B) Coincidental appearance of lethal sector and clonogen at the 1st division, corresponding to Figure 9-B. (C) Coincidental appearance of lethal sector and clonogen at the 2nd division. (D) Corresponding to Figure 9-A. (E) Appearance of only one clongen at the 3rd division, leading to small colony formation. (F) Appearance of three clonogens and one lethal sector (the predominance of clonogenic sectoring). (G) No appearance of either lethal sector or clonogen (the doubling of latent cell) at the 1st division. Figures (%) represent relative colony-size assuming no cell loss in the clonogenic progeny during 9–10 postirradiation generations (Reproduced from Figure 3 of the reference [9])

Another significant change of cell fate by split doses of α particles was the enhanced mitotic death that was noticed both in low and high doses (0.6 and 1.2 Gy). Though data were not shown here, I previously found the difference in the cell-cycle dependence of division delay between X rays and α particles; the largest delay at G1/ S boundary and the smallest delay at late S/ early G2 for X rays, whereas a smaller delay in G1 than S phase with maximum at late S/ G2 except for late G2 past G2 checkpoint beyond which no delay in division was recognized similarly as for X rays. Furthermore, division delay time was shortened by two split doses of X rays especially with regards to radioresistant late S cells, but such recovery was not found with α particles [7]. Using laser scanning cytometry, Goto *et al.* [22] reported that, relative to X rays, exposure to iron ions (LET 1,000 keV/μm) resulted in a substantially greater inhibition of S-phase progression as a result of a higher frequency of DNA replication-blocking clustered DNA damage. These findings suggested that an exposure of the 2nd dose to the cells arrested in S phase by the 1st dose of α particles might enhance inhibition of DNA replication, and the cells deficient in DNA replication might die when entered mitosis.

Lethal Sectoring

When an irradiated cell forms a colony consisting of more than 50 cells within 1–2 weeks, the cell has been generally judged as a survivor. Optimum time for colony-formation will be 9–10 generations. During this period, an unirradiated cell forms a colony of about 500–1,000 cells. If a small colony of about 50 cells derives from a single clonogenic progenitor (clonogen), it should appear in the progeny until the 3rd or 4th postirradiation division, since an exponential proliferation for 6 generations (2^6= 64) is required for the number of cells per colony to exceed the critical value of 50. Lethal sectoring is the process for survival in which lethal damage remaining in irradiated cells is eliminated during the postirradiation divisions as offspring without reproductive integrity (lethal sector) with the accompanying appearance of clonogen that may harbor nonlethal but no lethal damage. This process tended to occur during the postirradiation 1st–3rd divisions after exposure to α particles, and the 1st–4th divisions for X irradiation.

The colony-forming process was explored in detail by analyzing the pedigrees of HeLa cells surviving α (0.45 Gy) and X irradiation (3 Gy) which were constructed by means of long-term (~150 h, corresponding to 7–8 postirradiation generations) time-lapse observations (Figure 9). Although the observation period was shorter than optimum time for colony-formation (9–10 generations), the judgment on 'an irradiated cell was survivor or nonsurvivor' and 'the offspring was clonogen or lethal sector' was done successfully by taking the cell fate and the change in cell-cycle time into consideration. In these pedigrees, lethal sector was indicated as asterisk (✻), and clonogen was indicated as open symbol (○: stable type) or open triangle (△: unstable type). At first glance, pedigree (B) is simple and pedigree (A) is complicated. In pedigree (B), lethal sectoring was completed by the postirradiation 1st division; two sister cells appeared, the one retaining the reproductive integrity with a low incidence of delayed cell death (stable clonogen; two death and 60 normal divisions → 3.2% [2/62]) in the progeny, and the other whose descendants were dying after division (lethal sector). In the pedigree (A) lethal sectoring did not take place but the clonogen appeared at the 1st division. The process that was accompanied with the appearance of clonogen instead of lethal sector was defined as clonogenic sectoring. This clonogen divided further at the next division into stable and unstable ones (incidence of delayed cell death: >20%). As for the upper half of this pedigree, two successive lethal sectorings occurred at the 2nd and 3rd divisions.

The pattern of lethal sectoring is illustrated on the basis of total 54 pedigrees of survivors (Figure 10). Survivors are divided into LS negative and positive subgroups depending on the absence and presence of lethal sectoring in the pedigrees. No lethal damage is remaining in the former because of the repair of the damage during division delay at irradiated generation. The LS negative survivor by itself is clonogen (A). Here the survivor or its offspring harboring lethal damage yet retaining the reproductive integrity is defined as a latent cell (indicated by closed circle: ●). In this sense, the LS positive survivor is the latent cell. Four combinations of sister cells are expected after division of the latent cell; (I) lethal sector and clonogen, (II) lethal sector and latent cell, (III) clonogen and latent cell, (IV) two latent cells. Consequently, many patterns of lethal sectoring are possible when the postirradiation 1st–3rd divisions are taken into account (B–G). The process of lethal sectoring persists until the latent cell generates the lethal sector and the clonogen coincidentally (e.g., until the 2nd division in C). (E) is when only one clonogen appeared at the 3rd division and formed a colony consisting of slightly more than 50 cells until the 9th generation. (F) shows that 3 clonogens and one lethal sector appeared, demonstrating the predominance of clonogenic sectoring.

The number of pedigrees available for survivor groups was 31 and 23 for α and X irradiation, respectively. Lethal sectoring was found in 25 pedigrees (25/31: 81%) (α-particle LS positive) but not in 6 pedigrees (6/31: 19%) (α-particle LS negative) for the former, and it was found in 15 pedigrees (15/23: 65%) (X-ray LS positive) but not in 8 pedigrees (8/23: 35%) (X-ray LS negative) for the latter. Although□definite conclusion could not be drawn with the limited number of pedigrees available, the following differences seemed to be noted in the process of lethal sectoring between α-particle and X-ray survivors; 1) the process completed earlier for α-particle (until the postirradiation 3rd divisions) than for X-ray survivors (until the 4th divisions), 2) clonogenic and lethal sectoring occurred similarly in α-particle survivors, but the former predominated over the latter in X-ray survivors, 3) Mean number per pedigree was similar with respect to lethal sector (~1.5) both in α-particle and X-ray survivors, but the number of stable type clonogen was larger in X-ray (~2) than α-particle survivors (~1). These facts imply that the colony-size of the X-ray survivors is larger than that of α-particle survivors when compared at the same survival. This is consistent with the report of Walker *et al.* [23] that heavy ions caused a dose- and LET-dependent reduction in colony-size.

Delayed Cell Death

The incidence of delayed cell death in the clonogenic progeny was examined separately for LS positive and negative survivors (Table 2). A comparison between LS positive α-particle and X-ray survivors demonstrates that the mean incidence of dying cells is about two fold higher in the former (16.3% [194/1,190]) than in the latter (8.3% [90/1,079]). For α particles the value obtained with LS negative survivors (14.3% [45/324]) was similar to that obtained with LS positive survivors (16.3%). However, for X rays the value of LS negative survivors

(4.7% [16/337]) was significantly lower than that of LS positive survivors (8.3%) and was close to that of unirradiated control cells (3.6% [8/222]). These results clearly show that genomic instability was induced in LS positive but not in negative survivors by X rays, whereas the instability was induced both in LS positive and negative survivors by α particles, suggesting an intrinsic difference in the mechanisms of genomic instability induction between these two radiations.

Table 2. Incidence[a] of delayed cell death in the clonogenic progeny of LS positive or negative[b] surviving HeLa cells

Subgroups	α particles (0.45 Gy)	X rays (3 Gy)
LS positive	16.3% (194/1,190)[c, d]	8.3% (90/1,079)[c, e]
LS negative	14.3% (45/314)[d]	4.7% (16/337)[e, f]
Unirradiated cells	3.6% (8/222)[f]	

[a][(Number of abnormal cells)/ (Number of normal divisions and abnormal cells)] × 100 (%). Normal division means a bipolar one. Abnormal cells correspond to the ones undergoing death or abnormal division. [b]See Figure 10. [c]Highly significant difference ($p < 0.01$ by χ^2-test) between α particle LS positive and X ray LS positive survivors. [d]No significant difference ($p > 0.05$) between α particle LS positive and negative survivors. [e]Significant difference ($p < 0.05$) between X ray LS positive and negative survivors. [f]No significant difference ($p > 0.05$) between X ray LS negative survivors and unirradiated cells (Reproduced from Table 4 of the reference [9]).

Thus the incidence of delayed cell death was about two fold higher for α than X irradiation. The RBE of α particles for the induction of delayed cell death was estimated to be 9–14, when the RBE for cell killing (D_0 ratio: 4.4) or the ratio of exposed dose (X rays [3 Gy]/ α particles [0.45 Gy] = 6.7) was taken into consideration. This value is similar to maximum RBE for mutation induction (Kiefer [24]).

Delayed Division Delay and a Shortening of Cell Cycle Time

A novel phenomenon "delayed division delay (DDD)" was recognized with the progeny during the postirradiation 1st to 3rd generations (Figure 11). Cells irradiated with α particles (0.45 Gy) (A, B) or X rays (3 Gy) (C, D) showed a remarkable division delay at irradiated generation. The increase in cell cycle time from preirradiation control (pre-α irradiation: ~17 h, pre-X irradiation: ~19 h) was 12–16 h and ~10 h (29–33 h and ~30 h as actual cell cycle time) after α and X irradiation, respectively. The cell cycle time at the 1st generation recovered to a normal level in (A), (B) and (D), but it did not recover because of DDD in one of the sister cells born at the 1st division as shown in (C). DDD also occurred at the 2nd generation (A, D) and the 3rd generation (B). Generally, DDD is similar to division delay at irradiated generation for α irradiation, but it is much shorter than the latter in the case of X irradiation. DDD was found with the latent cells just prior to lethal sectoring (A [30 with star marks], C and D [22 with star marks]) and with the clonogens immediately after clonogenic sectoring (A [32 with star marks]) or the completion of lethal sectoring (B [33 with star marks]).

A prolonged cell cycle time caused by DDD is shown for the individual cases (Table 3). The mean DDD was much longer in α (10.8 h) than X irradiated cells (4.0 h). The mean frequency of DDD was ~30% per pedigree and ~10% per clonogen or latent cell for both radiations when detectable minor cell cycle change (1–3 h) was taken into account for its identification. It seemed that DDD was found with clonogen (~15% [6/40]) more predominantly than with latent cell (~7% [2/28]) for α irradiation and the converse was the case for X irradiation. In the present study, DDD could not identified in the progeny later than the 3rd generation after α irradiation because of limited observation period (~150 h), and so possibly α-particle-induced DDD can happen at later generations.

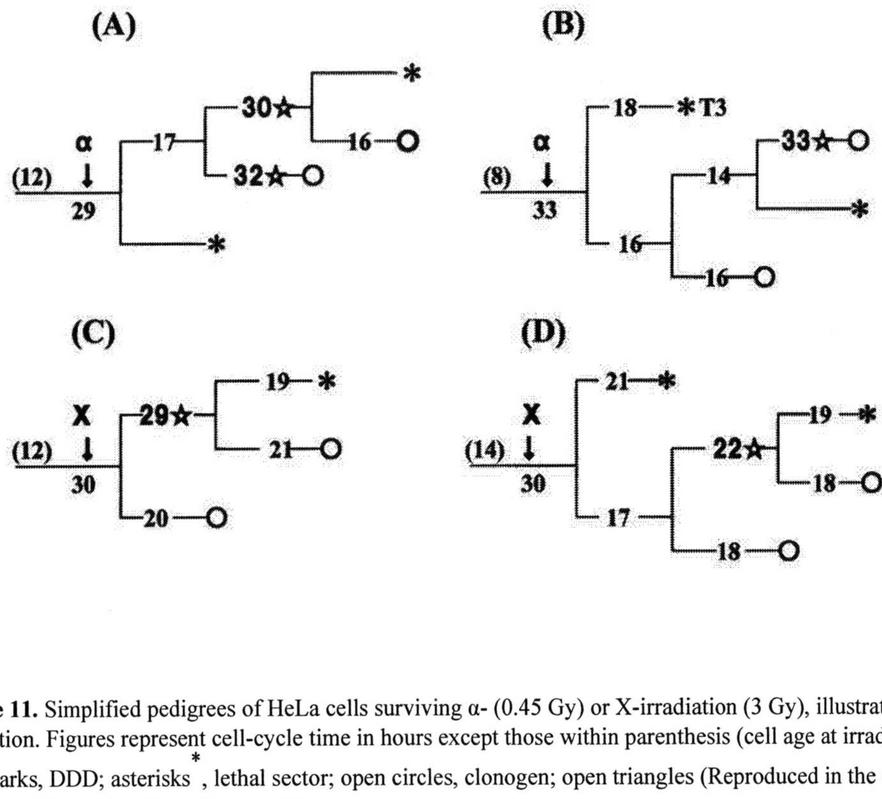

Figure 11. Simplified pedigrees of HeLa cells surviving α- (0.45 Gy) or X-irradiation (3 Gy), illustrating X-irradiation. Figures represent cell-cycle time in hours except those within parenthesis (cell age at irradiation). Star marks, DDD; asterisks *, lethal sector; open circles, clonogen; open triangles (Reproduced in the edited form from Figure 4 of the reference [9]).

Contrary to DDD, a significant shortening (~5 h) of cell cycle time corresponding to the time difference between preirradiation (19.4±1.3 h) and postirradiation 1st generation (mean±SD: 14.7±1.1 h) was found with the progeny of X-ray LS negative survivors. The shortened time interval was roughly identical to division delay at irradiated generation (mean 5.4 h). Some cellular components synthesized in an extra amount during division delay might enhance subsequent cell cycle progression in the progeny of the cells harboring no lethal nor nonlethal damage which would have a reverse influence.

Discussion

Features of the Damage Induced by α Particles Relative to X rays

Traversals of high-LET α particles (~125 keV/μm) through the higher-order structure of chromatin fibers may result in the clustering of damage on the chromosome (Goodhead [25]) and the formation of clustered DNA damage (CD) (Rydberg [26]) that is difficult to repair correctly (Rydberg *et al.* [27]), whereas low-LET X rays produce isolated sublethal (SLD) [19] or potentially lethal damage (PLD) (Phillips and Tolmach [28]) that can be repaired correctly by dose splitting or delayed plating. The fact that average 2.6 hit of α particles are required to give rise to one lethal event (Figure 2) indicates the probability of cell death by a single hit of α particles to nucleus of HeLa cells to be about 0.3 (~70% survival). Moreover, a high incidence of delayed cell death in the clonogenic progeny of α-particle survivors harboring no lethal damage (α-particle LS negative survivor) (Table 2) suggests that α particles are very efficient in inducing nonlethal damage, and a single hit of α particle to cell nucleus is sufficient to induce genomic instability in the progeny. However, no induced incidence of delayed cell death in the clonogenic progeny of X-ray LS negative survivors indicates complete repair of the damage (no production of lethal nor nonlethal damage) during division delay at irradiated generation.

Table 3. Prolonged cell-cycle times caused by delayed division delay (DDD) in HeLa cells during 1st–3rd generations after α or X irradiation

	Cell-cycle time (Value at gen. 0)[a]	
	α particles (0.45 Gy)	X rays (3 Gy)
Gen. 1	$22^{\dagger}$ (24.5), $24^{\dagger}$ (23)	$23.5^{\ddagger}$ (36.5), $25^{\dagger}$ (28) $29^{\ddagger}$ (30)
Gen. 2	$23^{\dagger}$ (18), $23^{\dagger\#}$ (26.5) $28^{\dagger\#}$ (20), $30^{\ddagger}$ (30) $32^{\dagger}$ (29), $36^{\ddagger}$ (30)	$20^{\dagger\#}$ (21), $21.5^{\dagger}$ (25) $22^{\ddagger}$ (27.5), $22.5^{\ddagger}$ (30)
Gen. 3	$26^{\dagger\#}$ (35), $32.5^{\dagger}$ (26) $33^{\dagger}$ (33)	
Mean±SD	28.1±4.6 (Control[b]: 17.3±1.5)	23.4±2.7 (Control[b]: 19.4±1.3)
Mean DDD[c]	10.8	4.0
Frequency	35% (11/31)[d], 12% (8/68)[e]	30% (7/23)[d], 10% (6/61)[e]

[a] Cell-cycle time at irradiated generation.

[b] Cell-cycle time of unirradiated control cells.

[c] Difference in mean cell-cycle time between irradiated and unirradiated control cells.

[d] Mean frequency of DDD/pedigree: [(Number of pedigrees in which DDD was found)/(Total number of pedigrees examined)] × 100 (%).

[e] Mean frequency of DDD/clonogen or latent cell, determined for LS positive survivors by the analysis of simplified pedigrees such as shown in Figure 10 −11.

† DDD with clonogen. ‡ DDD with latent cell. # DDD in LS negative survivors (Other DDD was found in LS positive survivors). Figures are represented in hours except those for frequency of DDD (Reproduced in the edited form from Table 6 of the reference [9]).

Radiation-induced lethal damage is manifested in the form of cell abnormality leading to death. Incomplete division followed by fusion mostly between sister cells is caused by dicentric chromosome; a typical 2 hit-type chromosome damage that is predominantly induced by low-LET radiation such as X or γ rays. Cell abnormality characteristic for high-LET radiation such as α particles is interphase or premitotic death; cell death before reaching postirradiation 1st mitosis. This type of death seems to be induced by unrejoined chromosome breaks which arrest the cells at G2 checkpoint (Qin and Li [29]) for a long time and commit them to undergo apoptosis. Significant fraction (37%) of HeLa cells irradiated with 1.2 Gy of α particles (~1% survival dose) underwent interphase death (Table 1). More detailed study with CHO cells exposed to high-LET heavy-ions suggests the requirement of multiple (>10) traversals of heavy-ions (e.g., Ne-ions) through cell nucleus for the induction of interphase death in the majority (63%) of cells [8]. Remarkable decrease in the frequency of interphase death in HeLa cells by two split doses of α particles (1.2 Gy → 2 × 0.6 Gy with 3 h interval) supports this possibility. On the other hand, split dose experiment at low dose of α particles (0.6 Gy; ~10% survival) revealed the reverse split dose effect on cell killing which correlated with the increase in incomplete division (Figure 8). A similar experiment on genomic instability remains to be done.

Mechanisms of Radiation-induced Genomic Instability

Although the greater potentiality of delayed cell death as an endpoint of genomic instability (GI) was recognized after exposure to α particles compared to X rays, the intrinsic difference was also found between these two radiations. GI could be induced by α particles both in LS positive and negative survivors; however, it was inducible in LS positive but not in negative survivors in the case of X rays (Table 2). This fact suggests the different mechanisms involved in the induction of GI by these two radiations. I would like to propose the fixation of unrepaired PLD (including sublethal damage) -mediated mechanism for X rays and the misrepaired CD-mediated one for α particles. In Figure 12, the presumptive pathways from primary radiation damage to the induction of LS and GI are shown. Here it is postulated that the fixation of remaining unrepaired PLD by a higher-order conformational change of DNA/chromatin fibers during the cell cycle (e.g., chromosome condensation during the postirradiation 1st mitosis) gives rise to both lethal and nonlethal damage and that the latter damage is responsible for GI induction after X irradiation (LS positive [GI+]). However, minor contributions of misrepaired PLD to the production of these damage can not be denied. When PLD is completely repaired, both lethal and nonlethal damage is not formed; there is no induction of genomic instability in X-ray LS negative survivors (LS negative [GI−]). On the other hand, both lethal and nonlethal latent damage will be produced at irradiated generation, mainly by misrepair and partly by the fixation of unrepaired CD after α-particle exposure (LS positive [GI+]). On the assumption that nonlethal damage is induced more predominantly than lethal damage by α particles and a single hit of α -particle to cell nucleus is sufficient to induce nonlethal damage, this damage can be produced in the cells harboring no lethal damage (LS negative [GI+]). Thus genomic instability is inducible both in LS positive and negative survivors for α particles, whereas it is inducible in LS positive but not in negative survivors for X rays. This explanation agrees well with the finding of Roy *et al.* [30] that delayed cell death was suppressed through PLD recovery by the delayed plating of X-irradiated human embryo cells. It is also supported by the finding [23] that the cross section for the effect of an LET-dependent reduction of colony-size was 100–300 μm^2. This was at least as large as the cell nuclear area and up to 3 times as large as the cross section for cell killing. A higher potentiality for GI induction by high-LET radiation will be one of the causes for a heavy-ion induced reduction of colony-size.

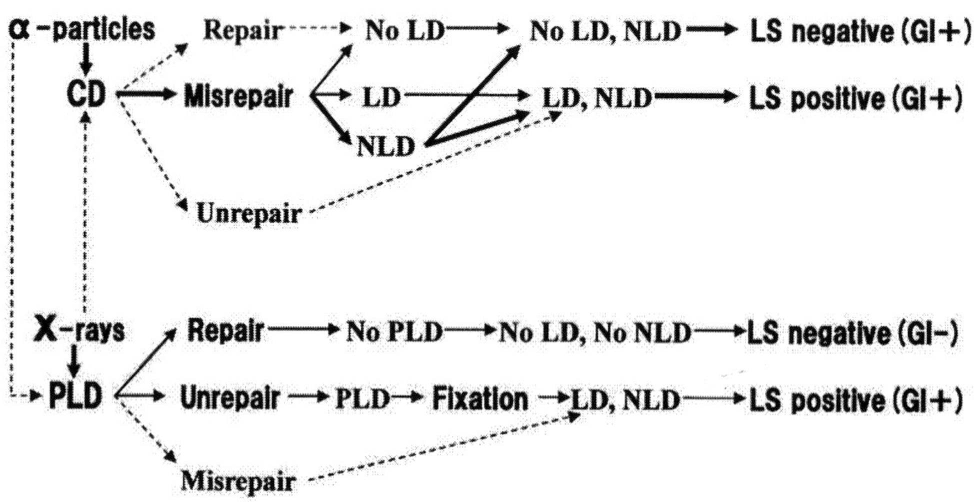

Figure 12. The presumptive pathways from primary radiation damage (CD, PLD) to the appearance of LS negative and positive survivors accompanied by the induction or no induction of genomic instability. See Figure 10 for LS negative and positive. CD, clustered DNA damage; PLD, potentially lethal damage; LD, lethal latent damage; NLD, nonlethal latent damage responsible for the induction of genomic instability; Fixation, PLD fixation by a higher-order conformational change of DNA/chromatin fibers during the cell-cycle (e.g., chromosome condensation during the postirradiation 1st mitosis); (GI-), no induction of genomic instability; (GI+), induction of genomic instability. Bold and dotted arrows indicate predominant and minor pathway, respectively (Reproduced from Figure 5 of the reference [9]).

Mechanisms of Delayed Division Delay

This is the first report demonstrating the occurrence of delayed division delay (DDD) in the progeny of cells irradiated with ionizing radiation, although Thompson and Humphrey [31] previously reported on the second-division delay that occurred mainly because of DNA replication block by pyrimidine dimers at the postirradiation 1st generation in the progeny of late S/G2 mouse L-P59 cells exposed to UV. Since cell cycle checkpoints arrest the cell cycle upon the occurrence of DNA double-strand breaks (DSB) in irradiated cells [29], DDD serves as an indicator of delayed DNA DSB. According to the suggestion of Suzuki *et al.* [32] that DNA DSB may create nonlethal potentially unstable chromosome regions (PUCR) at the break-rejoining sites through DNA DSB repair, α particles are assumed to generate clustered PUCR via CD, causing a persistent and pronounced DDD. Delayed DNA DSB at PUCR sites with the accompanying reactivation of DNA damage responding gene products (p53, NBS1, CHK2) [33] may trigger DDD. However, the mechanisms for delayed DNA DSB remains to be elucidated.

At present, there is no evidence suggesting an intracellular presence of the enzyme system that specifically breaks the fragile misrejoining sites of DNA double-strand structure. Therefore, I would like to propose a hypothesis that delayed DNA DSB is induced by a higher-order conformational change of DNA/chromatin fibers during the cell cycle, especially during late G2/M transition, based on the following results obtained from the PLD fixation experiments using tsBN2 mutant (Sasaki and Nishimoto [34]) in which premature chromosome condensation (PCC) occurs irrespective of the cell cycle at nonpermissive temperature (40℃) (Figure 13). Although the incubation at 40℃ for 3 h alone did not lower the survival of unirradiated tsBN2 cells, the cell killing effect of X rays was enhanced by this treatment immediately after irradiation for the cells synchronized at various stages of the cell cycle except those at around the late G2/M phase, where chromosome condensation occurred at a permissive temperature (33.5℃). This suggests that the enhanced interaction between PLDs (isolated DNA DSBs) by chromosome condensation may fix PLD to lethal damage. Therefore, we can not deny the possibility that fragile misrejoining sites may change to DNA DSB by chance during late G2/M transition.

SUGGESTIONS FOR THE FUTURE STUDY

Time-lapse Photography

Present study indicates the usefulness of DDD as a new endpoint of radiation-induced genomic instability, but a long-term time-lapse observation of irradiated cells is required to detect DDD. Since the field size of photograph taken with 35mm camera under low magnification (×10 objective and ×5 ocular lens) is about 1.5 mm^2 and large enough to include about 500 HeLa cells, the colony-forming process for about 10 surviving cells can be followed by reviewing time-lapse photographic records on cell fates during at least 7 postirradiation generations. However, the cells must be cultured under the condition where the inhibition of cell proliferation by cell to cell contact is avoided.

More important information may be obtained by combining a time-lapse photography with immunostaining. For instance, immunostaining of the cells with anti-phosphorylated H2AX antibody immediately after the ending of time-lapse observations will provide some information concerning the relation of lethal sectoring and genomic instability with DNA damage remaining in the progeny of surviving cells, since Suzuki *et al.* [35] observed that phosphorylated histone H2AX foci remained in interphase cells even 96 h after exposure to X rays.

Radiation Source

A point α source of ^{241}Am is easy to handle and its α particles (LET~125 keV/μm at air path length of 12 mm) are very effective in inducing DDD and delayed cell death, therefore are recommended to be used in the study on genomic instability using cultured cells as materials.

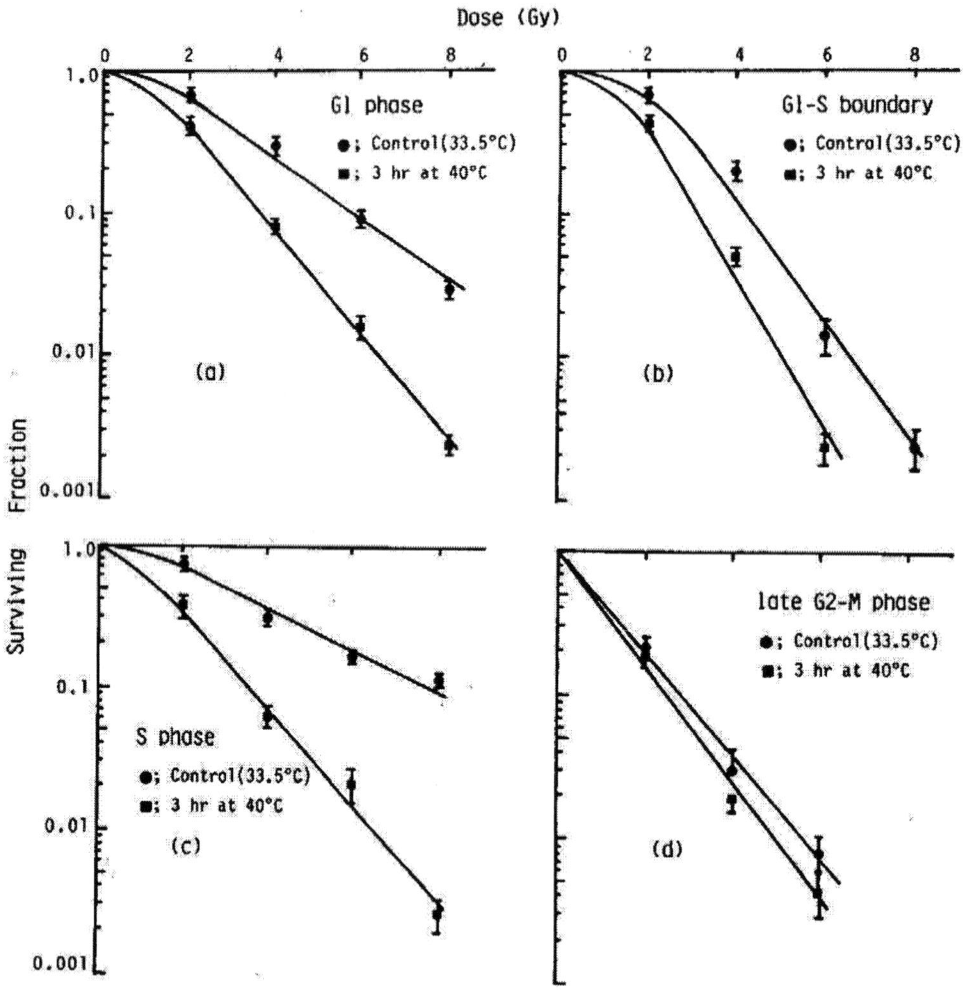

Figure 13. Effect of postirradiation incubation at 40℃ (3 h) on X ray dose-survival curves of synchronized tsBN2 cells. (a): Cells synchronized in G1 phase (3 h after release from G0 phase) were exposed to X rays, incubated at 40℃ for 3 h. (b) Cells synchronized at G-S boundary by 2.5 mM hydroxyurea were exposed to X rays and incubated at 40℃ for 3 h after replacement of hydroxyurea. (c): Cells synchronized in S phase (3 hours after release from G1-S boundary) were exposed to X rays, and incubated at 40℃ for 3 h. (d): Asynchronous cells were exposed to X rays and incubated for 2 h at the permissive temperature (33.5 ℃). The cells at prophase to anaphase were harvested using mitotic selection, plated, and incubated at 40℃ for 3 h. (■): Cells incubated at 40℃ for 3 h after X irradiation. (●): Control cells which were incubated at 33.5 ℃ after X irradiation (Reproduced from Figure 7 of the reference [34]).

Recently, the study on mutagenic or carcinogenic effect of accelerated high-LET Fe-ions on astronauts during space flight has been becoming increasingly important. Presence of non-hit cells, identified with a time-lapse observation (Mehnati *et al.*[36]), among hit cells is one of the features characteristic for exposure to very high-LET Fe-ions (Figure 14). Bystander effect may occur in non-hit cell (Little [37]). This effect will be ascertained if heavy-ion's hit or non-hit to individual cells can be judged accurately e.g. by using CR-39 plastics, on which cells are growing, before beginning a time-lapse observation of cell fates.

Selection of the Cells Undergoing Delayed Division Delay

Delayed DNA DSB should arrest the cells at the checkpoint located in G2 until the rejoining of DNA DSB is completed. During delay time that is more than 10 h for α irradiation, the cells should become significantly larger than the G2 phase cells undergoing no DDD. The fraction of the cells undergoing DDD may be small, but can be increased by selecting larger cells using a pertinent method. Immunostaing with anti-phosphorylated H2AX antibody may be useful in the recognition of the cells undergoing DDD.

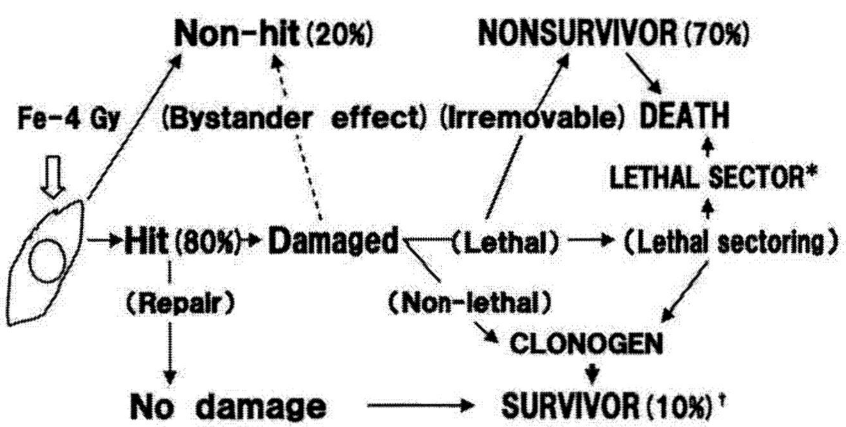

Figure 14. Fates of CHO cells exposed to 4 Gy Fe-ions (2,000 keV/µ□). *Offspring without reproductive integrity; †Surviving fraction corrected by excluding non-hit fraction (20%) from survivors and irradiated cells (Reproduced from Figure 6 of the reference [36]).

Conclusion

Both lethal and nonlethal damage may be induced by the same primary damage; CD for α irradiation and PLD in the case of X irradiation. Lethal damage tended to be manifested as the cell cycle arrest-dependent interphase death after exposure to α particles and as an incomplete division followed by fusion in the case of X irradiation. Lethal damage remaining in the progeny of surviving cells is eliminated as lethal sector during the posirradiation 1st–4th divisions. Nonlethal damage remaining in the clonogenic progeny is responsible for the induction of genomic instability. This damage is supposed to correspond to the fragile sites on DNA double-strand structure created by misrejoining of DNA DSB through the misrepair of CD and the fixation of unrepaired PLD for α and X irradiation, respectively. Delayed DNA DSB at the fragile sites, assumedly induced by a conformational change of DNA/chromatin fibers during mitosis, should trigger DDD in the progeny of surviving cells. DDD is much longer in α- (mean: ~11 h) than X-irradiated cells (~4 h), correlating well with a higher incidence of delayed cell death in the progeny of α-particle survivors. Contrary to DDD, a significant shortening (~5 h) of cell cycle time at the postirradiation 1st generation was found with the progeny of X-ray survivors harboring no lethal nor nonlethal damage.

Acknowledgments

I thank Radiation Research Society and Japan Radiation Research Society for permitting me to reproduce figures and tables of our papers published in Radiation Research and Journal of Radiation Research, respectively.

References

[1] Marin, G. and Bender, M. A. (1966) Radiation-induced mammalian cell death: Lapse-time cinemicrographic observations. *Exp. Cell Res. 43*: 413–423.

[2] Sasaki, H. and Hayashi, H. (1979) A nuclear thread bridging the sister cells prior to radiation-induced cell fusion. *Radiat. Res. 77*: 577–585.

[3] Hurwitz, C. and Tolmach, L. J. (1969) Time lapse cinemicrographic studies of X-irradiated HeLa S3 cells. I. Cell progression and cell disintegration. *Biophysical J. 9*: 607–633.

[4] Thompson, L. H. and Suit, H. D. (1969) Proliferation kinetics of X-irradiated mouse L cells studied with time-lapse photography. II. *Int. J. Radiat.Biol. 15*: 347–362.

[5] James, A. P. and Werner, M. M. (1966) Radiation-induced lethal sectoring in yeast. *Radiat. Res. 29*: 523–536.

[6] James, A. P., Werner, M. M., Saunders, A. S. and Harris, M. A. (1968) Persistence of X-ray induced lethal sectoring in yeast. *Radiat. Res. 34*: 475–487.

[7] Sasaki, H. (1984) Cell killing and division delay in asynchronous and synchronized HeLa cells irradiated with alpha particles or x rays. *Radiat. Res. 99*: 311–323.

[8] Sasaki, H., Yatagai, F., Kanai, T., Furusawa, Y., Hanaoka, F., Zhu, W.-G. and Mehnati, P. (1997) Dependence of induction of interphase death of Chinese hamster ovary cells exposed to accelerated heavy ions on linear energy transfer. *Radiat. Res. 148*: 449–454.

[9] Sasaki, H. (2004) Lethal sectoring, genomic instability, and delayed division delay in HeLa S3 cells surviving alpha- or X-irradiation. *J. Radiat. Res. 45*: 497–508.

[10] Deering, R. A. and Rice, R. (1962) Heavy ion irradiation of HeLa cells. *Radiat. Res. 17*: 774–786.

[11] Ngo, F. Q. H., Blakely, E. A. and Tobias, C. A. (1981) Sequential exposures of mammalian cells to low and high-LET radiations. *Radiat. Res. 87*: 59–78.

[12] Barendsen, G. W., Koot, C. J., van Kersen, G. R., Bewley, D. K., Field, S. B. and Parnell, C. J. (1966) The effect of oxygen on impairment of the proliferative capacity of human cells in culture by ionizing radiations of different LET. *Int. J. Radiat. Biol. 10*: 317–327.

[13] Hall, E. J., Gross, W., Dvorak, R. F., Keller, A. M. and Rossi, H. H. (1972) Survival curves and age response functions for Chinese hamster cells exposed to X-rays or high-LET alpha-particles. *Radiat. Res. 52*: 88–98.

[14] Keller, A. M. and Rossi, H. H. (1972) The theory of dual radiation action. *Curr. Top. Radiat. Res.* Q. *8*: 85–158.

[15] Konishi, T., Takeyasu, A., Yasuda, N., Natsume, T., Nakajima, H., Matsumoto, K., Asuka, T., Sato, Y., Furusawa, Y. and Hieda, K. (2005) Number of Fe ion traversals through a cell nucleus for mammalian cell inactivation near the Bragg peak. *J. Radiat. Res. 46*: 415–424.

[16] Hopwood, L. E. and Tolmach, L. J. (1971) Deficient DNA synthesis and mitotic death in X-irradiated HeLa cells. *Radiat. Res. 46*: 70–84.

[17] Hurwitz, C. and Tolmach, L. J. (1969) Time-lapse cinemicrographic studies of X-irradiated HeLa S3 cell. II. *Cell fusion. Biophysical J. 9*: 1131–1143.

[18] Dodson, H., Bourke, E., Jeffers, L. J., Vangnarelli, P., Sonoda, E., Takeda, S., Earnshaw, W. C., Merdes, A. and Morrison, C. (2004) Centrosome amplification induced by DNA damage occurs during a prolonged G2 phase and involves ATM. *EMBO J. 23*: 3864–3873.

[19] Elkind, M. M. and Sutton, H. (1960) Radiation response of mammalian cells grown in culture. I. Repair of x-ray damage in surviving Chinese hamster cells. *Radiat. Res. 13*: 556–593.

[20] Hill, C. K., Carnes, B. A., Han, A. and Elkind, M. M. (1985) Neoplastic transformation is enhanced by multiple low doses of fission-spectrum neutrons. *Radiat. Res. 102*: 404–410.

[21] Hieber, L., Ponsel, G., Roos, H., Fenn, S., Fromke, E. and Kellerer, A. M. (1987) Absence of □dose-rate effect in the transformation of C3H 10T1/2 cells by α-particles. *Int. J. Radiat. Biol. 52*: 859–869.

[22] Goto, S., Watanabe, M. and Yatagai, F. (2002) Delayed cell cycle progression in human lymphoblastoid cells after exposure to high-LET radiation correlates with extremely localized DNA damage. *Radiat. Res. 158*: 678–686.

[23] Walker, J. T., Todd, P. and Walker, O. A. (2002) Heritable non-lethal damage to cultured human cells irradiated with heavy ions. *J. Radiat. Res. 43* (Suppl.): S187–S191.

[24] Kiefer, J. (2002) Mutagenic effects of heavy charged particles. *J. Radiat. Res. 43* (Suppl.): S21–S25.

[25] Goodhead, D. T. (1992) Track structure consideration in low and low dose rate effects of ionizing radiation. *Adv. in Radiat. Biol. 16*: 7–44.

[26] Rydberg, B. (1996) Cluster of DNA damage induced by ionizing radiation: Formation of short DNA fragments. II. Experimental detection. *Radiat. Res. 145*: 200–209.

[27] Rydberg, B., Cooper, B., Cooper, P. K., Holley, W. R. and Chatterjee, A. (2005) Dose-dependent misrejoining of radiation-induced DNA double-strand breaks in human fibroblasts: Experimental and theoretical study for high- and low-LET radiation. *Radiat. Res. 163*: 526–534.

[28] Phillips, R. A. and Tolmach, L. J. (1966) Repair of potentially lethal damage in x-irradiated HeLa cells. *Radiat. Res. 29*: 413–432.

[29] Qin, J. and Li, L. (2003) Molecular anatomy of the DNA damage and replication checkpoints. *Radiat. Res. 159*: 139–148.

[30] Roy, K., Kodama, S., Suzuki, K. and Watanabe, M. (1999) Delayed cell death, giant cell formation and chromosome instability induced by X-irradiation in human embryo cell. *J. Radiat. Res. 40*: 311–322.

[31] Thompson, L. H. and Humphrey, R. M. (1970) Proliferation kinetics of mouse L-P59 cells irradiated with ultraviolet light: A time-lapse photographic study. *Radiat. Res. 41*: 183–201.

[32] Suzuki, K., Ojima, M., Kodama, S. and Watanabe, M. (2003) Radiation-induced DNA damage and delayed induced genomic instability. Oncogene *22*: 6988–6993.

[33] Suzuki, K., Yokoyama, S., Waseda, S., Kodama, S. and Watanabe, M. (2003) Delayed reactivation of p53 in the progeny of cells surviving ionizing radiation. *Cancer Res. 63*: 936–941.

[34] Sasaki, H. and Nishimoto, T. (1987) Chromosome condensation may enhance X-ray-related cell lethality in a temperature-sensitive mutant (tsBN2) of baby hamster kidney cells (BHK21). *Radiat. Res. 109*: 407–418.

[35] Suzuki, M., Suzuki, K., Kodama, S. and Watanabe, M. (2006) Phosphorylated histone H2AX foci persist on rejoined mitotic chromosomes in normal human diploid cells exposed to ionizing radiation. *Radiat. Res. 165*: 269–276.

[36] Mehnati, P., Morimoto, S., Yatagai, F., Furusawa, Y., Kobayashi, Y., Wada, S., Kanai, T., Hanaoka, F. and Sasaki, H. (2005) Exploration of 'over kill effect' of high-LET Ar- or Fe-ions by evaluating the fraction of non-hit cell and interphase death. *J. Radiat. Res. 46*: 343–350.

[37] Little, J. B. (2003) Genomic instability and bystander effects: a historical perspective. *Oncogene* 22: 6978–6987.

In: New Research on Genomic Instability
Editor: Eleanor J. Glascow, pp. 207-260

ISBN: 1-60021-320-0

Chapter VIII

How a Cell Defends Itself against Genomic Instability Caused by Chromium

Nelly Sapojnikova**, *Tamar Kartvelishvili*, *Marina Abuladze and Nino Asatiani

Department of Physics of Biological Systems, Institute of Physics, Tbilisi, Georgia

Abstract

Chromium (Cr) compounds are regarded as a complex genotoxin. Soluble and insoluble hexavalent chromium salts have been demonstrated to induce morphological and neoplastic transformation in human and animal cells. Due to its rapid uptake by cellular anion transporters, soluble Cr(VI) has been considered as the primary carcinogenic form of chromium. Within a cell Cr(VI) reduces to lower oxidation states Cr(V/IV/III), accompanied by ROS (reactive oxygen species) production. Cr(VI) reduced forms, along with ROS, affect many cellular functions by damaging nucleic acids (as Cr-DNA adducts, DNA strand breaks, DNA-DNA and DNA-protein crosslinks), oxidizing proteins and causing lipid peroxidation. Cells have evolved many ways to either avoid or repair these injurious consequences. In cells, the toxic capacity of Cr(VI) can be decreased by: 1) activation of DNA damage checkpoints, which cause transient growth arrest to repair and prevent the transmission of damaged chromosomes; 2) antioxidant defense systems, which can neutralize the ROS generated by Cr(VI) action; 3) apoptosis as a highly organized physiological mechanism of eliminating heavily damaged cells. The majority of real life exposure to Cr(VI) is low and long-term exposure to this ecologically negative factor, which is permanently present in the atmosphere, soil and water. The human epithelial-like cells were used as a model of different possible

* E-mails: nellys@iphac.ge, tkartvelishvili@iphac.ge, (tkartvelishvili@hotmail.com), marina.abuladze@ caucasus.net, nasatiani@iphac.ge

hazardous routes of Cr(VI) exposure (e.g., ingestion, inhalation, and dermal contacts). In this chapter dose-dependent involvement of defense mechanisms in response to low level and long-term Cr(VI) treatment are analyzed in dynamic. Low level Cr(VI) doses were assigned at the boundary of cell growth disorders and apoptosis for the studied cell line. The apoptogenic and nonapoptogenic low doses initially produced ROS to a comparable level but affected differently DNA damage-induced cell cycle checkpoints and the antioxidant defense system (including glutathione cycle, catalase, superoxide dismutase). The nonapoptogenic concentration increased activity of the antioxidant system, and caused temporal cell cycle arrest elevating S phase cell number followed by the increasing of the G_2/M cell compartment that resulted in transient growth arrest. The detected imbalance of the activity of the glutathione-dependent system at the cell treatment with the toxic apoptogenic Cr(VI) concentration correlated with the permanent growth arrest at the S phase of the cell cycle, lowered DNA synthesis, irreversibly inhibited proliferation and development of apoptosis. The small compartment of viable cells that had escaped from apoptosis retained the ability to proliferate, and the resultant cell population was enriched for polyploid multinuclear cells.

Key words: chromium, cell cycle, apoptosis, reactive oxygen species, antioxidant defense system

1. Introduction

Human Health Risk and Exposure Assessment of Chromium (VI)

Exposure to certain metals is associated with an increased risk of cancer. Based on epidemiological data arsenic, chromium, nickel, beryllium and cadmium are confirmed human carcinogens, while lead, cobalt, antimony and iron are considered potential carcinogens [1-3]. Excess occupational exposure to metals especially in mining and industrial settings can be a major cause of metal related cancer [1].

Chromium is one among the metals recognized as genotoxic, carcinogenic, and mutagenic in humans, rodents, and a variety of experimental systems [4-7]. Its uniqueness comes from its strict assignment of carcinogenicity to one level of oxidation, namely Cr(VI) as opposed to the physiologically essential Cr(III), as part of the glucose tolerance factor [8-10].

Chromium (Cr) is ubiquitous in the environment occurring naturally in soils, rocks, and living organisms. Cr exists primarily in two stable valence forms: hexavalent (Cr(VI)) and trivalent (Cr(III)). The latter is the most environmentally stable valence state of chromium, and occurs naturally in most environmental media as a water-insoluble form. In contrast, Cr(VI) rarely occurs naturally and is produced by anthropogenic sources, such as the chrome plating industry, stainless steel welding, chrome pigmenting, and leather tanning.

Health consequences of inhalation exposure to Cr(VI) include allergic contact dermatitis, inflammation, airway hypersensitivity reactions, parenchymal diseases, pulmonary fibrosis, bronchitis, emphysema, bronchial asthma and respiratory cancers [11-14]. The most common sites for metal related cancers are the lungs and skin. Cr(III) has extremely low or no toxicity via all routs of human exposure and does not pose a cancer hazard, because of its low capability of passing through cell membranes via simple diffusion or pinocytosis.

The chromate oxyanion $[CrO_4]^{2-}$, the dominant form of Cr(VI) in neutral aqueous solutions (at physiological pH), can cross cellular membranes via the surface anion transport system ($SO_4{}^{2-}$ and $HPO_4{}^{2-}$ channels) due to structural similarity to sulfate and phosphate anions, and is biologically active [15-17]. At physiological conditions Cr(VI) is nonreactive toward DNA in vitro because of mutual electrostatic repulsion and requires reductive activation to cause biological damage. In contrast to Cr(VI), Cr(III) can bind to DNA in a cell free system. Cr(III) forms six coordinated complexes with molecules containing O, N or S atoms. Cr(III) acts as a hard Lewis acid and strongly prefers coordination to negatively charged oxygen groups [18]. DNA having an abundance of phosphate, nitrogen and oxygen donor groups is an ideal target for metals, especially transition metals due to their capacity to form coordinated complexes.

Reduction of Cr(VI) inside a cell results in the production of unstable high reactive Cr(V,IV), stable Cr(III) [19-21] and a wide range of reactive oxygen species (ROS). The reactive intermediates of Cr(VI) and ROS interact with different cellular components and organelles eventually leading to toxicity. Cr(VI) can produce DNA damage from either an oxidative pathway or a metal-binding pathway that results in a wide variety of DNA lesions, including damages to chromosomes. The neoplastic transformation of cells is connected with serious alterations of their genetic code. Any molecule that can bind with constituents of cell nuclei may affect the genetic code. In addition, oxidative damages to protein and lipids may play an important role in chromium carcinogenesis [22,23].

Generation of ROS and Cr(VI) Redox Processing

Oxygen is an inevitable component of aerobic life. Incomplete reduction of oxygen to water during normal aerobic metabolism generates ROS. The main ROS such as superoxide anion ($O_2^{\cdot-}$), singlet oxygen ($^1O_2$), hydrogen peroxide (H_2O_2), and nitric oxide (NO), which reacts with superoxide anion producing different types of reactive nitrogen species (RNS), are very transient species and play an important role in many physiological and pathological processes.

One of the most popular theories to explain oxygen toxicity has been the superoxide theory, which proposes that oxygen toxicity is due to overproduction of superoxide anion [24].

The mitochondrion is the organelle in eukaryotic cells responsible for aerobic respiration and it is the most common source of ROS. In normal cells, 1-2% of electrons carried by the mitochondrial electron transport chain leak from this pathway and pass directly to oxygen generating superoxide anion, which can be a source of other ROS by developing different types of chain reactions [25]. Besides mitochondrion, ROS are generated by irradiation, metal-catalyzed reaction, inflammatory cell activation (neutrophils and macrophages), and chemical reduction of oxygen by intracellular oxidases, peroxidases, oxygenases [24,26]. Superoxide rapidly dismutates to H_2O_2 by superoxide dismutase (SOD), or reduces Fe(III) to

Fe(II), releasing the iron from storage sites. H_2O_2 is detoxified by catalase to H_2O and O_2, or in the presence of transition metals (e.g. Fe(II) or Cu(I) ions) is reduced to the hydroxyl radical (•OH) by Fenton and Haber-Weiss reactions [24].

$O_2^{\cdot-}$ and H_2O_2 are unavoidable by-products of aerobic metabolism. Most biomolecules resist univalent redox reactions and are unreactive with $O_2^{\cdot-}$, which is a single electron oxidant of only moderate strength. One way in which $O_2^{\cdot-}$ is believed to cause toxicity is through its participation in •OH radical production, representing an extremely powerful oxidant.

ROS can function as beneficial (e.g. defense response against infection) or dangerous agents. Accumulating evidence suggests that ROS production in non-phagocyting cells participates in normal physiological signaling by growth factors and cytokines [27-29]. Excess ROS are harmful because they react with and modify all classes of cellular macromolecules causing behavioral abnormalities, cytotoxicity and mutagenic damage. Cell exposure to environmental oxidants, such UV and ionizing radiation, heavy metals, redox active chemicals, hypoxia, hyperoxia increases ROS production that shifts cell redox status to a more oxidized state known as an oxidative stress.

According to the uptake-reduction model of Cr(VI) carcinogenicity [15], once inside cells Cr(VI) reduces to the lower oxidation states Cr(V,IV,III) by different nonspecific intracellular reductants, such as ascorbic acid, glutathione (GSH), cysteine, riboflavin, carbohydrates, NADH, NAD(P)H, nucleotides, glutathione reductase (GR) [14,30-34]. During Cr(VI) reduction a wide variety of reactive intermediates, such as oxygen-, carbon-, and sulfur- centered radicals form from complexing ligands and may be a source of cell disorder [35,36].

The general proposed mechanism of oxygen activation by Cr(VI) is provided by Fenton-like and Haber-Weiss reactions including Cr(VI/V), Cr(V/IV), Cr(IV/III) and Cr(III/II) redox pairs or redox couples [22,37,38]. Substrate of the mentioned reactions are $O_2^{\cdot-}$ and H_2O_2 (two common metabolic products, that do not react with DNA), which realize redox cycling of chromium anions, producing a powerful DNA-damaging hydroxyl radical, resulting in the Haber-Weiss reaction.

$$Me^{n+} + H_2O_2 \rightarrow Me^{(n+1)+} + \bullet OH + OH^- \text{ (Fenton-like reaction)}$$
$$Me^{(n+1)+} + O_2^{\cdot-} \rightarrow Me^{n+} + O_2$$

The balance of these two reactions is:

$$H_2O_2 + O_2^{\cdot-} \xrightarrow{Me^{n+}/Me^{(n+1)+}} O_2 + \bullet OH + OH^- \text{ (Haber-Weiss reaction)}$$

It is believed that Cr(III) can also maintain redox activity in vivo through Cr(III)/Cr(V) [39], or Cr(III)/Cr(II) redox pairs [40], re-activating the radical cascade.

Various molecules, such as amino acids [41], proteins [42], DNA and RNA [43] form complexes with Cr(III). In addition, elevation of ROS in excess generates pro-oxidant conditions damaging DNA, RNA, proteins and lipid components. All of these factors may be a source of the pathogenesis of many human diseases and aging [36].

Antioxidant Defense Systems

Oxidative stress is defined as an imbalance between the production and removal of reactive oxygen species [24]. The effect of ROS is balanced by antioxidant systems, which provide either direct or indirect protection to cells against adverse effects at different biological sites. The cellular protective antiradical mechanisms consist of multiple interacting enzymatic and nonenzymatic antioxidants. As the main products of chromium(VI) reduction are superoxide anion, hydrogen peroxide and hydroxyl radical, among antioxidants we have chosen the main components responsible for the transformation of the mentioned reactive oxygen species, and capable to take part in Cr(VI) reduction as well.

Superoxide Dismutase

Superoxide dismutases (SODs) are present in a variety of aerobic organisms, but absent in the most anaerobs. The wide distribution of superoxide dismutases among aerobic organisms points to the special role of these enzymes [44]. It is proposed to play a central role in protecting cells against harmful effects of superoxide anion radicals. Their sole function is to remove the superoxide formed via the univalent oxygen reduction and thus to protect the cells against oxygen toxicity.

SODs ameliorate oxygen toxicity by dismuting $O_2^{\cdot-}$

$$O_2^{\cdot-} + O_2^{\cdot-} + H^+ \rightarrow H_2O_2 + O_2$$

In humans SOD generally exists in two forms inside the cells: Cu,ZnSOD in the cytoplasm and MnSOD exclusively in the inner mitochondrial space. Besides, there is an extracellular SOD (Ec-SOD) [45,46].

SOD isoforms differ in the nature of the active metal center, amino acid content, and numbers of subunits. SOD catalyses the superoxide dismutation by alternate reduction and reoxidation of the transition metal ion at the active site [47,48]

Mitochondrial MnSOD is a homotetramer (~96 kDa), containing one manganese Mn(III) per subunit [36]. Cu,ZnSOD is a homodimeric enzyme (~32 kDa) containing one copper Cu(II) per subunit joined to the buried Zn(II) by a bridging histidyl imidasolate group [49].

It should be noted that MnSOD does not react with H_2O_2, while Cu,ZnSOD can interact with H_2O_2, that causes Cu,ZnSOD inactivation. Extracellular Ec-SOD is a secretory, tetrameric Cu, Zn containing glycoprotein, explaining SOD activity in extracellular fluids such as plasma and lymph [50].

Catalase

Hydrogen peroxide, one of the major reactive oxygen species is formed mainly from dismutation of superoxide usually generated from mitochondrial transport chain. Although H_2O_2 can be a product of some oxidases, such as glycolate oxidase, urate oxidase or flavoprotein dehydrogenases localized in peroxisomes [24].

Pathological conditions, which increase the rate of H_2O_2 production, will lead to the accumulation of H_2O_2 in cytosol and mitochondria. Being uncharged and not very reactive, it can easily diffuse between cell compartments. Elevation of H_2O_2 levels can be a cause of harmful consequences, such as depletion of ATP, GSH, NADPH pools, and induction of mitochondrial permeability, disrupting the mitochondrial membrane potential that can trigger the apoptotic pathway of cell death [51].

Safe disposal of H_2O_2 in cells is carried out by GPx and catalase. The former functions in both cytosol and mitochondria, the latter only in peroxisomes.

Catalase is a homotetramer (~240 kDa). Each protein subunit contains a ferric Fe(III) heme group bound to its active site [48]. It decomposes of H_2O_2 to H_2O and O_2.

$$2H_2O_2 \rightarrow 2H_2O + O_2$$

H_2O_2 decomposition proceeds by a two step reaction involving H_2O_2 reduction to H_2O, followed by H_2O_2 oxidation to O_2. Owing to the oxidant capacity of the oxidized ferric group, generated at the first step of H_2O_2 decomposition, catalase can also catalyze peroxidase-type reactions, such as an oxidation of alcohols (methanol, ethanol) to aldehydes, and formic acid to CO_2 [24].

Glutathione Cycle

Glutathione (GSH) is the main low molecular weight thiol present in most mammalian cells (generally in the millimolar range) and involved in bioreductive reactions, maintenance of enzyme activity and protection against oxidative stress [52-54]. There is much evidence connecting metal toxicity to oxidative modification of thiol groups in low molecular compounds and proteins. The thiol groups are central components of redox-sensitive proteins and control pathways [55].

Most of the cellular GSH (about 90%) is disposed in the cytosol, with the remainder in many organelles (including the mitochondria, nuclear matrix, and peroxisomes) [56]. GSH exists in two major forms: reduced GSH and oxidized GSSG.

γ-Glu-Cys-Gly **(GSH)**   γ-Glu-Cys-Gly **(GSSG)**
|
γ- Glu-Cys-Gly

GSH is synthesized in cells by the consecutive action of two cytosolic enzymes γ-glutamate cysteine ligase (also referred to as glutamyl cysteine synthetase) and GSH synthetase. The former is a rate-limiting enzyme and can be regulated by feedback inhibition of synthesis by GSH.

GSH has a potent electron-donating capacity as indicated by the high negative redox potential of the GSH/GSSG redox couple. The reducing power of GSH is a measure of its free radical scavenging, electron-donating, and sulfhydryl-donating capacity. The ratio of GSH/GSSG plays an important role in regulating the cellular redox status, since it is the most abundant thiol-disulfide redox buffer in the cell. The changes of the cell redox-state may have a serious impact on cellular function - a higher GSH/GSSG ratio (high redox potential) is connected with cell proliferation, while a lower GSH/GSSG ratio (low redox potential) is associated with cell differentiation [53,54].

GSH plays an important role in a number of critical cellular processes. The reactions in which GSH is involved can be distinguished as reversible and irreversible. The reversible loss of GSH represents a GSH transformation into GSSG and/or mixed low molecular and protein disulfides, which are finally converted into GSSG. Glutathione oxidation up to GSSG proceeds via reduction of different substrates at the GSH expense. The main enzymes taking part in GSH oxidation to GSSG can be conditionally divided into three groups, namely: enzymes reducing hydroperoxides [57,58], disulfides [59,60], and other substrates (e.g. ribonucleotides to deoxyribonucleotides, dehydroascorbate to ascorbate) [61,62].

GSSG reduction to GSH is a key event in GSH/GSSG balance. The recycling of GSSG to GSH is accomplished mainly by glutathione reductase (GR).

$$\text{GSSG} + \text{NADPH} + \text{H}^+ \rightarrow 2\text{GSH} + \text{NADP}^+$$

GR is a dimeric flavoprotein (~100-120 kDa) consisting of two protein subunits. Each of them contains one active center. The enzyme is mainly located in the cytosol and partly in the mitochondria, microsomes and nuclei. The general biological function of GR is to support GSH high level.

Irreversible loss of GSH represents a GSH transformation into metabolites which could not be regenerated, thus the maintenance of the GSH pool can be only supported via its synthesis or resynthesis. GSH loss proceeds via two enzymes, glutathione-S-transferase, and γ-glutamyl transpeptidase. Among numerous reactions catalyzed by the former enzyme, only production of glutathione thioethers GSR leads to the irreversible loss of GSH [52,63]. The later enzyme hydrolyzes γ-link between glutamate and cysteine in GSH, initiating GSH catabolism [64]. The irreversible loss of GSH can be also accomplished via the formation of stable complexes as Me-GSH and GSH-Me-DNA in the process of transition metals reduction [34,65]. The GSH-Me-DNA conjugates make an essential contribution to the metal-induced genotoxicity. The influence of intracellular GSH level and some products of its metabolism on the processes connected to the activation of the signal transduction receptors were considered in [64,66-68].

Among enzymes of the GSH reducing systems, only GPx catalyzes H2O2 reduction. Se-containing glutathione peroxidases (GPx) catalyze the decomposition of H2O2 and organic hydroperoxides at the expense of GSH [57,58]. GPx removes H2O2 (or ROOH) by coupling its reduction to H2O (or alcohol) with oxidation of reduced GSH.

$$2\text{GSH} + H_2O_2 \rightarrow \text{GSSG} + 2H_2O$$

$$2\text{GSH} + \text{ROOH} \rightarrow \text{GSSG} + \text{ROH} + H_2O$$

Four isoforms of GPx have been identified in mammalian cells. The most characterized of the GPx is GPx1. This form reduces hydrogen peroxide and almost all organic hydroperoxides (except hydroperoxides of phospholipids). GPx1 (~80 kDa) is a homotetrameric protein, each subunint contains one selenium atom as selenocysteine (the amino acid cysteine in which the sulfur atom are replaced by selenium) in the active site. GPx1 is widely distributed in different cell types. Other GPx forms are specifically arranged and poorly detected in the most nonspecific tissue [48].

Unlike SOD and catalase, GPx requires several secondary enzymes (GR and G-6-PDH) and cofactors (GSH, NADPH and glucose-6-phosphate) to function at high efficiency. Glucose-6-phosphate dehydrogenase and glucose-6-phosphate do not act on ROS directly, but they regulate the NADPH level [69] and thus indirectly enable GPx to function. GPx in addition to affecting GSH/GSSG, controls the cellular content of H2O2 (or other organic hydroperoxides) and NADPH/NADP+, so it may regulate the cellular redox balance [70].

GPx shares H2O2 as the substrate with catalase. Cooperation of GPx and catalase depends on different factors. One of them is the subcellular compartenmalization of the enzymes. Whereas catalase is localized in the peroxisomes, GPx is found in the cytosol and in the matrix of mitochondria (about 10% of total). As mitochondria lack catalase, H2O2 arising from mitochondria (which is mostly subjected to oxidative stress) should be mainly decomposed by GPx. The second factor defining the GPx and catalase cooperation is the intracellular level of H2O2. GPx has much higher affinity to H2O2 than catalase. That is why GPx is a major source of protection against low levels of oxidative stress (low level of H2O2), whereas catalase becomes more significant at the severe oxidative stress (high level of H2O2) [24]. The third factor is the type of cells. In liver GPx (along with GSH cycle) can resist to H2O2, but catalase is a main executor of H2O2 in lung and eye. Besides, defense from H2O2 depends on enzyme availability in cells, e.g. the brain has no catalase, and thus stress arising from H2O2 will be coped by GPx.

The levels of the intracellular antioxidant enzymes are under genetic control. This regulation is mostly studied in bacteria. It is shown that elevation of intracellular concentrations of O2-. and H2O2 coincide with activation of enzymes, such as catalase, SOD, and peroxidase due to the upregulation of gene transcription [71-73]. These genes knockout bacteria become hypersensitive to damage by O2 [74].

The early study [75] has estimated the influence of Cr(VI) on the expression of a number of oxidative stress genes in normal human lung LL24 and human lung carcinoma A549 cells. None of analyzed genes (catalase, GPx1, GR, MnSOD, Cu,ZnSOD, HO1) were induced in A549 cells using a wide range of Cr(VI) concentrations (2.5-200 μM). In LL24 cells, only stress-inducible heme oxygenase (HO1) was expressed at a low, nontoxic concentration of Cr(VI). Genechip technology used to analyze differential gene expression shows that definite redox genes are induced by Cr(VI) action in A549 cells. Several fold expressions of Cu,ZnSOD, and GPx caused by Cr(VI) action indicate that the cellular antioxidant defense systems are activated in response to Cr(VI) to decompose induced O2-. and H2O2 [76].

Thus antioxidant defense systems provide the endogenous barrier against ROS. Now, the role of antioxidant defense systems is considered in the context of the fine-tuning of ROS levels in the redox regulation of the cell cycle and apoptosis [77].

Effect of Chromium (VI) and Chromium-induced ROS on Signal Transduction

Toxicants can interact with the normal cell signaling systems leading to impaired cellular functions and ultimately to toxicity. Impaired cell functions cover a wide range of responses including loss of viability, apoptosis, proliferation, alterations of gene expression, and mutations. Toxicological dose-response behaviors depend on toxic compound concentrations, persistence of the impact and the state of the intracellular signal motifs (nuclear transcriptional receptors, kinase/phosphatase cascades, G-coupled protein receptors, etc.). Early transient responses may be linked with the activation or deactivation of biological switches regulating signaling pathways [78,79]. In moving from one phenotypic state to another the cell components have to change in concert to achieve pleiotrophic alterations in cell characteristics and maintain homeostasis within a range of conditions. Diminished adaptability to stress or diminished physiological function can result in disease or cell death. Recent advances in cell signaling research provide the evidence that multiple sets of signal transducing molecules are preorganized and sequestered in distinct compartments within the cell. Compartmentalization spatially organizes signal units into hierarchical interconnected sets of molecules through their participation in different classes of interacting units [80,81].

Phosphorylation-dependent signaling pathways are involved in cellular responses to H2O2 and chromium impact, activating nuclear transcription factors such as NF-kB and AP-1.

All eukaryotic cells possess complex mitogen-activated protein kinases (MAPK) pathways, which coordinately regulate diverse cellular activities and mediate mitogen- and stress-activated signals. Signaling through the MAPK pathways culminates in the activation of a variety of transcription factors modulating gene expression. To date, five distinct groups of MAPK have been characterized in mammals, and the most extensively studied are extracellular signal-regulated kinases (ERK 1/2), c-Jun N-terminal kinases (JNK), and p38 kinases. MAPKs can be activated by a wide variety of different stimuli, but in general ERK1 and ERK2 are preferentially activated in response to growth factors and other stimuli involved in cell proliferation and differentiation processes, while the JNK and the p38 kinases are more responsive to stress stimuli ranging from, osmotic shock and ionizing radiation, to proinflammatory cytokines. Each family of MAPKs is composed of a set of evolutionary conserved sequentially acting kinases: a MAPK, a MAPK kinase (MAPKK), and a MAPKK kinase (MAPKKK), and they are activated through phosphorylation by the mitogen-activated protein kinase (MAPK) kinase (MEK). Despite the diversity in function and upstream signals MAPKs are strongly activated in a highly conserved mechanism that involves phosphorylation on both the Thr and Tyr residue catalyzed by a MAPK kinase. There are different MAPK kinases specific for each subgroup of MAPKs. One well-known function of MAPKs is to regulate gene expression through phosphorylation of downstream transcription factors [82-86]. The lifespan of activated MAPKs and cell types expressing them can determine diverse gene expression patterns that result in different physiological consequences. Thus, transient JNK and p38 activation by tumor necroses factor α (TNFα) is a survival factor, while a persistent induction results in apoptosis [87,88]. Transient ERK activation leads to proliferation, while a persistent activation mediates growth arrest or differentiation signaling [89,90]. Although JNK has been considered primarily a stress-responsive protein kinase, recent evidence supports the suggestion that JNK signaling is involved in cell growth and proliferation of unstressed cells. The JNK inhibition in a KB-3 human carcinoma cell line resulted in an expanded cell cycle with delays in both S and G2 phases [91].

Chromium effects on signal transduction pathways may be direct, through the interaction of Cr(VI) and its intermediates with proteins, or indirect through the formation of ROS such as hydroxyl radicals, thiyl radicals, superoxide radicals, and hydrogen peroxide [92]. Multiple possible damage sites define the character of cellular responses to chromium exposure. Using cultured mammalian cells it was demonstrated that Cr(VI) did not remain bound at the cell surface, but passed through the cell membrane, and was dispersed throughout the cell in the first hours of Cr(VI) treatment [39]. Chromium enters into the cells even at low concentration as it was shown in the dose-dependent uptake study, using the human lung epithelial cell line [37].

The imbalance of the antioxidant defense systems of the cell, simultaneously with ROS production, modulates different signaling processes [23,28,29,93]. Even mild intracellular redox destabilization caused experimentally by diamide without ROS generation, inhibited proliferation of human colonic CaCo2 cells. The suppression of cell proliferative activity was responsive to GSH/GSSG status, and the cells were arrested at G1-S transition and G2/M phase [94].

Biological molecules involved in cell signaling and gene regulations are very sensitive to the redox status of the cell [48]. The reduced/oxidized ratio provides the means of dynamically regulating intracellular redox state and is the one mechanism by which cells may transduce oxidative stress into the inducible expression of a wide variety of genes implicated in cellular processes such as growth, survival, death and mutations. Tumour necrosis factor receptors, CD95/Fas, and protein tyrosine kinases, involved in signal transduction are characterized by several cysteine-rich regions that are susceptible to redox modification [95]. Various studies have demonstrated that H2O2 may be an important second messenger in the cell signaling that activates different members of signaling modules dependent on cell type, doses and impact persistence [96-98]. The rapid formation and disappearance of ROS also can serve as 'on and off' switches for cellular signaling [92]. Hydrogen peroxide is believed to activate MAPK signaling by deactivation of protein tyrosine phosphatases [99-101]. H2O2 activates nuclear transcription factors such as AP-1 signalling through mechanisms that require phosphorylation of cJun following activation of JNK kinase [102]. Low level oxidative stress that lacked apoptosis induction in U937 human lymphoid cells, caused activation of p38 cascade and mitotic arrest [103]. Sub-lethal doses of Cr(VI) activated JNK, ERK, and p38 in human lung adenocarcinoma cell line CL3 through oxidative stress, and the induction was mediated by diverse redox mechanisms [104].

Exposure to chromium markedly activated JNK and p38 and moderately activated ERK in a dose- and time-dependent manner. Co-administering Cr(VI) with oxidative modulators shows that p38 positively correlated with oxidative stress, while JNK activity could be enhanced by either a quencher or activator of redox reactions. Activation of JNK, p38, and ERK by a very high dose of chromium (500 μM) had also been reported in human bronchial epithelial cells [90]. Selective activation of JNK by hydrogen peroxide required stimulation of Src kinase activity [105]. The diverse hydrogen peroxide and chromium impact was detected at the low noncytotoxic Cr(VI) exposures that did not cause global oxidative stress or apoptosis, but increased ROS in A549 human lung carcinoma cells, and selectively activated JNK relative to other MAP kinases [106]. JNK activation in response to Cr(VI) and H2O2 (1 mM) shared requirements for intracellular thiol oxidation, activation of Src family kinases, and p130cas. But Cr(VI) did not mimic hydrogen peroxide and directly activated specific Src family kinases (predominantly Fyn). Several organic forms of Cr(III) induced lymphocyte apoptosis through the induction of Src family members such as p561ck, p59fyn, and p53/56lyn [107]. The cross talk is proposed between redox-mediated signaling and the phosphorylative cascade of MAP kinases pathways [108].

Genotoxic Lesions in Cellular DNA

DNA damage is considered the most serious cellular modifications as DNA is not synthesized de novo but copied and hence DNA damages, which have mutagenic and cytotoxic effects, causes genomic instability and a remarkable threat to proper functioning of cells.

In eukaryotes, genomic DNA is packed with histones to form nucleosomes, which in turn condense to form a re compact chromatin structure [109]. Even during DNA synthesis nucleosomes are rapidly generated on newly replicated DNA not leaving DNA naked for a long time [110]. Nonetheless the primary structure is under the constant injurious action of oxidants from physiological processes, such as mitochondrial respiration, and other biochemical reactions. And only the highly coordinated action of the cellular repair machine makes DNA a very stable molecule.

Oxidative stress from pathophysiological conditions such as xenobiotic metabolism, and inflammation greatly accelerates DNA damages. The main kinds of DNA primary structure damages are i) base- or deoxyribose lesions; ii) strand breaks; iii) cross-links. From all wide spectra of ROS, superoxide or hydrogene peroxide at physiological levels do not react with any of the DNA structural components. By contrast, highly reactive hydroxyl radicals and singlet oxygen are very harmful for the DNA structure. The most powerful •OH radical attacks sugars and all four DNA bases. 1O2 selectively attacks guanine. To pose their damaging effect •OH radicals should be formed in the vicinity of DNA, as they cannot diffuse.

The oxidative concept in chromium genotoxicity proposes that products of Cr(VI) reduction - ROS, and various possible Cr(V/IV/III) coordinated complexes with cellular reductants, in the vicinity of DNA in vivo can catalyze redox reactions, whose products oxidize DNA. Such products of DNA oxidative damage include modified bases, abasic sites, single and double breaks in the phospho-sugar backbone of DNA and base adducts of carbon-centered radicals including DNA-cross-linked nuclear proteins (e.g. thymine to tyrosine) and DNA-DNA crosslinks [111,112].

It was demonstrated that Cr(VI) may lead to an increase of 8-oxodG levels. This lesion is the principal form of oxidative DNA damage and a key biomarker relevant to carcinogenesis. The increase of 8-oxodG was observed after Cr(VI) exposure of cells in culture and in vivo in rat lung DNA after inhalation exposure to chromium [113,114].

The use of the alkaline single cell electrophoresis ('comet') assay and alkaline elution studies for the evaluation of single and double strand breaks as well as alkali labile sites (e.g. apurinic-apyrimidinic sites) in DNA after exposure to Cr(VI) of cultured cells and experimental animals showed the increase of such lesions in dose- and time-dependent manner [115,116]. Moreover, increased DNA strand breaks in peripheral lymphocytes were observed in chrome-plating workers correlating with the elevated level of chromium lymphocyte concentrations [117].

Intracellular reduction of Cr(VI) has been known to lead to the formation of stable ternary adducts with DNA. Thus cysteine-Cr(III)-DNA and glutathione-Cr(III)-DNA are the major type of DNA modifications observed after Cr(VI) exposure of CHO cells [34]. Glutathione forms a complex with Cr(VI) in a rapid equilibrium step, followed by a slow reduction of Cr(VI). Cr(VI) produces a large amount of Cr-DNA adducts in the presence of glutathione [118]. Glutathione-Cr(VI) complexes in the vicinity of DNA can produce radicals during Cr(VI) reduction to Cr(V) and Cr(III) and as a result ternary adducts can be generated, causing several kind of DNA lesions.

Different kinds of DNA-protein cross-links (DPS) can be formed under Cr(VI) reduction. One of these types is covalently linked DNA and proteins, resulting from free radical attack on chromatin. The other type is the result of Cr-DNA adducts interaction with proteins. Cross-links of the second type can be distinguished from the first type by the dissociation with increased ionic strength solutions or metal chelators. DNA-protein crosslinks are potentially bulky and helix distorting and have the potential to block the progression of translocating protein complexes.

DNA-DNA interstrand cross-links (DDS) were detected in chromate-treated normal human lung cells. The observed DDS are EDTA (metal chelator agent) sensitive which points to Cr(III) as an agent connecting two strands of DNA [119]. Cr(III) in vitro chelates DNA via guanine N7 and nearest PO2 group [120]. This Cr(III) feature may be responsible for guanine-guanine DNA interstrand cross-links induced by chromate in living cells [6]. DDS as well as DPS may induce double-strand breaks during replication presumably due to replication fork collapse and consequent nuclease attack.

Chromium can cause chromatin destabilizing structural effects by DNA conformation modifications. In the cell free system it was shown that covalent attachment of chromium to DNA induces alterations in DNA structure, which are recognized by high mobility group proteins HMGB1 and HMGB2 (formerly known as HMG1 and HMG2). Moreover these proteins recognize chromium damaged DNA recovered from Cr(VI)-treated cells in dose-dependent manner [121]. It is known that DNA binding by HMGB1/2 is DNA conformation dependent [122]. A characteristic feature of HMG proteins is their ability to recognize single-stranded DNA and a variety of bent and unwound DNA structures, including four-way junctions, cruciform DNA, cis-platin modified DNA, B-Z DNA junctions, and stem loops [123,124]. HMGB1 has been shown to play a role in enhancing the activating properties of p53 [125]. p53 has been recognized as a guardian against cellular stressors. It is a transcription factor, and exerts its protective effects via transcription activation of a great number of its downstream genes, such as a cell cycle-control gene p21 (with possibility of repair) and an apoptosis-inducing gene Bax [126].

DNA-Repairing Systems in Response to Chromium Genotoxicity

The cells use various repair mechanisms adapted to the different types of DNA lesion [127]. Repair of eukaryotic DNA damages proceeds in the contex of packed chromatin structure. Histone-DNA interactions within nucleosomes can be loosened or remodeled to access repair machinery proteins to DNA by chromatin modifying and remodeling enzymes [128]. Obviously, chromatin structure obstructs DNA both from damaging action as well as from the detection of DNA lesions and their subsequent repair.

The main guardians against DNA damages due to cellular metabolism are involved in the broad spectrum of DNA lesions induced by Cr(VI). DNA repair pathways can be divided into: direct repair, base excision repair (BER), nucleotide excision repair (NER), repair of DNA double-strand breaks (dsDNA), which includes homologous recombination (HR) and nonhomologous end joining (NHEJ) mechanisms.

One of the direct repair mechanisms consists in the activation of photolyases. They repair UV-induced photoproducts. Photolyase works without recruitment of any other proteins. The human homolog of the Drosophila (6-4) photolyase gene is induced 5.17 fold in the human lung type II epithelial cell line (A549) by 300 μM potassium chromate after 2 h of exposure. This is first evidence of induction of photolyase in response to agent other than UV light [76].

The base excision repair (BER) mechanism is responsible for repair in DNA lesions that do not strongly disturb the DNA backbone, e.g. modified DNA bases, abasic sites. OGG1 DNA glycosilase/β-lyase is the BER enzyme that is involved in recognition and removal of 8-oxodG [129,130].

The exposure to 100 μM sodium chromate for 16 h, and to more than 25 μM for 28 h downregulates the expression of the repair glycosylase for 8-oxodG (hOGG1) in A549 cells. However, exposure to 200 μM H_2O_2 has not shown an inhibition of hOGG1 expression by RT-RCR and Western blotting. It is proposed that the mechanism of hOGG1 inhibition by Cr(VI) is not mediated by hydrogen peroxide as a by-product of its reduction, rather by the possible inhibition of the key transcription machinery proteins' interactions by coordination with Cr(III) [114]. The 8-oxodG repair activity in crude tissue extract from rat lung after inhalation exposure to hexavalent chromium has been monitored in time-and dose-dependent manner. The results point to a decrease of the repair activity in 2 weeks of exposure, and recovery at 3 weeks of repeated exposure for both low and high exposure groups [131].

NER machinery recognizes and repairs the broad spectrum of bulky DNA-distorting lesions. DNA interstrand cross-links, and DNA-protein/amino acid ternary adducts are substrates for NER. NER is a complex, multifunctional and multicomponent system. In mammalian cells, two proteins XPA-RPA (essential for global genome repair) and XPC-HR-23B (essential for transcription-coupled repair) initiate the recognition of damage. The following step is recruitment of multiprotein complex (XPG, XPB, XPD etc.), which is a part of the RNA polymerase II transcription machinery. XPB and XPD are helicases producing a 'bubble' structure containing dsDNA/ssDNA junctions at both ends of damaged oligonucleotide fragment [132]. A 'bubble' is a substrate for cleavage by two structure-specific endonucleases XPF/ERCC and XPG [133,134]. The involvement of NER in the protection of cells against chromate toxicity has been recently shown by Patierno et al. [135] and Zhitkovich et al. [136]. The used approach consists in comparing normal and NER-deficient cells in the removal of Cr-DNA adduct. Cells were exposed to 1-5 μM potassium chromate for 3 h and after 7 h Cr-induced lesions were successfully removed in NER-proficient cells. However, NER-deficient XPA, XPC and XPF human cells were severely impaired in the removal of Cr-DNA adducts. [136].

NER-deficient XPD and ERCC4/XPF cells were considered in parallel with complementation by XPD and ERCC4/XPF proteins. Complementation does reverse the effect of the compromised Cr-induced DNA lesion repair at low doses of Cr(VI) exposure [135].

The same approach was used to represent the HR mechanism for preventing Cr(VI)-induced DNA double strand breaks, which can be the result of a collapsed DNA replication fork forming upon DNA-protein and DNA-DNA cross-links, or from oxidative destruction of deoxyribose residues [137].

Cells are constantly exposed to oxidants from metabolic and other biochemical reactions as well as external factors, suggesting that intracellular antioxidants and DNA repair systems could be important determinants for low levels of DNA damage.

EXPERIMENTAL

Cell Culture and Chromium Treatment

The L-41 cell line is an established human epithelial-like cell culture (Research Center of Medical Genetics, Russian Academy of Medical Science, Moscow, Russia) that was derived from J-96 cell line originally obtained from a patient with monocytic cell leukemia [138]. The L-41 cell line was secondary, routinely established and subcloned by viral transformation (Coxsackie B-3), and was characterized as epithelial-like monolayer culture with high proliferative capacity and adherent cell growth [139]. The L-41 cells were maintained as adherent cells in Eagle's culture medium supplemented with 10% donor calf serum, 2 mM L-glutamine, 100 units of penicillin/ml, and 100 μg of streptomycin/ml at 37°C in a 5% CO_2 incubator. Cells were harvested with trypsin (0.25%)/Na_2EDTA solution. The cells seeded at the concentration $1x10^5$/ml were confluent in 3 days.

Cr(VI) as potassium chromate was added to the logarithmic growth phase at 48 h of growth (80% of confluence). The concentration range and the exposure time depended on the experimental procedure.

Cytotoxicity Assay

Cytotoxicity was studied by the MTT colorimetric assay based on the ability of live and not dead cells to reduce a yellow tetrazolium-based compound 3-(4,5-dimethylthiazol-2-yl)-2,5-diphenyltetrazolium bromide (MTT) to blue formazan product by cell dehydrogenases [140]. The MTT assay has been widely used for cytotoxicity monitoring in drug sensitivity testing [141,142]. The MTT assay was carried out as described by Mosmann [140] with some modifications. Briefly, cells at a concentration of $3x10^4$ per well (200 μl) were seeded in 96-well plates and were grown 48 h before exposure to different Cr(VI) concentrations at the range from 2 to 200 μM for the indicated time periods. Following chromate treatment cells were washed with PBS and the complete growth media without Cr(VI) was administered to the culture. The cell survival was detected in 48 h after the beginning of Cr(VI) treatment. 20 µl of the MTT solution (5 mg/ml) was added to each culture well and the plates incubated for an additional 4 h at 37°C. At the end of the incubation with MTT the medium was removed and the colored formazan precipitate was solubilized with DMSO (100 µl per well). After 10-15 min cell survival and background was determined by absorbance at 560 nm and 620 nm respectively, using the scanning multiwell spectrophotometer (Anthos Labtec Instruments, Austria). The MTT assay was carried out in five replicate wells. The data are presented as % of control value.

Cell Growth Analysis

Cells were seeded at a concentration of $1x10^5$/ml in culture tubes and were grown for 48 h to achieve 80% of confluence (sub-confluent culture) before they were exposed to 0, 2, 5, 10 and 20 µM potassium chromate. After Cr(VI) treatment cells were harvested with trypsin/EDTA solution and suspended in the complete growth media. The total cell number was determined by hemocytometer in five replicate culture tubes for each concentration and each time point. Dead cells were excluded by trypan blue staining assay. For mitotic index estimation and morphological analysis cells were grown on cover slips in 24-well plates. After fixation with ethanol, glacial acetic acid (3:1) cells were stained with Giemsa and analyzed by microscope.

The mitotic index (Mi) was scored as the number of mitosis among 1,000 cells growing on cover slips and expressed as a percentage.

For long-term chromium treatment, cells were seeded at $1x10^5$/ml using T25 flasks and treated with 2 and 20 µM Cr(VI) at 48 h of growth for 24 h. After treatment (24 h) cells were rinsed with PBS and fresh complete growth media without chromate was added to the culture. The cells were collected by trypsinization at 72 h, counted by hemocytometer, reseeded at $3x10^3$ in triplicate 60 mm^2 dishes and incubated for 7 days. Then cells were harvested by trypsin/EDTA solution, counted, and cell morphology was analyzed by fluorescent microscope at acridine orange staining, and DNA content estimated by flow cytometry using FACScan (Becton Dickinson, USA).

Cell Cycle Analysis

DNA content of propidium iodide (PI) stained cells was measured by flow cytometry, and separation into phases of the cell cycle was based on the PI fluorescence. PI intercalates into double-stranded nucleic acids, binds to DNA stoichiometrically and its fluorescence is enhanced upon binding [143]. The enzyme RNase was used to degrade double-stranded RNA. Cells after Cr(VI) treatment were harvested by trypsinization, cell concentration was adjusted to $1x10^6$ cells/ml for each sample, centrifuged at 600g for 5 min, washed in PB. The pellets were vigorously vortexed, fixed in ice-cold 70% ethanol added drop-by-drop to the vortexed cells and maintained overnight at 4°C. Cells were then stained with a fluorochrome solution containing 20 μg/ml PI in PBS, 50 μg/ml RNase A and 0.1% Triton X-100 for 30 min at 37°C and analyzed using FACScan (Becton Dickinson, USA). The DNA histogram showed the cell cycle distribution of the nonapoptotic cells. The apoptotic cells should be observed as a distinct sub-G_1 peak of the hypodiploid DNA. In the long-term Cr(VI) treatment study the increased DNA content characterized the enhanced ploidy of the cell samples. All results are expressed as the mean percent cells from each cell cycle compartment of three experiments.

Nonradioactive Assay for DNA Synthesis Measurement

The monitoring of DNA synthesis is an indirect parameter of cell proliferation and is suitable for the study of the regulation of DNA synthesis itself [144]. The pyrimidine analog BrdU (5-bromo-2^1-deoxyuridine) is incorporated into newly synthesized cellular DNA of proliferating cells. Incorporation of BrdU can be detected as early as 5 min after administration of the precursor into the cultures [145]. We used the colorimetric immunoassay for the quantification of DNA synthesis based on the measurement of BrdU incorporation during DNA synthesis (Roche Diagnostics GmbH, Germany). The assay was carried out according to the protocol provided by the manufacturer. Briefly, the cells were cultured in 96-well plates at the concentration of 3 x 10^4 per well (200 μl) for 48 h (sub-confluent culture) prior the Cr(VI) treatment. At the indicated time 20μl per well BrdU labeling solution (final concentration: 10 μM BrdU) was added to the culture and the cells were reincubated for additional 2 h (pulse of BrdU) at 37°C. After the labeling period BrdU was detected by the enzyme immunoassay (ELIZA assay) and quantified by measuring the absorbance at the 450 nm (reference wavelength is 690 nm) using the scanning multiwell spectrophotometer (Anthos Labtec Instruments, Austria). The absorbance values directly correlate to the amount of DNA synthesis. The BrdU assay was carried out in four replicate wells for each experimental point.

Apoptosis Measurement

Cell death measurement was performed using the Annexin V staining assay. The assay is a very sensitive probe, which identifies cells undergoing apoptosis, based on externalization of phosphatidylserine (PS) in the early/intermediate stages of apoptosis from the inside to the outside of the cell membrane [146]. The assay detects PS by using FITC-labeled Annexin V, which is a Ca^{++}-dependent phospholipids-binding protein and has a strong natural affinity for PS. Combining Annexin V-FITC staining with propidium iodide (PI), which binds straight to nucleic acids in the cells that have already lost their membrane integrity, provides the possibility to distinguish early apoptosis from secondary necrosis (late phase of apoptosis). The Annexin V-FITC/PI assay effectively detects apoptosis in adherent mammalian cells by flow cytometry [147]. The assay was carried out according to the protocol provided by the manufacturer (Sigma, USA). Briefly, after the chromate treatment cells were gently trypsinized, centrifuged at 600g for 5 min, washed in PBS and 10^6 cells were resuspended in the 200 μl of the binding buffer with 15 μl of fluorescent Annexin V and 1 μl of PI. The samples were incubated 10 min at 4°C prior the flow cytometry analysis. The data are presented as the percentage of Annexin V-positive cells (early apoptotic), Annexin V/PI-double positive cells (late apoptotic), PI-positive (necrotic cells) and Annexin V/PI double negative cells (viable cells).

For morphological detection of apoptosis, the cell population was analyzed by fluorescent microscopy after the staining with a mixture of acridine orange (AO) and ethidium bromide (EB) for the simultaneous detection of viable, nonviable and apoptotic cells [148]. AO intercalates into DNA of viable and nonviable cells and makes it appear green. It also binds to RNA, but because it can not intercalate, the RNA stains red-orange. EB is only taken by nonviable cells, intercalates into DNA making it appear orange. Thus, dead cells have bright orange chromatin, as EB overwhelms AO. Apoptotic nuclei can be detected by highly condensed chromatin and uniformly staining, and the overall nuclear brightness is less than that of a normal cell. The dye mix (100 μg/ml AO and 100 μg/m EB) was prepared in phosphate buffered saline and placed at the bottom of a glass tube, where 25 μl of cell suspension at $5x10^5$ – $5x10^6$ cells/ml in complete medium was added and gently mixed. The mixture was placed (10 μl) on a microscopic slide and examined with a 40x – 60x dry objective.

The enzymatic activity of caspase-3 was determined by using chromogenic substrate Ac-DEVD-pNA, containing the specific cleavage site (DEVD), linked to p-nitroanilide (Caspase-3 Assay kit, Sigma, USA). Cell lysate of untreated cells or cells treated with chromium (25 μg of the total protein) were mixed with chromogenic substrate and the absorption at 405 nm was taken as specific caspase activity.

DNA fragmentation into oligonucleosomes was used as a marker of apoptosis. DNA fragmentation was estimated by electrophoresis of extracted DNA on 1.5 % agarose gel in 1xTAE (40 mM Tris-acetate, 1mM EDTA) buffer, and visualized by ethidium bromide staining [149].

Measurement of ROS

The detection of ROS in living cells was carried out by the flow cytometry method using 2'-7'-dichlorofluorescein diacetate (DCFH-DA, Sigma, USA). Cell-permeant DCFH-DA is oxidized in live cells to its fluorescent derivative 2'-7'-dichlorofluorescein (DCF) in the presence of ROS (predominantly hydrogen peroxide and partially •OH and •NO radicals). ROS detection is proportional to an increase in the basal level of the probe [25]. Accumulation of DCF was measured by an increase of fluorescence at 530 nm and the mean fluorescent intensity (MFI) was used for estimation of intracellular level of ROS. DCFH-DA was added (10 μM final concentration) to L-41 cells (about $1x10^6$ cells) and the mixture was incubated for 30 min at 37°C. Hydrogen peroxide (10 mM) was used as a positive control. Then the incubation cells were subjected to flow cytometry analysis (FACScan, Becton Dickinson, USA).

Mitochondrial Membrane Potential Assessment

Mitochondrial potential in situ, is an important indicator of mitochondrial function and dysfunction. We have used a flow cytometry method for the quantitative measurement of mitochondrial potential utilizing the lipophilic cationic fluorescent probe 3,3'-dihexyloxacarbocyanine iodide - $DiOC_6$. Flow cytometry in conjunction with a fluorescent probe provides determination of the magnitude of mitochondrial membrane potential in a large number of individual cells of heterogeneous cell populations. The decrease of $DiOC_6$ uptake is observed upon the reduction of mitochondrial membrane potential. The assay requires a relatively small number of cells (10^3-10^5) that are equilibrated with very low dye concentrations (less than 1 μM) [150,151]. For the estimation of mitochondrial membrane potential 20 μl of 0.2 μM solution of $DiOC_6$ was added per $1x10^5$ cells for 15 min at 37°C and analyzed by flow cytometry using FACScan instrument (Becton Dickinson, USA). Excitation of $DiOC_6$ was at 488 nm, emission was measured at 530 nm.

Preparation of Crude Cell Extract

The catalase and SOD activities, protein expression were investigated in crude cell extract. Cells (~10^7) were harvested by centrifugation (3,000 rpm, 5 min, 4°C), rinsed twice in 50 mM phosphate buffer, pH 7.8. The rinsed cells were re-suspended in a definite volume of above-mentioned buffer 1:4 (w/v), sonicated five times for 10 sec bursts (44 kHz), centrifuged (14,000 rpm, 20 min, 4°C), and the soluble extract was used as a sample. The crude cell extract was obtained by the treatment of the cell pellet with lysis buffer M-PER® Mammalian Protein Extraction Reagent (Pierce, USA) following the manufacturer instructions as well. The crude cell extracts were standardized per microgram of total protein. Protein concentrations in the cell extract were determined using BCA protein assay reagent (Pierce, USA).

Glutathione Antioxidant System

Glutathione concentration was determined by using BIOXYTECH GSH/GSSG-412™ colorimetric assay (Oxis, USA). The assay uses Ellman's chromogenic reagent 5,5'-dithiobis-2-nitrobenzoic acid (DTNB), which reacts with GSH to form spectrophotometrically detectable product at 412 nm. GSSG existing in the sample is reduced to GSH by GR assay followed by the reaction with Ellman's reagent. The method allows determining GSH_t. The second reagent used in this assay 1-methyl-2-vinylpyridinium trifluoromethanesulfonate (M2VP) scavenges GSH, but does not interfere with GR assay. The remained GSSG is estimated by procedure described above. The method allows determining an origin GSSG content in the sample. Glutathione reductase (GR) activity was measured by using the BIOXYTECH GR-340™Assay (Oxis, USA). The assay is based on the oxidation of NADPH to $NADP^+$ by GR. The reduction of GSSG is determined indirectly by the measurement of the consumption of NADPH as demonstrated by a decrease in absorbance at 340 nm as a function of time, ε_{NADPH}=6620 $M^{-1}cm^{-1}$. Glutathione peroxidase (GPx) activity was determined by using BIOXYTECH GPx-340™ colorimetric assay for cellular glutathione peroxidase (Oxis, USA). The assay is based on the indirect measure of the GPx activity. Oxidized glutathione (GSSG) produced upon reduction of hydrogen peroxide by GPx is recycled to its reduced state by the enzyme glutathione reductase (GR). The oxidation of NADPH to $NADP^+$ was measured by procedure described above.

SOD-in Gel Assay

The technique involves photoreduction of nitro blue tetrazolium (NBT) for the determination of activity of superoxide dismutase following native PAGE (polyacrylamide gel electrophoresis). NBT substrate is incorporated into polyacrylamide gels by adding the superoxide anions inducing agent riboflavin. The protein corresponding to SOD can be then visualized as an achromatic zone through the inhibition of NBT reduction via SOD. Crude extract was mixed with the loading buffer in the absence of 2-mercaptoethanol and SDS and loaded onto the gel without heating. Following electrophoresis in 12.5% gels for 3 h at 30 mA, the gels were subjected to a two-step staining procedure described by Steinman [152]. The gels were soaked in the solution, containing 28 μM riboflavin, 28 mM TEMED in 36 mM potassium phosphate buffer, pH 7.0 for 20 min and then in 2.45 mM NBT for 10 min at 25°C.

The positions of the two isozymes of SOD, Cu,ZnSOD and MnSOD were identified by incubation of the cell lysate at 37°C with 2% (w/v) sodium dodecyl sulfate (SDS) for selective inactivation of MnSOD [153]. After incubation, the upper achromatic zone was identified as MnSOD and the lower as Cu,ZnSOD.

Catalase Activity

Catalase activity in the crude cell extract was determined by measuring the rate of H_2O_2 (10 mM) decomposition in 50 mM potassium phosphate buffer (pH 7.0) in the presence of the cell crude extract at 240 nm and 25°C, $\varepsilon_{H_2O_2}$=43.6 M^{-1} cm^{-1} [154].

Werstern Blot Analysis

Sodium dodecyl sulfate polyacrylamide gel electrophoresis (SDS-PAGE) was performed by the method of Laemmli [155], using 15 % gels. 50 μg of the total protein was run per lane. The resolved proteins were transferred to a nitrocellulose membrane. A western blot assay was performed using antibodies against Bax, actin and secondary anti-rabbit IgG (Sigma, USA). The reaction was with ECF substrate (Santa-Cruz, USA). Actin expression served as the loading control.

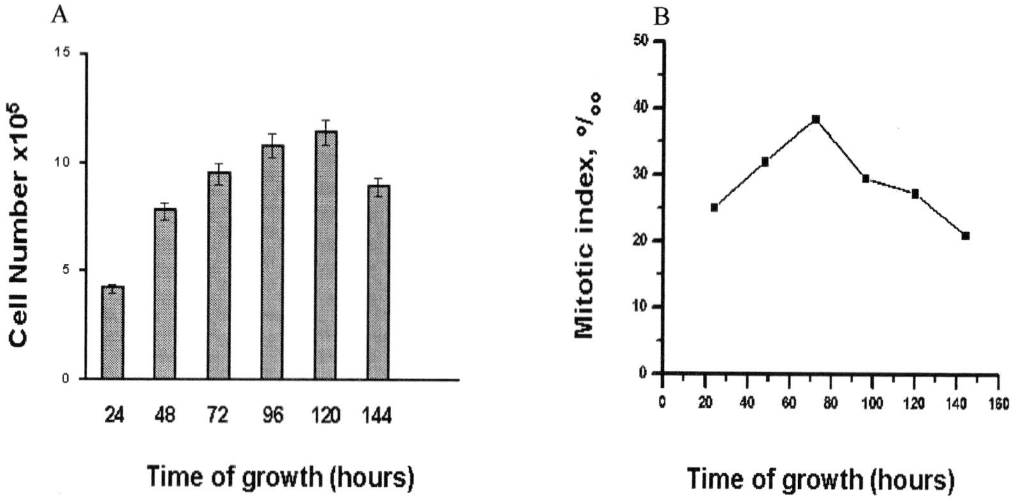

Figure 1. Time-dependent growth characteristics of human L-41 epithelial-like cultured cells. L-41 cells were grown in culture tubes at concentration of $3x10^5$ per tube for 144 h without medium replenishment. The viable cell number was determined by hemocytometer using trypan blue exclusion assay at different time points (from 24 h to 144 h). For mitotic index estimation cells were grown on cover slips in 24-well plates. After fixation in ethanol, glacial acetic acid (3:1) cells were stained with Giemsa and analysed by microscope. Mitotic index (Mi) was scored as the number of mitosis per 1000 cells. (A) Time-dependent cell growth estimated by the viable cell number. (B) Cell proliferation capacity estimated by the number of mitotic cells.

Statistical Analysis

Experiments were carried out at least in triplicate unless otherwise stated. All values were expressed as the mean ± SD and analyzed using one-way ANOVA with Scheffe's test. A p-value less than 0.05 was considered statistically significant.

RESULTS

Low-dose and Low-dose Rate Chromium Action on the Asynchronous Human Epithelial-like L-41 Cells

3.1.1. Cr(VI)-induced Cytotoxicity

The human epithelial-like L-41 cell line was used to study cell responses to Cr(VI) treatment. The studied cell line is characterized by high proliferate capacity. Mitotic index (Mi,$^o/_{oo}$) of L-41 cells reaches about 35-40 $^o/_{oo}$ at 48-72 h of cultivation. The sub-confluent state of attached cell population (80% of confluence) is achieved at 48 h. The number of viable cells gradually decreases after 120 h due to culture medium depletion (Figure 1). Flow cytometry analysis demonstrates a rather stable cell cycle phase distribution in the logarithmic growth phase (48 and 72 h of growth), and the polyploid cell fractions never exceed 3 % in control cultures (Figure 2). The distinct increase of the sub-G_1 peak of the hypodiploid DNA at the late growth stages (up 72 h) indicates to the presence of the apoptotic cells in the culture [143].

To establish the time-dose relationship of Cr(VI) toxicity for the studied cell line, cells were exposed to Cr(VI) as potassium chromate at the concentration range of 2 - 200 μM (2, 5, 10, 20, 50, 100, 150 and 200 μM). Chromium was administered to the sub-confluent cell culture for 2, 6 and 24 h and the MTT-cytotoxicity assay was carried out. As it was demonstrated (Figure 3A), the concentration of 2 μM did not reveal cytotoxicity by the MTT assessment during the studied exposure periods. The Cr(VI) concentrations above 5 μM affected the growth of L-41 cells. The concentration of 20 μM decreased cell viability to 50% of control value after 24 h exposure to chromate. The Cr(VI) concentrations of 50 μM up to 200 μM decreased the absorbance at 570 nm compared to control in dose- and time-dependent manner causing acute toxic effect already after 2 and 6 h of Cr (VI) exposure.

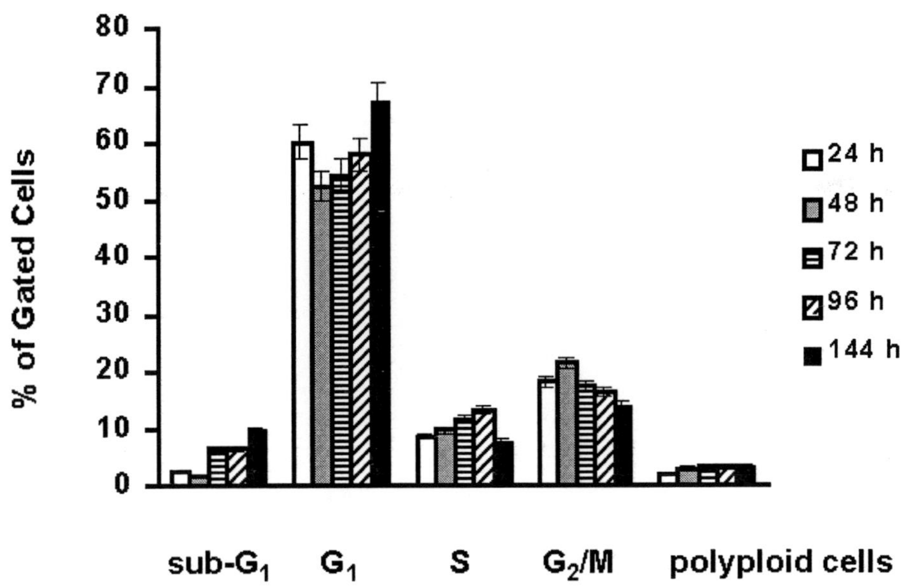

Figure 2. Time-dependent effect of L-41 cell growth without medium replenishment on cell cycle progression and polyploid cell formation.

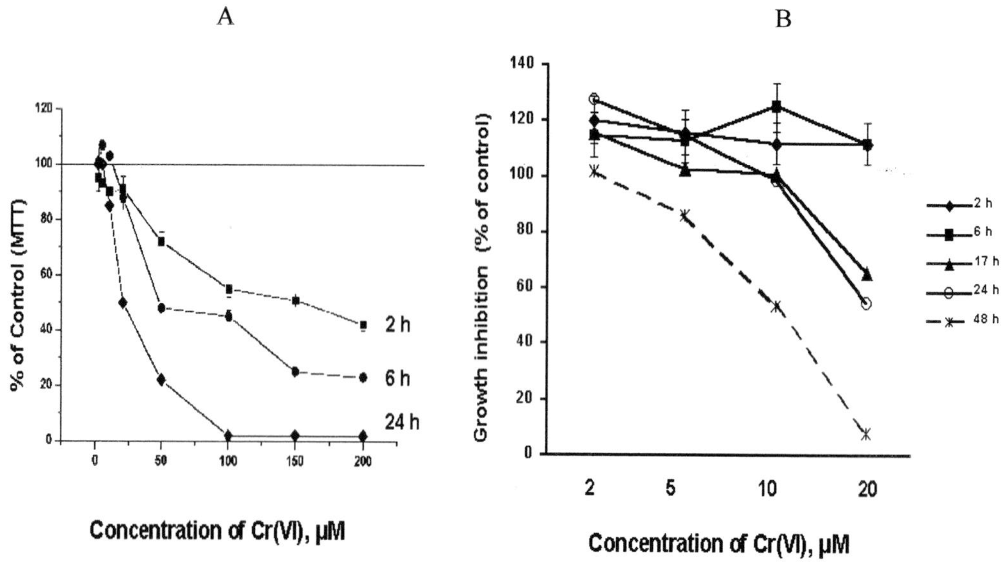

Figure. 3. Time- and dose-dependent Cr(VI)-induced cytotoxicity by MTT assessment. (A) L-41 cells were grown in 96-well plates (3 x 10^4 per well) for 48 h (sub-confluent monolayer) prior to chromium treatment at the concentration range of 2-200 μM for different time periods: 2, 6 and 24 h. (B) The L-41 cells were grown for 48 h of permanent Cr(VI) action (dashed line) at the concentration range 2-20 μM Cr(VI). After chromium treatment for different time periods (2, 6, 17, and 24 h) the cultures were rinsed with phosphate-buffered saline and replaced to fresh medium up to 48 h prior to the measurment (solid lines).

Besides, we have examined the time-course action of Cr(VI) on the cell viability by MTT-cytotoxicity assay in the concentration range of 2-20 μM after a 48 h exposure without medium replenishment and the cell recovery after Cr(VI) administration for 2, 6, 17, and 24 h followed by the cell growth in complete medium without chromium up to 48 h. It was revealed that the action of 5 and 10 μM Cr(VI) was successfully recovered even after 24 h of permanent cell exposure to chromium. The effect of 20 μM Cr(VI) was already irreversible after 17 h of cell treatment with chromium (Figure 3B).

Cr(VI) doses of 2 to 20 μM were used for further studies of cell responses to chromate treatment regarding 2 μM as nontoxic and 20 μM as toxic and the cutoff value of chromium concentration, above which Cr(VI) exposure resulted in acute toxicity.

3.1.2. Cell Growth Dynamics Study at Cr(VI) Exposure

To verify that Cr(VI) affects the cell growth and not the decreased metabolic rate of the cells by the MTT assessment of chromium toxicity, direct viable cell counting was also done by hemocytometer using trypan blue exclusion assay. The measurements had been performed at various time points from 6 to 96 h after chromium administration into the growth media at the Cr(VI) dose range of 1-20 μM. It was demonstrated (Figure 4) that low nontoxic concentration (2 μM) caused the cell population growth perturbation at the initial period of Cr(VI) treatment (up to 48 h) that was followed by the restoration of the population cell number up to control level (72 h). The growth retardation at the low nontoxic concentrations of Cr(VI) was not accompanied by the loss of cell viability as was assessed by trypan blue staining. The apoptotic cells also were not registered morphologically.

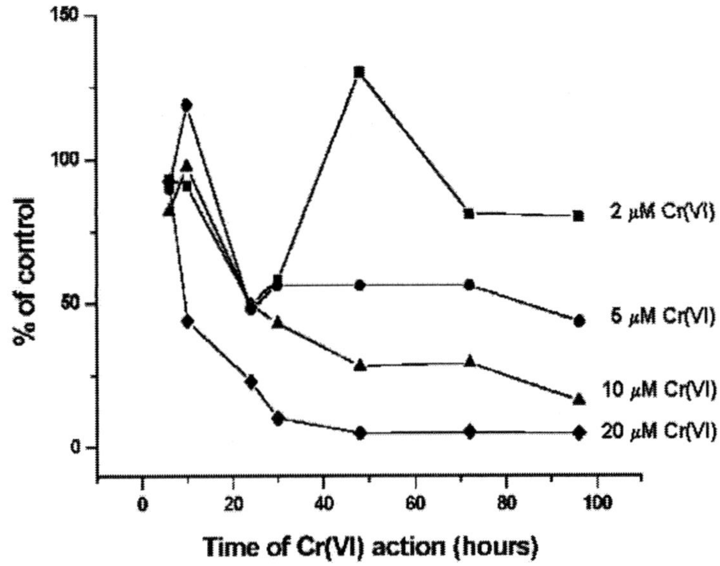

Figure 4. Time- and dose-dependent Cr(VI)-induced cell growth inhibition. L-41 cells were grown in culture tubes at concentration of 3 x 10^5 per tube for 48 h (sub-confluent monolayer) before Cr(VI) exposure. The total cell number was determined by hemocytometer in five replicate culture tubes at different time points after chromium introduction into the growth media. Dead cells were excluded by trypan blue staining.

The Cr(VI) dose of 5 μM (Figure 4) analogically disturbed the cell growth, but the resultant cell population number at 96 h was significantly decreased relevant to control (to 40 %).

The Cr(VI) dose of 10 μM also revealed the time-dependent growth inhibition, reducing cell population number to 10% of control at 96 h.

The toxic concentration of 20 μM irreversibly decreased the cell number already at 48 h, enhancing cell detachment from the monolayer. The Cr(VI) concentration range that caused loss of cell viability paralleled the results obtained with the MTT assay. The direct cell number study at different time points of Cr(VI) exposure permitted discrimination of the studied Cr(VI) concentrations as the nontoxic dose (2 μM) that caused the transient growth arrest and the toxic one (20 μM) resulted in cell death.

3.1.3. Apoptosis as the Defense Mechanism against the Damaging Chromium Action

The time-course action of 2 and 20 μM Cr(VI) was analyzed by microscope up to 96 h. It was demonstrated that 2 μM Cr(VI) did not affect cell viability and induce of apoptosis (Figure 5A). The time-dependent loss of cell viability at the cell exposure to 20 μM Cr(VI) was accompanied by the increased level of apoptotic cells. The cell treatment with 20 μM Cr(VI) (Figure 5B) increased apoptotic cells fraction determined by morphological changes consistent with apoptosis up to 20-30% of cell number at 24 h, and up to 80-90% at 48 h of Cr(VI) action..

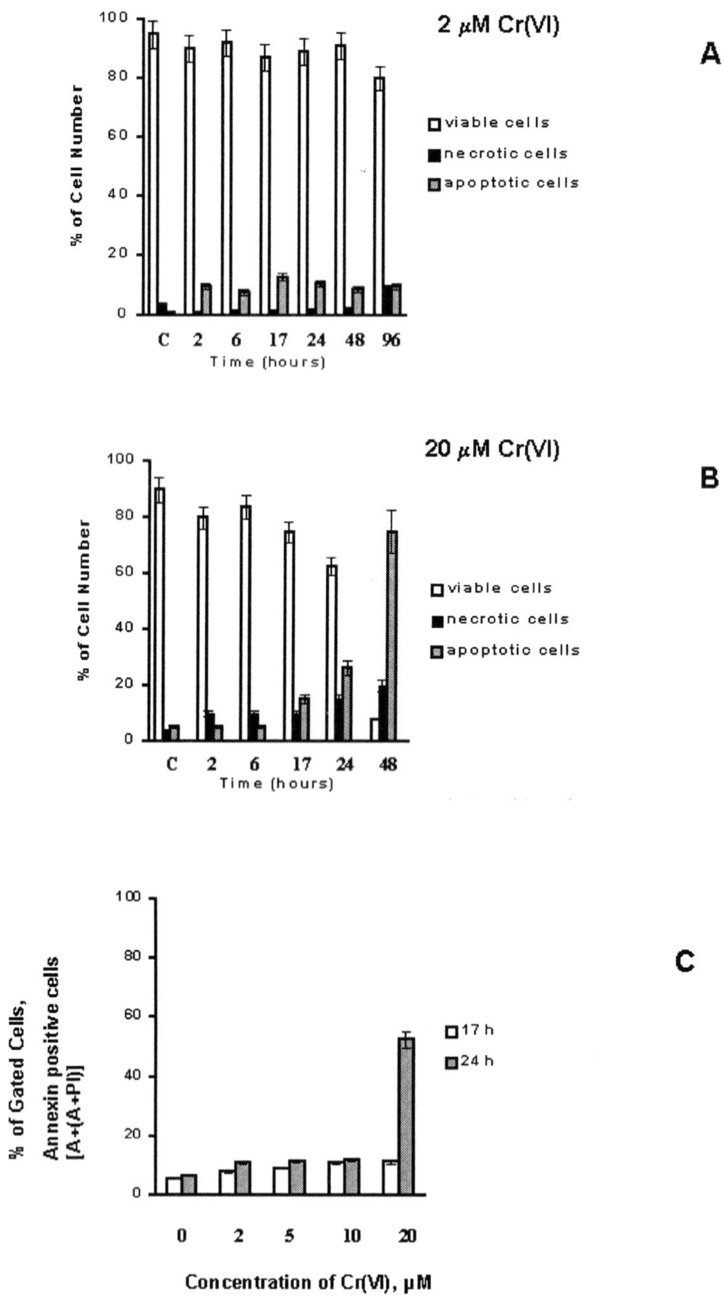

Figure 5. Effect of Cr(VI) on cell survival after treatment of L-41 cells with various chromium concentrations for different time periods. (A,B) Viability of L-41 cells under 2 and 20 μM Cr(VI) action respectively was determined by morphological analysis as described in Experimental. The data presented are mean values ± SD from five separate sets of experiments and significant difference ($p<0.05$) was obtained between the groups of cells under 20 μM Cr(VI) action for 17 h, 24 h, and 48 h. (C) Flow cytometry assessment of apoptosis at AnnexinV/PI staining. L-41 cells were grown for 48 h in 25 cm^2 culture flasks (sub-confluent monolayer) prior to 17 and 24 h of Cr(VI) treatment at concentration range of 2-20 μM. The data are presented as % of gated cells at flow cytometry analysis: A - AnnexinV- positive cells (early apoptosis); A+PI -AnnexinV/PI- double positive cells (late apoptosis). Results are the mean of three experiments ± SD.

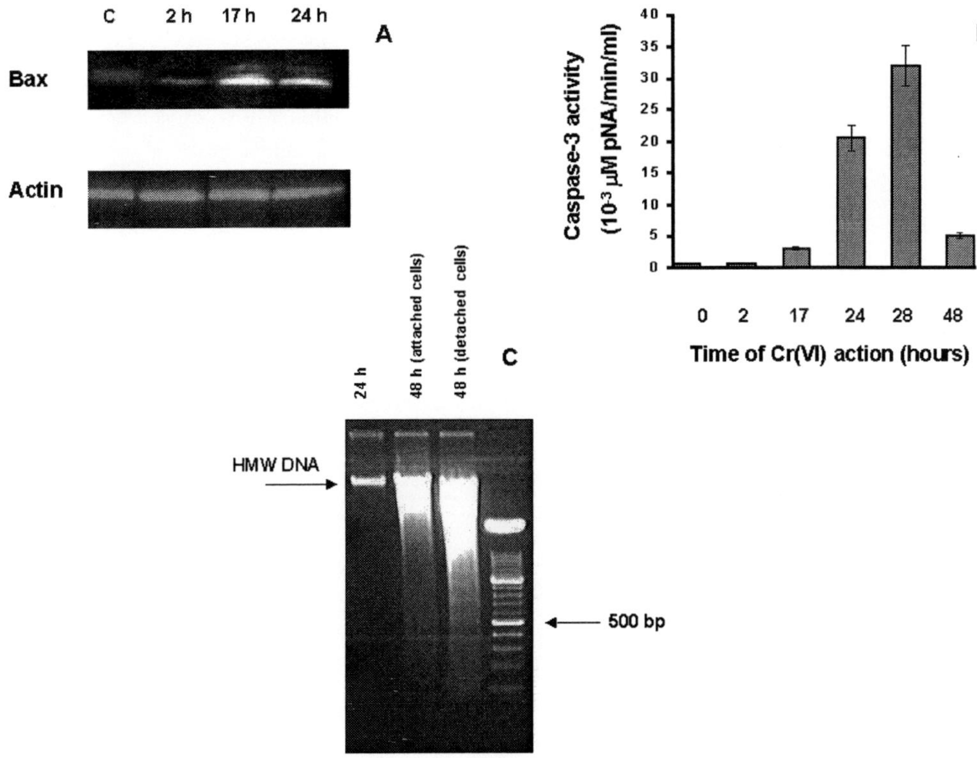

Figure 6. Apoptosis analysis of the L-41 cells under 20 μM Cr(VI) action. (A) Expression of Bax protein in L-41 cells after 20 μM Cr (VI) treatment for the appropriate times. Immunoblots were probed with anti-Bax antibody and developed by using chemilumeniscent detection. Actin expression served as an internal control for protein loading. (B) Caspase-3 activity after chromium treatment over a time course of 2- 48 h. Caspase-3 activity expressed by mean ± SD (n=3) was largely increased when L-41 cells were exposed to 20 μM Cr(VI) for 24 h ($p<0.05$). (C) DNA extracted from Cr(VI)-treated (20 μM) L-41 cells for 24 h and 48 h. After treatment for 48 h attached and detached L-41 cells display the classical laddering degradation of DNA. 100 bp DNA ladder (Bio-Rad, USA) was used as a DNA marker.

The flow cytometry analysis of AnnexinV/PI stained cells (Figure 5C) also demonstrated that 24 h exposure to Cr(VI) at the concentration of 20 μM caused the enhanced cell death by the mode of apoptosis, as the fraction of AnnexinV-positive cells increased up to 50% relevant to control.

The other apoptotic marker, namely, over-expression of Bax, a protein promoting apoptosis was estimated by Western blot analysis. The expression of Bax in L-41 cells before and after various times of Cr(VI)-treatment are presented in Figure 6A. The results indicate an increase in Bax protein from 17 h of 20 μM Cr(VI) treatment. This protein is under transcriptional control of p53, which is DNA-damage sensitive. Bax translocates to the mitochondria and induces cell death by mitochondrial dysfunction, by favoring the instability of mitochondrial membrane [156].

The chromium-induced apoptotic cell death under 20 μM chromium was further confirmed by caspase-3 activation. Caspase-3 is an executioner caspase, and is indispensable for apoptotic chromatin condensation and DNA fragmentation [157]. The activity of caspase-3 was examined in time-dependent manner (Figure 6B). The results have revealed that caspase-3 activity begins to increase at 17 h and increases significantly at 24 h of the permanent incubation of L-41 cells with chromium.

The Bax protein increase and the activation of caspase-3 from 17 h of 20 μM Cr(VI) treatment coincided with the irreversibility of chromium impact beginning from 17 h as it had been demonstrated by the MTT assay (Figure 3B).

DNA fragmentation caused the appearance of a 48 h of exposure of L-41 cells to 20 μM Cr(VI) in still attached as well as detached apoptotic cells (Figure 6C).

Our data are consistent with published observations regarding the cell death via apoptosis in the toxicity of Cr(VI) [158,159]. Cells treated with Cr(VI) exhibit apoptotic features, depending on cell line, dose and exposure time; 300 μM Cr(VI) after 3 h of exposure causes apoptosis in 24 h in human lung tumor A549 cells [159], a 24 h exposure to 12.5 μM Cr(VI) is apoptogenic for chronic myelogenous leukemic K562 cells, 0.6 μM Cr(VI) is toxic for cultured J774A.1 murine macrophage cells after 48 h [158].

3.1.4. Effect of Cr(VI) on DNA Synthesis

The evaluation of DNA synthesis was performed by the measurement of BrdU incorporation into the newly synthesized DNA during its 2 h pulse incubation with cells using the colorimetric immunoassay. The measurement of the absorbance at 450 nm generally quantifies the level of cell proliferation and depends on the population capacity to proliferate, the changed cell number under the stress treatment, and the level of DNA synthesis itself. The impact of Cr(VI) exposure on the BrdU incorporation at the concentration range of 2 to 100 μM (2, 5, 10, 20 and 100 μM) was detected at 2, 4, 6, 24 and 48 h after chromium administration into the culture media. The low nontoxic dose of chromium (2 μM) that caused the transient growth arrest did not affect the DNA synthesis compared to the control over the studied period according the parameter of BrdU incorporation. The toxic apoptogenic concentration (20 μM) inhibited DNA replication within the first hours of the Cr(VI) treatment (Figure 7). Thus, the BrdU incorporation decreased in 2 h by 10% and in 6 h by 50%. At this time point (6 h) the 10%-20% decrease was also detected for the intermediate low toxic concentrations of 5 and 10 μM. The acute toxic dose of 100 μM permanently inhibited the BrdU incorporation to 50% after 6 h of Cr(VI) exposure. The dose-dependent decrease of DNA synthesis at 6 h of Cr(VI) exposure up to 20 μM could not be influenced by the diminished cell number in the population as 6 h of Cr(VI) treatment at the dose range of 2-20 μM did not reduce it (Figures 3A,4). In our model system the inhibition of DNA replication increased gradually up to 6 h and later restored after 24 h in the case of mild toxic concentrations (5 and 10 μM).

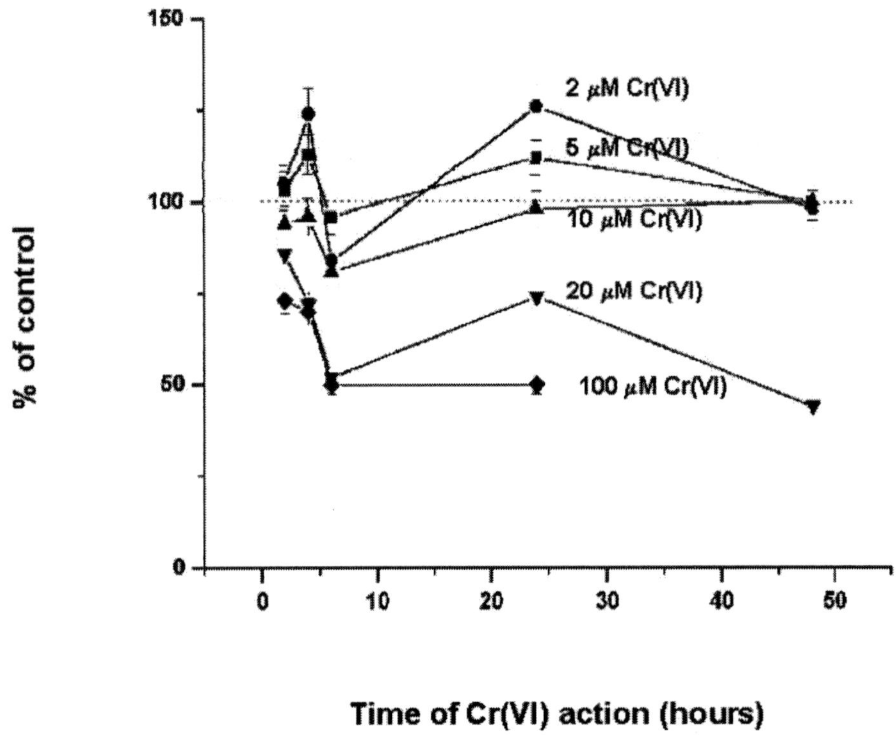

Figure 7. Time- and dose-dependent Cr(VI) effect on DNA synthesis. DNA synthesis was estimated by the colorimetric immunoassay of BrdU incorporation into the newly synthesized DNA at 2 h pulse incubation with cells. Effect of Cr(VI) exposure at concentration range of 2 - 100 μM. L-41 cells were grown in 96-well plates at the concentration of $3x10^4$ per well (200 μl) for 48 h prior to Cr(VI) treatment. Each treatment group was carried out in five replicate wells. Similar results were obtained in three independent sets of experiments. Data are presented as a percentage of the control. Standard deviation of mean (mean±SD) is indicated as error bars.

In normal human lung fibroblasts treated with low doses of Cr(VI), resulting in greater than 90% survival, a dose-dependent inhibition of ^{3}H-thymidine incorporation into DNA was observed immediately after treatment and the blocking DNA replication process was guanine-specific as a result of guanine-guanine DNA interstrand cross-links [6]. It was demonstrated that chromium-DNA adducts were of short half life averaging 7 h for 4 human cell lines [136]. Adducts removal rate was proportional to the initial genotoxic insult, and at low level damage approximately 50% repair was observed after 4 h. The mild toxic concentrations caused the restoration of DNA synthesis after 24 h in Cr-treated human fibroblast and lung epithelial cells that was in concordance with our results.

3.1.5. Effect of Cr(VI) on Cell Cycle Progression

DNA damage may cause cell cycle delay presumably to provide an opportunity for a cell to repair the lesions before replication. Multiple cyclin-dependent kinases (Cdks) regulate the G_1, S and G_2/M phases of the cell cycle, and Cdks are themselves regulated and cooperate with other protein kinases. Checkpoints are mostly related to DNA damage and mitotic progression. Arrest in G_1 prevents aberrant replication of damaged DNA and arrest in G_2

allows cells to avoid segregation of defective chromosomes. Massive DNA damage or deprivation of nucleotides forces a cell to be arrested in S phase [160-164]. It is generally accepted that cell cycle progression alters in response to Cr(VI) exposure influencing the growth of the cultured cell population. Thus, the chromate treatments inhibited cell growth by specific blocking of the cell progression through S phase of the cell cycle in dose-dependent character [6].

It had been demonstrated that human lung fibroblasts (HLF) in response to Cr(VI) concentrations of 6 μM for 24 h, followed by release into complete medium, induced an initial G_1 arrest, and subsequent S phase delay, while 10 μM exposure induced S phase arrest in HLF cells 24-48 h later after chromium removal. Low to moderate levels of DNA damage resulted in transient cell cycle checkpoint arrest in HLF cell culture, and higher concentrations caused irrepairable DNA damage and cell cycle blockade from which the majority of cells could not emerge and regain replication potential leading to apoptosis or to terminal growth arrest [165]. In the human lung epithelial cell culture A549 Cr(VI) at low concentrations was able to cause cell growth arrest at G_2/M in dose-dependent manner while higher concentrations (more than 25 μM) induced apoptosis, and ROS generated by Cr(VI)-stimulated cells were involved in Cr(VI)-induced cell growth arrest [166]. Also it had been demonstrated that mild intracellular redox imbalance itself could inhibit cell proliferation of CaCo2 (the human colon cancer cell line) arresting cells at G_1/S and G_2/M phases [94].

In the current study the applied dynamic flow cytometry analysis provided the ability to follow the changes of the cell cycle phase distribution along with chromium exposure. To examine whether the transient loss of cell number at low level Cr(VI) exposure was due to the inhibition of cell cycle progression in L-41 cells, the DNA levels at various phases of cell cycle was measured using flow cytometry analysis. The DNA histogram shows the relative number of cells in G_1/G_0, S and G_2/M phases of the cell cycle. The dose-dependent modulation of cell cycle progression at chromium exposure for 24 h is presented in Figure 8. The cell treatment with Cr(VI) at the low concentration of up to 20 μM increased the cell population of S phase relevant to control from 4.3% to 10.4%, 12.8%, and 15.3% subsequently for 2, 5 and 10 μM. The toxic dose of 20 μM increased the S phase cell number to 9.7% and decreased the cell number in G_2/M cells from 12.6% (control) to 6.1%. The cell increase in G_2/M phase was detected only in the case of nontoxic concentration of 2 μM from 12.6% to 17%. The obtained data gave basically static information for the single time point of chromium exposure detecting only the fact of the disorders in the cell cycle phase distribution at stress conditions.

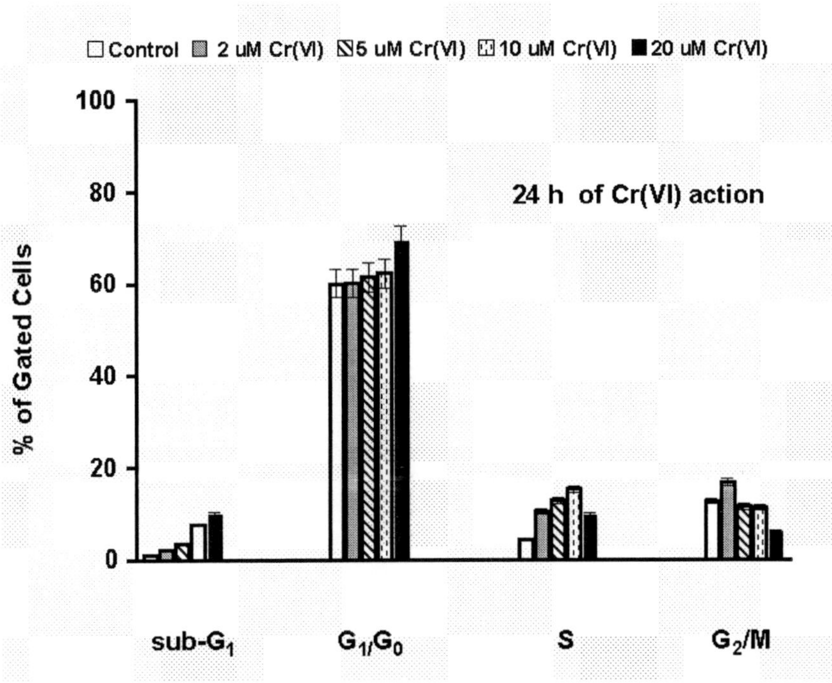

Figure 8. Dose-dependent effect of Cr(VI) exposure for 24 h on cell cycle progression. L-41 cells were plated in a 24 well plate at $1x10^5$/ml concentration. Cells were grown for 48 h (sub-confluent monolayer) and exposed to Cr(VI) for 24 h at concentrations 2, 5, 10 and 20 μM. After Cr(VI) treatment cells were trypsinized, cell concentration adjusted to $1x10^6$/ml for each sample, and routinely prepared for flow cytometry analysis using propidium iodide. All results are expressed as the mean of percentage of the cells from each cell cycle compartment of three experiments (mean ± SD).

The recording of DNA histogram at Cr(VI) treatment with time permits following the temporal changes in the cell cycle phase distribution at the transient growth arrest of nonapoptotic cells and at the irreversible growth inhibition of cells committed to apoptosis. Cell cycle phase distribution in response to cell exposure to the nontoxic Cr(VI) concentration of 2 μM and the apoptogenic concentration of 20 μM in the time course (Figure 9) considerably differed. The cell exposure to 2 μM Cr(VI) for 10 h produced a temporary cell number increase of S phase fraction (1.7 fold of control), at 24 h cells were increased both in S (1.5 fold of control) and G_2/M (1.35 fold of control) compartments, later (30 h) cell cycle distribution did not differ from control. In the long-term culture (96-144 h) the fraction of G_2/M cells increased in the time course for about the same extent as had been shown for S phase in 10 h of Cr(VI) exposure. It may be proposed that the nontoxic concentration initially arrested cells in S phase and then slowed down cell movement through the cell cycle. The toxic apoptogenic concentration of 20 μM induced within 2 h significant cell arrest in S phase (about 2.5 fold of control). The cell arrest in G_2/M phase was not detected at the studied time-points, moreover this cell fraction decreased to 40% of control after 30 h of treatment. Mitotic cells were not found in the population after 17 h of Cr(VI) exposure to 20 μM (Figure 10), which confirmed the persistent cell delay in S phase compartment.

A B

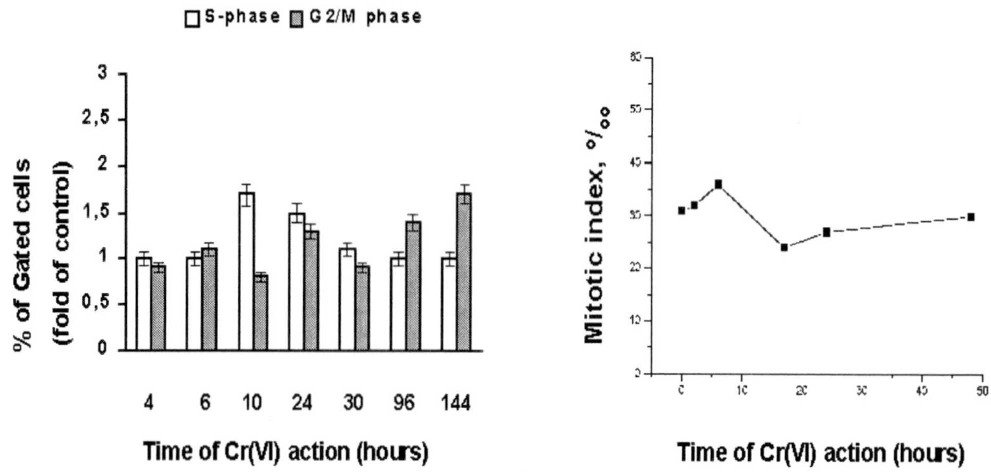

Figure 9. Time-course effect of Cr(VI) exposure on cell cycle progression. (A) Effect of 2 μM chromium treatment on cell cycle phase distribution. Data are presented as percentage of the cells in S and G_2/M phases (fold of control) of three independent experiments. Standard deviation of mean (mean±SD) is indicated by error bars. (B) The dynamics of mitotic index (Mi, $^{o}/_{oo}$).

A B

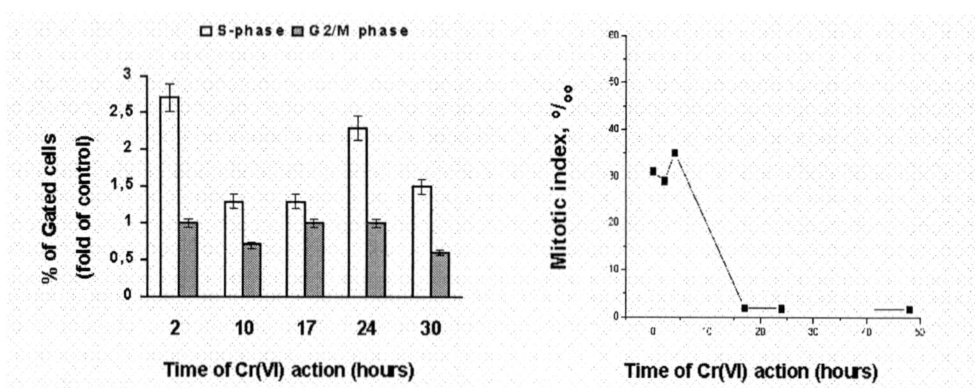

Figure 10. Time-course effect of Cr(VI) exposure on cell cycle progression. (A) Effect of 20 μM chromium treatment on cell cycle phase distribution. Data are presented as percentage of the cells in S and G_2/M phases (fold of control) of three independent experiments. Standard deviation of mean (mean±SD) is indicated by error bars. (B) The dynamics of mitotic index (Mi, $^{o}/_{oo}$).

It may be supposed, that transient cell growth arrests at the nontoxic chromium concentration (2 μM) impact was related to the temporal inhibition of the cell cycle progression. The initial increase of S phase cell number was followed by the increase of G_2/M compartment. The toxic Cr(VI) dose of 20 μM irreversibly delayed cells in S phase of the cell cycle and was accompanied by the inhibition of DNA synthesis (Figure 7). The S phase arrest and the inhibition of DNA synthesis were detected at 2 h of chromium exposure, whereas the reduced growth and apoptosis were found later after 24 h of Cr(VI) treatment. The time scale for the onset of apoptosis was more protracted than for the induction of disturbances in cell cycle progression. The toxic concentration caused permanent S phase delay of certain biphasic character with two peaks in 2 h and 24 h after chromium administration into the growth media. It may be supposed that the first increase (2 h) was connected with the initial inhibition of DNA synthesis, and the second one reflected the failed DNA repair process and coincided with the development of apoptosis. Previously we have demonstrated that 20 μM caused the increase of caspase-3 activity in 24 h of Cr(VI) exposure. The microscopic analysis provided the detection of the disappearance of mitotic cells that also indicated a profound cell cycle arrest (Figure 10B). The activated caspase-3 during apoptosis may be implicated in proteolysis of a number of cell cycle regulatory or machinery proteins, such as mitotic spindle proteins and proteins of mitotic checkpoint system [163,167-170].

The nontoxic concentration (2 μM) Cr(VI)) affected the cell cycle phase distribution later after 10 h of Cr(VI) exposure and temporarily increased the S phase cell number followed subsequently by the transient cell increase in G_2/M compartment.

The current study suggests that the chromium doses that caused temporary cytostasis of the cell population, allowed the retention of the coordinated cell reactions to stress as demonstrated by the transient cell cycle and growth arrest in response to Cr(VI) treatment.

The toxic concentrations upregulated the cell defense systems, irreversibly caused growth arrest, and massive cell death via apoptosis, but some cells that survived after 24 h of chromium treatment, could enter additional cell cycles.

3.1.6. Effect of Cr (VI) on Long-term Cell Survival

Long-term survival and regained replicative potential as an outcome of a genotoxin exposure is estimated by a standard clonogenic assay. Cr(VI) exposure caused a dose-dependent decrease in clonogenicity in various cell culture lines [65,104,165]. In normal human fibroblasts (BJ) immortalized by human telomerase gene transfection (BJ-hTERT) cumulative cellular fates over time (apoptosis, terminal growth arrest or clonogenic survival) depended on the extent of DNA damage within population exposed to a narrow range of increasing doses of Cr(VI) [165]. In this study we applied flow cytometry and morphological analysis to characterize the survived sub-populations after long-term chromium exposure. The cell survival under Cr(VI) exposure to 2 μM and 20 μM was examined in 10 days after 24 h chromium action. The Cr(VI) treatment (24 h) was followed by the cell growth in complete media without chromium for 72 h, then cells were trypsinized, reseeded at the concentration of 10^3/ml for additional 7 days and applied to cell morphology (fluorescent microscope, AO staining) and flow cytometry analysis. The toxic concentration of 20 μM caused extensive cell death via apoptosis in 72 h after chromium removing from the growth media (data not shown), as the majority of cell population retained the intact DNA-damage responses. The low levels of surviving cells were able to continue growth. The resulting cell population (7 days after reseeding) was characterized by the high degree of polyploid giant multinuclear and apoptotic cells (Figure 11). The cells with the high level of DNA content increased in 10 fold compared to control. The examined increase of polyploid cells was not detected in the case of the nontoxic concentration of 2 μM. The survived cell compartment after 24 h toxic chromium treatment escaped from highly regulated elimination of damaged cells by apoptosis and terminal growth arrest. The disorders in control of apoptosis and cell cycle checkpoint pathways may permit the survival or the growth of cells with genomic abnormalities, enhancing the chance of malignant transformation [170-172].

3.2. Intracellular Antioxidant Defense System of the Asynchronous Human Epithelial-like L-41 Cells as an Important Determinant for Low Levels of DNA Damage

3.2.1 Effect of Cr(VI) on ROS Generation

The detected disruption of the cell cycle progression and the inhibition of BrdU incorporation in the first hours of Cr(VI) treatment of L-41 cell culture could potentially be the result of the enhanced ROS production that affects signal transduction pathways associated with growth regulation and development of apoptosis [28,173]. The effect of the nontoxic and the apoptogenic concentrations of chromium on ROS production were detected by the fluorescent DCF assay. The applied flow cytometry method in conjunction with DCFH-DA probe enables the production of total ROS at the single cell level to be quantified by enhanced fluorescence of oxidized compound DCF indiscriminating the contribution of any individual ROS component [174]. It generally determines hydrogen peroxide because of its high stability and detectable steady-state level and partially of •OH, •NO radical levels [25].

The measurement of intrinsic cellular ROS level at different time points of growth L-41 cell culture (days) was assessed using the fluorescent DCF assay to characterize unspecific intracellular oxidative stress due to long term incubation of the cell culture without medium replenishment (up to 144 h). The data are presented in Figure 12A as MFI (mean fluorescent intensity) of gated cells (%). The MFI for each time point was rather variable but did not significantly differed from each other from 48 h indicating to the high adaptive capacity of L-41 cell culture.

The data of the measurement of Cr(VI) induced ROS are presented in Figure 12B as percent of MFI compared to control. An initial rapid increase of ROS was detected after 2 h exposure to 2 and 20 μM of Cr(VI) subsequently for 72% and 90%. The enhanced production of ROS was determined also at 24 h retaining its dose-dependent manner - 2 μM and 20 μM caused about 50% and 100% increase of MFI relevant to control. The change of ROS level under 2 μM of Cr(VI) action was normalized at 48 and 96 h, whereas incubation with 20 μM Cr(VI) resulted in a significant second increase of MFI at 48 h of Cr(VI) treatment up to 4.0 fold of control. The initial ROS elevation could be regarded as the consequence of the direct impact of the intracellular chromium reduction. Exposure to the toxic concentration of 20 μM

Cr(VI) resulted in cell death via apoptosis and the detected second increase of ROS production revealed the establishment of apoptotic process.

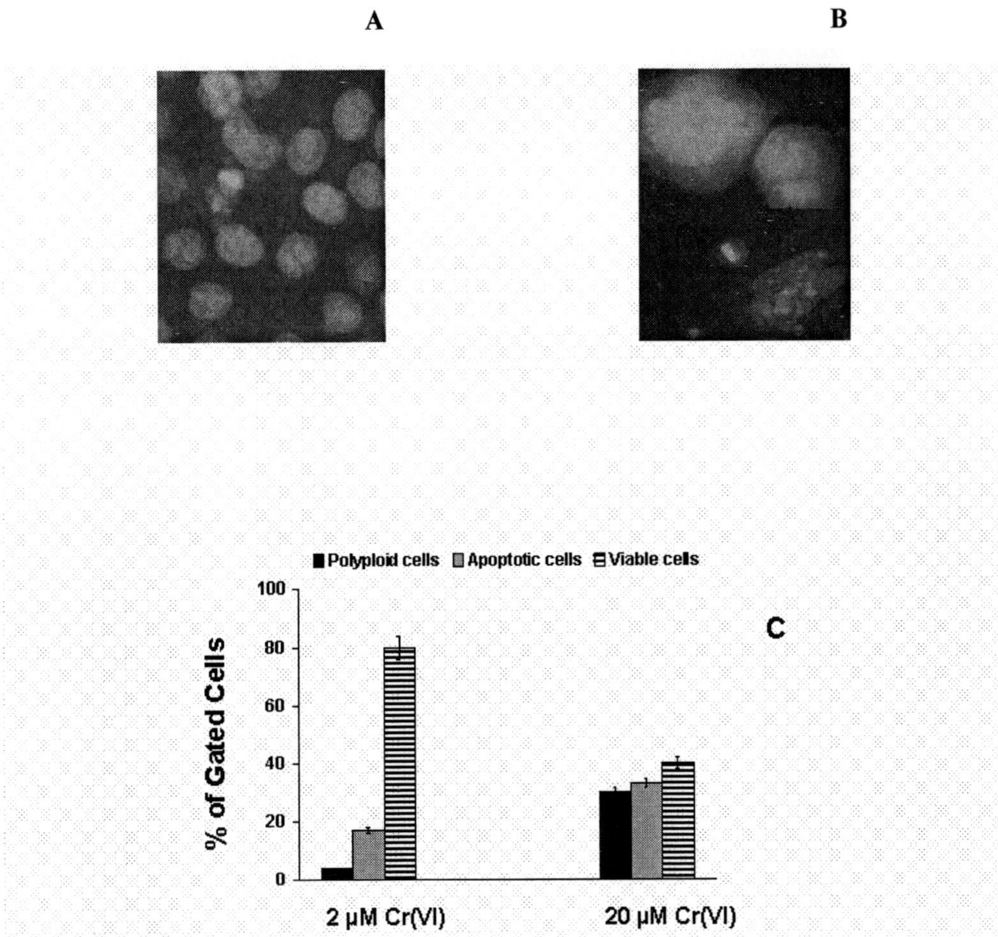

Figure 11. Long-term cell survival after 24 h of Cr(VI) treatment with the nontoxic and the toxic concentrations. L-41 cells were grown for 48 h in 24 well plates (1x10^5/ml). The Cr(VI) treatment (24 h) was followed by the cell growth in fresh medium without chromium for 72 h, then cells were trypsinized and reseeded at the concentration of 10^3/ml for additional 7 days (growth without medium replenishment). Cell population was morphologically analyzed by fluorescent microscope at acridine orange staining (100 μg/ml, prepared in phosphate buffered saline). Cells were grown on cover slips for viewing under microscope (magnification 10x90). (A) Cells treated with 20 μM Cr(VI) for 24 h. (B) Cells grown for 10 days after 24 h treatment of 20 μM Cr(VI) concentration. Shown are the scans of two different photos to demonstrate the giant and mitotic cells in the cell population. (C) Comparative panel of polyploid and viable cells detected by flow cytometry and apoptotic cells detected by fluorescent microscope analysis in resulted cell population (10 days) after 24 h of Cr(VI) treatment with 20 μM. Data are presented as percentage of the control. Standard deviation of mean values (mean ± SD) from three separate sets of experiments is demonstrated as error bars.

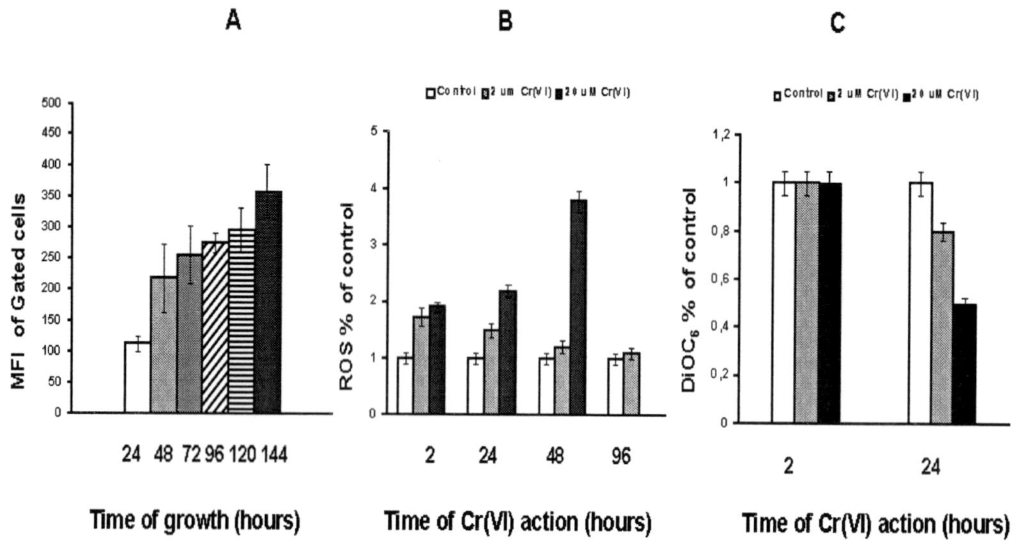

Figure 12. ROS level and mitochondrial damage in control untreated and Cr(VI)-treated L-41 cells in dynamics. (A) Time-dependent effect of L-41 cells growth without medium replenishment on ROS level. ROS were detected by FACScan using DCFH-DA. The mean fluorescent intensity (MFI) was used for ROS estimation. (B) ROS level in Cr(VI)-treated L-41 cells with nontoxic (2 μM) and the toxic (20 μM) concentrations. Data are presented as fold of the control. (C) The mitochondrial membrane potential was examined by FACScan with $DiOC_6$ staining. Data are presented as percentage of the control. The data presented are mean values ± SD from three separate sets of experiments.

The changes of mitochondrial membrane function may play a proapoptotic role, its membrane permeabilization is an early event of apoptosis [51,151], and the mitochondrion is the most common source of ROS during apoptosis [25]. Using a flow cytometry method in conjunction with fluorescent probe $DiOC_6$ we determined the magnitude of mitochondrial membrane potential of the same cell samples at 2 and 24 h of Cr(VI) treatment with concentrations of 2 and 20 μM. The results presented in Figure 12C demonstrated that 2 h exposure with 2 and 20 μM Cr(VI) did not influence the mitochondrial membrane potential, whereas 24 h exposure decreased it in dose-dependent manner for about 20% and 50% in the case of 2 and 20 μM of Cr(VI) respectevely. The ROS generation at 2 h of Cr(VI) exposure could not be connected with the changes of the mitochondrial membrane function assessed by $DiOC_6$ staining. The decrease of mitochondrial membrane potential to 50% relevant to control at 24 h exposure of 20 μM Cr(VI) coincided with 50% increase of Annexin V-positive cells (Figure 5C) and reflected the increased fraction of the apoptotic cells.

Earlier the ability of Cr(VI) to generate $O_2^{\cdot-}$ and H_2O_2 within 30 min treatment had been observed by intracellular staining by hydroethidine, a special fluorescent dye for $O_2^{\cdot-}$, and DCFH-DA using a laser scanning confocal microscope in human lung epithelial cells A549 [159]. The rapid dose-dependent formation of ROS upon Cr(VI) stimulation of A549 cells had been demonstrated by electron spin resonance (ESR) trapping measurements [37,166]

In our study the initial ROS production was generated in a dose-dependent manner at the narrow range of the examined low Cr(VI) doses exposed to L-41 cells (from 2 μM to 20 μM), and the discrepancy in the intensity of fluorescence between 2 and 20 μM did not exceed

20%. The concentration-response relationship of ROS intracellular production in A549 cells after 30 min treatment with potassium dichromate revealed the more pronounced differences in fluorescence intensity at higher Cr(VI) concentrations from 12.5 μM up to 800 μM [175].

3.2.2. Characterization of Endogenous Antioxidant System of L-41 Cells in Time-dependent Manner

The cell possesses a very important capacity to accomplish the nonspecific reactions in response to the impact of normal and pathological surroundings. The biological sense and character of the reactions include mobilization of functional reserves to support cell homeostasis. The cell antioxidant defense system is an important marker of the cell adaptive capacity and thus, plays a substantial role in realization of adaptive mechanisms. Control of the cell redox status, as a measure of cell resistance, can favor the development of the means to defend it against negative factors.

Induction experiments show that bacteria pre-incubated with pure O_2 or with $O_2^{-\cdot}$ generating compounds become more resistant to the toxic effects of ROS [44]. Similarly, mammalian cells and/or animals exposed to the brief hyperoxia or hypoxia, which is connected to ROS elevation, are resistant to oxidative stress due to elevation of antioxidant enzyme activities [176,177].

Thus, in order to characterize the antioxidant defense system of L-41 cells, from the outset we have studied the most important components of antioxidant defense system during long-term cell growth in culture medium without Cr(VI) introduction (Figure 13).

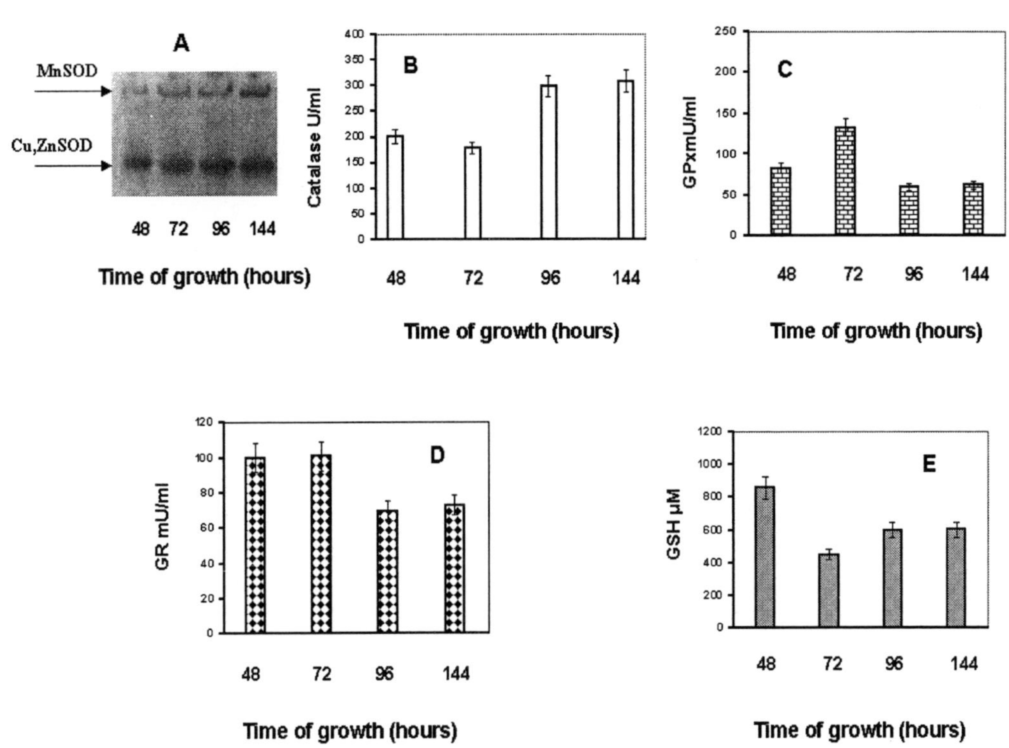

Figure 13. Antioxidant defense system of L-41 cells grown without medium replenishment (controls). The data presented are mean values ± SD from three separate sets of experiments

SOD activity in L-41 cells during the long-term cell growth without chromium action was determined by the intensity of achromatic zones in native polyacrylamide gel. The achromatic zones of SOD positions in the gels are presented in the invert images. The increase of Cu,ZnSOD activity at 72 h of the cell growth and both the MnSOD and Cu,ZnSOD activities at 96 h and 144 h point to the elevation of H_2O_2 level caused by superoxide dismutation. Although catalase and GPx share the substrate H_2O_2, catalase has much lower affinity for H_2O_2 than GPx [24]. GPx is thought to be a very sensitive enzyme to the cell intracellular changes and is a major source of protection against low level of H_2O_2, whereas catalase becomes more significant at the high level of H_2O_2. Thus there are definite correlations between the slightly intensified Cu,ZnSOD activity and slightly heightened GPx (but not catalase) at 72 h of the cell growth and between significantly increased MnSOD and Cu,ZnSOD activities and catalase (but not GPx) level at the 96 h and 144 h of the cell growth (Figure 13 B,C). The amelioration of H_2O_2 by GPx is connected to the inclusion of the components of GSH dependent cycle, GSH and GR. At the late stages of growth (96 h and 144 h) activities of GPx and GR decreased, but balance between GPx and GR were preserved (Figure 13 C,D). At this condition GSH concentration is stable, which shows system homeostasis is supported (Figure 13E).

The key antioxidant enzymes can be interchangeable to support intracellular antiradical defense. The inhibition of one of them can cause the induction synthesis of the others. Besides, relatively freely mobile low molecular antioxidants appear in the foreground to support homeostasis.

3.2.3. Antioxidant Responses during Nontoxic Chromium Action

2 μM chromium (VI) as potassium chromate was administered into the culture medium of L-41 cells grown for 48 h and effect of Cr(VI) long-term action on SOD, catalase, GPx, and GR activities was studied (Figure 14). Enzyme activities were assessed at the early (2 h, 4 h, and 24 h) as well as the late stages (48 h and 96 h) of Cr(VI) action. The total cell growth proceeded up to 144 h as in controls. As it follows from Figure 14A activities of both MnSOD and Cu,ZnSOD (2 h and 4 h) sharply increased compared to control. Simultaneously activity of GPx (but not catalase) augmented that correlate with not strong, but the equal diminution of GR activities (Figure 14B,C,D).

It seems at these conditions 24 h of 2 μM Cr(VI) action is a critical point for activities of studied antioxidant enzymes. Namely, Cu,ZnSOD diminished essentially in parallel with essential increase of catalase activity. At the same time GPx reached its maximal level and GR began to augment slightly. As in mammalian cells GSH content is higher than H_2O_2 concentration, any excess of peroxides intensifies the reaction rate and GSSG accumulation.

At the 2 μM Cr(VI) long-term exposure (48 h and 96 h) catalase and GPx responses are distinct. Catalase kept its high level, but GPx level decreased approaching its control levels. Thus both enzymes are in line with corresponding controls (Figure 14B,C).

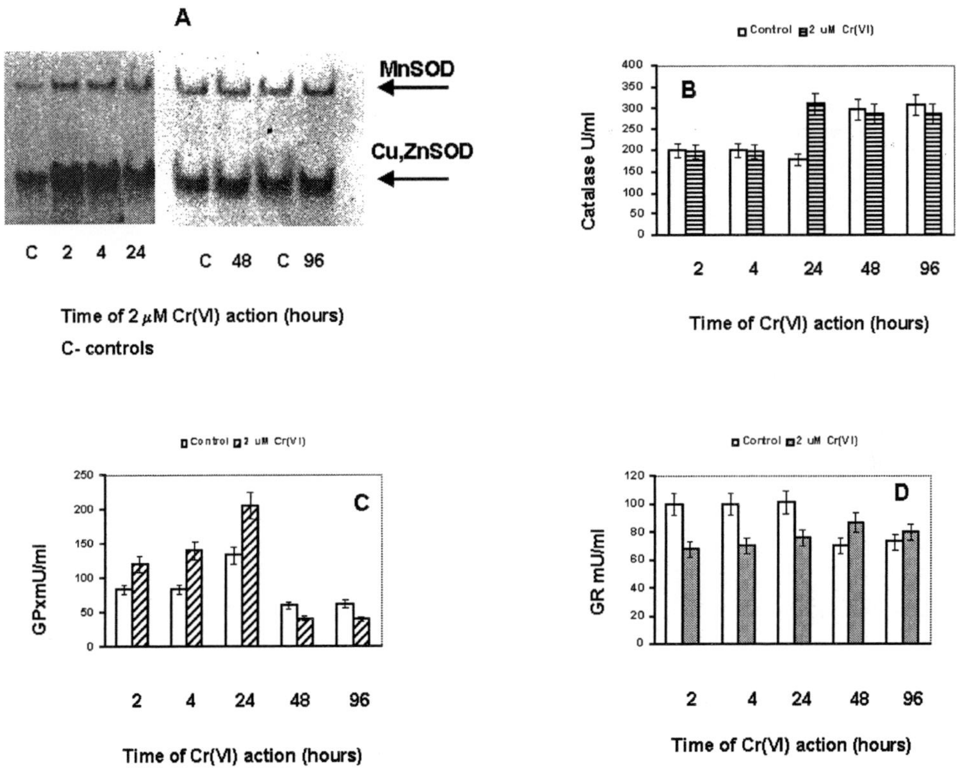

Figure 14. Effect of nontoxic 2 μM Cr(VI) on antioxidant enzymes activities of L-41 cells. Cr(VI) was added at the 48 h of cell growth (80% of confluence), and the cells were grown under permanent Cr(VI) action without medium replenishment. The data presented are mean values ± SD from three separate sets of experiments.

Simultaneous increase of both enzyme activities (24 h of 2 μM Cr(VI) action) points to the intensification of the oxidative stress accompanied by high generation of H_2O_2, specifically as this process is preceded by $O_2^{-\cdot}$ active dismutation by SODs at the initial steps of Cr(VI) action. Further the SOD activities are in line with corresponding control levels. It should be considered that besides H_2O_2 decomposition, GPx removes organic peroxides. At the same time catalase revealing high specificity to H_2O_2 in oxidized condition can display peroxidatic activity.

GPx functioning needs reduced GSH, which is either synthesized or reduced by GR. The main biological sense of GR is to maintain high GSH/GSSG ratio by reduction of GSSG to GSH, precisely supporting the reversible reactions of GSH cycling. As GSH accomplishes its main functions in the reduced form, a high activity of GR is especially necessary at the oxidative stress [178]. As it was mentioned above (Figure 14 C,D) from 24 h of Cr(VI) action GR activity slightly elevated (GPx was maximal) and maintained this level at late stages of Cr(VI) action (48 h and 96 h). This, in turn, prevented depression of GSH function supporting the redox state of cells. As it follows from Figure 15 the maximal loss of GSH corresponds to maximal level of GSSG (24 h of Cr(VI) action). After that the levels of GSH and GSSG approach their corresponding controls. Thus, 24 h is critical for this system followed by the restoration of cell redox balance. GSH/GSSG ratio, which is the most important indicator of the cellular redox status, is restored. However, the changes of GSH in response to Cr(VI) are cell type dependent [178]. The elevated content of reduced GSH testifies to the evolution of adaptive cell response to the oxidative stress.

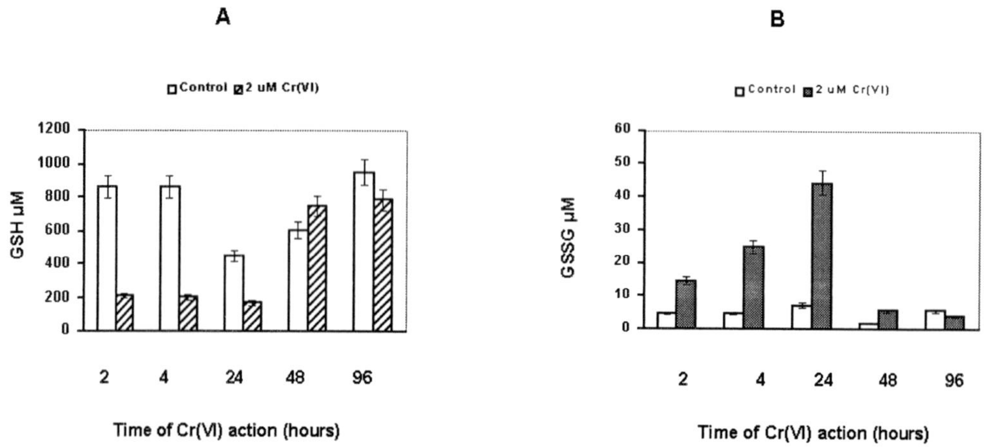

Figure 15. The changes of GSH (A) and GSSG (B) levels (in µM) in L-41 cells grown under 2 µM Cr(VI). Cr(VI) was added at the 48 h of cell growth (80% of confluence), and the cells were grown under permanent Cr(VI) action. The data presented are mean values ± SD from three separate sets of experiments

Thus, low-dose, nontoxic 2 µM Cr(VI) long-term action causes the weak oxidative burst in L-41 cells at 24 h of the permanent cell incubation, which is already over in the next 24 h. The key enzymes involved in the antioxidant defense (SODs, GPx, catalase) react in a compensatory manner. Antioxidant defense systems cope with oxidative stress reversing dynamical balance between ROS and antioxidants, retaining cell homeostasis.

3.2.4. Antioxidant Responses during Toxic Chromium Action

It is known, that GSH takes part not only in the main line of defense against ROS, but it is crucially required for redox-control of disulfide bonds during the fulfillment of such important protein functions, as enzymatic activity, functioning of membrane structures and different types of the active transport [62,179,180].

The reduction of protein disulfides into thiols represents an important process of the vital cell functions accomplished by thioredoxin and glutathione systems [68]. The glutathione-dependent defense system, supporting cell buffer capacity, plays a special role in regulating toxic stress resistance. In this connection, the changes of GSH level at 20 µM Cr(VI) action are of special interest. As it follows from Figure 16A GSH activity decreases at 2 h, then it increases at 4 h under the toxic Cr(VI) concentration. The repeated decrease of GSH level was observed at 24 h of Cr(VI) action. It should be noted that GSH maintains corresponding control levels at 4 h and 24 h. It is known, that different unfavorable factors (carcinogens, cytotoxins, and possibly transition metal ions, etc) initially reduce GSH level, and a cell responds to this by GSH overproduction [181,182]. The data presented in Figure 16B testify to this effect. At initial steps of Cr(VI) action (2 h and 4 h) GSSG concentration did not change significantly and was undetectable at 24 h (Figure 16B), while GSH level after sharp decrease at 2 h, sharply elevated and maintained control levels at 4 h and 24 h (Figure 16A).

Subsequent action of Cr(VI) after 48 h resulted in the complete exhaustion of GSH pool accompanied by 80% apoptosis. A depletion of intracellular GSH has been reported to occur in several different apoptotic systems [183,184]. The increased concentration of GSSG was not detectable in human Jurkat T lymphocytes induced to undergo apoptosis under anti-FAS/APO-1 antibody [184]. In our system, the increase of GSSG was not observed in response to lowering GSH levels in the cells undergoing apoptosis under toxic Cr(VI) and it was detected under the nontoxic concentration at the early time of its action. This may not only reflect the different extent of the oxidative stress depending on the chosen Cr(VI) concentrations (2 μM and 20 μM), but the different character of GSH metabolism.

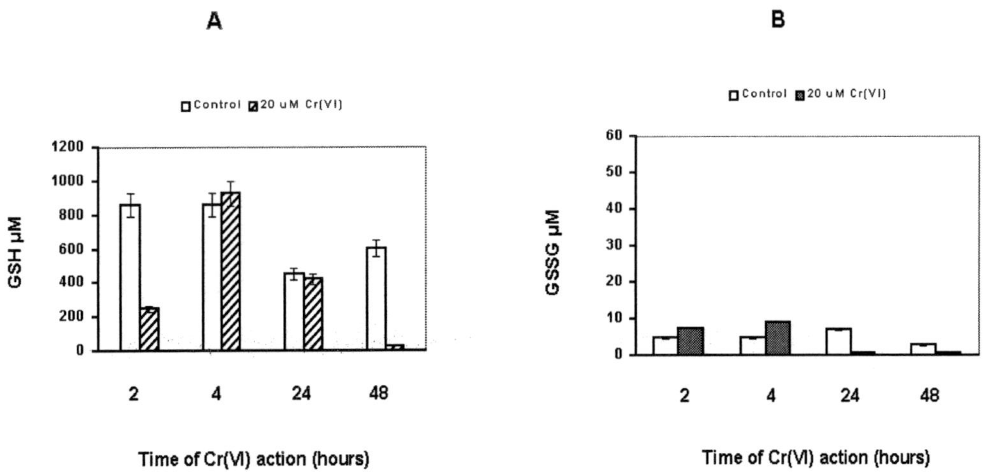

Figure 16. The changes of GSH (A) and GSSG (B) levels (in μM) in L-41 cells grown under 20 μM Cr(VI). Cr(VI) was added at the 48 h of cell growth (80% of confluence), and the cells were grown under permanent Cr(VI) action. The data presented are mean values ± SD from three separate sets of experiments.

Figure 17 points to activity changes of SOD, catalase, GPx, and GR at 20 μM Cr(VI) exposure. Cr(VI) was introduced into the L-41 cell culture medium at 48 h of the cell growth (80% of confluence).

The changes of enzyme activities observed at 20 μM Cr(VI) (Figure 17) and 2 μM Cr(VI) (Figure 14) are considerably different. Time course of 20 μM Cr(VI) action was analyzed for 48 h, since this time point was already connected with 80-90% apoptosis determined by morphological changes and DNA ladder analysis. In the initial steps of toxic Cr(VI) treatment (2 h and 4 h) the suppression of GSH dependent enzyme system was observed, namely GPx and GR were essentially attenuated in 2 h of Cr(VI) action, followed by a slight tendency to increase (4 h), although both enzyme activities were still low compared to controls. The subsequent development of Cr(VI) action (24 h) entails a considerable diminution of the both enzyme activities (especially GR). The more prolonged Cr(VI) action (48 h, 80-90% apoptosis) caused a complete inactivation of the both enzymes (Figure 17C,D).

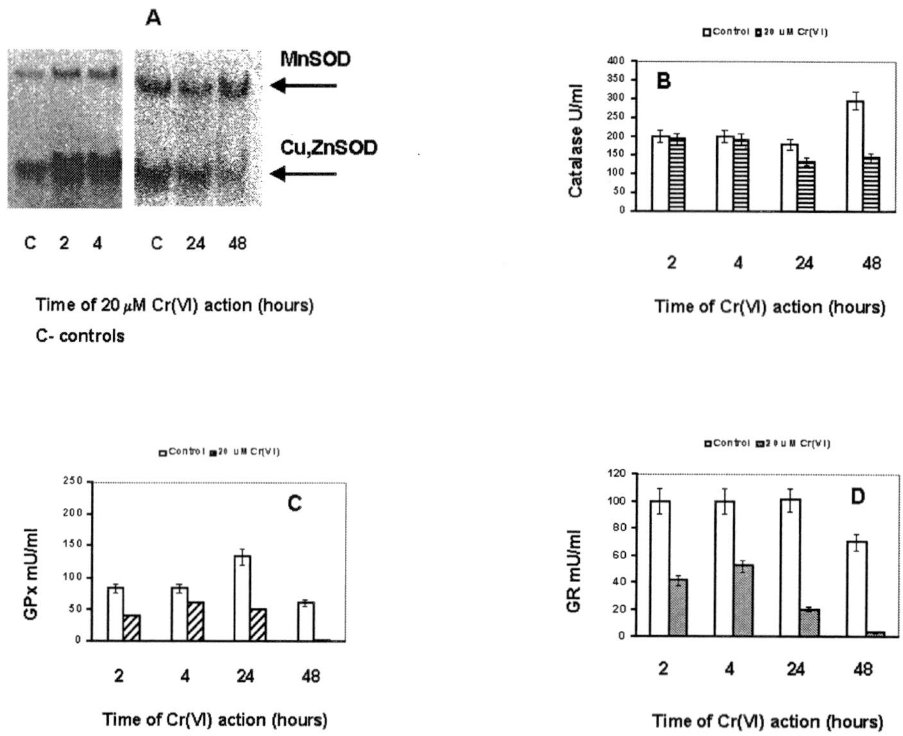

Figure 17. Effect of toxic 20 μM Cr(VI) on antioxidant enzymes activities of L-41 cells. Cr(VI) was added at the 48 h of cell growth (80% of confluence), and the cells were grown under permanent Cr(VI) without medium replenishment. The data presented are mean values ± SD from three separate sets of experiments.

It is generally known that GR operation is connected with enzymes oxidizing GSH and its reversible cycling is only connected with GR activity [62]. Thus, it can be supposed that GR is less effective already from the initial steps of 20 μM Cr(VI) treatment. Inhibition of GR appears to be an important aspect of the toxicity of Cr(VI) [185].

SOD and catalase responses to toxic Cr(VI) exposure were considerably distinguished from GSH dependent enzyme responses. It follows from Figure 17A that Mn- and Cu,ZnSOD activities sharply reinforced (2 h and 4 h). Cr(VI) long-term action attenuates their activities up to control levels (24 h) followed by the essential decrease of Cu,ZnSOD activity (48 h).

In these experimental conditions, catalase proved to be the most stable enzyme. Its activity diminished only at 24 h of Cr(VI) action and becomes two-fold lower compared to control at 48 h of Cr(VI) action (Figure 17D).

These results agree with the data, testifying the more effective resistance of SOD and catalase to the Cr(VI) toxic action [186]. In addition, it seems that these conditions entail a disturbance of adjusted interchangeable enzymatic antioxidants, which can cause an activation of pathways implicating cells to the death by the apoptotic mode [77].

It should be noted, that Cr(VI) toxic concentrations can change activity and/or concentration of intracellular enzymatic and nonenzymatic antioxidants due to reactions typical for the oxidative stress as well as due to their direct participation in the Cr(VI) reducing processes [31,187].

Conclusion

The major cellular targets of Cr(VI) toxicity are lung epithelial cells and lung fibroblasts. However human Cr(VI) intoxication is also associated with hepatotoxicity, nephrotoxicity cardiotoxicity and immunotoxicity [188]. Toxicity of Cr(VI) has been also demonstrated for the various eukaryotic cells in culture, such as gastric mucosa cells, peripheral blood lymphocytes [189], human lung epithelial and fibroblast cell lines [7,76,102,159,165], human lymphoblast cell line (Jurkat cells) [190], and murine macrophage J774A.1 cells [158].

Chromium(VI) inside a cell acts as a multipotent agent, activating various biochemical pathways. Any biomacromolecules can be its targets. Cr-induced DNA damages can result in arrest or induction of transcription, induction of signal transduction pathways, replication errors, and genomic instability. The key transcription machinery proteins interactions can be inhibited by coordination with Cr(III) (the final cellular metabolite of hexavalent chromium), particularly it was observed for one of the repair enzyme [114]. Inhibition of the expression of DNA repair system components by carcinogenic metals is considered as the possible mechanism of their carcinogenicity [191].

When a cell needs to defend itself against injurious Cr(VI) action, stress-response pathways, transient or terminal growth arrest, and apoptosis are activated in reply. The boundary of cell cycle disorders and apoptosis induced by Cr(VI) are cell type dependent. As the main products of Cr(VI) reduction are ROS and the harmful redox imbalance can be developed in a cell, the antioxidant defense system is considered among the main stress-response pathways.

In this chapter we continue to consider the narrow range (2-20 μM) of low-dose and low-dose long-term Cr(VI) action on human epithelial-like L-41 cells in dynamics. The particular emphasis is made on the cell population defense responses at the whole cell level. The initial delay in growth under nontoxic Cr(VI) concentrations for the studied model Cr(VI) correlates with transient cell cycle arrest in S and G_2/M phases. The cell cycle analysis at various time points of chromium(VI) impact demonstrates that the temporal increase of the S phase cell number was subsequently followed by the transient cell increase in G_2/M compartment. This indirectly points to the occurrence of DNA damage, as a cell proliferation may be stopped by checkpoint signals as long as the repair is going on. After the processing of the lesions is accomplished, the cells re-enter the cell cycle and divide.

The toxic concentrations for the studied model Cr(VI) irreversibly caused the growth arrest and massive cell death via apoptosis. The level of DNA synthesis was diminished in the first hours of chromium(VI) impact as assessed by BrdU incorporation. The time-dependent S phase arrest was of certain biphasic character with two peaks in 2 and 24 h after Cr(VI) exposure. It may be supposed that the early S phase delay (2 h) was connected with the initial inhibition of DNA synthesis, and the second one (24 h) reflected the failed DNA repair process. The failure to repair DNA lesions leads to genomic instability and thus the defense programmed cell death (apoptosis) may be induced. The serious irreversible destabilization of chromatin structure has been observed after 24 h of 20 μM Cr(VI) action in our previous study by calorimetry method that could be the result of the accumulation of DNA strand breaks leading to the diminished thermostability of chromatin. Besides, the appearance of the new weak high temperature endotherm points to the formation of another type of chromium-mediated DNA damages as DNA-DNA and/or DNA-protein crosslinks, stabilizing a small part of chromatin structure [192]. The majority of L-41 cell population retained the intact DNA-damage responses (the terminal growth arrest or apoptosis), but some cells survived after 24 h of chromium treatment and entered additional cell cycles. The obtained cell population was characterized by a high level of apoptosis and enriched by polyploid cells that reflected the disorders in control of cell cycle checkpoint pathways and enhanced the possibility of carcinogenesis.

One of the crucial biomarkers of cell adaptation to toxicants is the response of the antioxidant defense system, which plays the special role in regulating toxic stress resistance. In our previous study an overall capacity of antioxidant system, which includes all enzymatic and nonenzymatic antioxidant factors represented in the crude cell extract of L-41 cells, was established by ESR spin-trapping technique in the model Fenton reaction. Cells exposed to 2 μM Cr(VI) revealed a high suppressing capacity against formation of the hydroxyl radical, similar to the control untreated cells. The intracellular antioxidant capacity under toxic chromium concentration (20 μM) revealed less suppressing capacity as compared with the nontoxic concentration [193,194]. In this chapter the changes of the main antioxidant system components are considered individually with and without Cr(VI) action in dynamics. It was found, that the glutathione-dependent system is fine-tuning system in the narrow range of the low Cr(VI) concentrations. The particular marker of the toxic Cr(VI) action is the depletion of GSH-dependent antioxidant defense system. Inhibition of GR is an important aspect of the Cr(VI) toxicity in the studied system.

ACKNOWLEDGMENT

The work was supported partially by the International Science and Technology Center (ISTC), grant G-349.

REFERENCES

[1] Steenland, K., Loomis, D., Shy, C. and Simonsen, N. (1996). Review of occupational lung carcinogens. *Am. J. Ind. Med.,* 29, 474-490.

[2] Fu, H. and Boffetta, P. (1995). Cancer and occupational exposure to inorganic lead compounds: a meta-analysis of published data. *Occup. Environ. Med.,* 52, 73-81.

[3] NTP National Toxicology Program Report on Carcinogens. US Department of Health and Human Services, Public Health Service, 2002.

[4] Tsapakos, M. J., Hampton, T. H., Sinclair, P. R., Sinclair, J. F., Bement, W. J. and Wetterhahn, K. E. (1983). The carcinogen chromate causes DNA damage and inhibits drug mediated induction of porphyrin accumulation and glucuronidation in chick embryo hepatocytes. *Carcinogenesis,* 4, 959-966.

[5] Xu, J., Wise, J. P. and Patierno, S. R. (1992). DNA damage induced by carcinogenic lead chromate particles in cultured mammalian cells. *Mut. Res.,* 280, 129-136.

[6] Xu, J., Bubley, G. J., Detrick, B., Blankenship, L. J. and Patierno, S. R. (1996). Chromium (VI) treatment of normal human lung cells results in guanine-specific DNA polymerase arrest, DNA-DNA cross-links and S-phase blockage of cell cycle. *Carcinogenesis,* 17, 1511-1517.

[7] Ye, J., Zhang, X., Young, H. A. and Shi, X. (1995). Chromium (VI)-induced nuclear factor-kB activation in intact cells via free radical reactions. *Carcinogenesis*, 16, 2401-2405.

[8] Costa, M. (2003). Potential hazards of hexavalent chromate in our drinking water. *Toxicol. Appl. Pharmacol.,* 188, 1-5.

[9] Bal, W. and Kasprzak, K. (2002). Induction of oxidative DNA damage by carcinogenic metals. *Toxicol. Lett.,* 127, 55-62.

[10] Anderson, R A. (1998). Chromium, glucose intolerance and diabetes. *J. Am. Coll. Nutr.,* 17, 548-555.

[11] Sorahan, T., Burges, D. C., Hamilton, L. and Harrington, J. M. (1998). Lung cancer mortality in nickel/chromium platers, 1946-1995. *Occup. Environ. Med.,* 55, 235-242.

[12] Gibb, H. J., Lees, P. S. J., Pinsky, P. F. and Rooncy, B. C. (2000). Lung cancer among workers in chromium chemical production. *Am. J. Ind. Med.,* 38, 115-126.

[13] Zhitkovich, A., Voitkun, V., Klutz, T. and Costa, M. (1998). Utilization of DNA-protein cross-links as a biomarker of chromium exposure. *Enviorn. Health Perspect.,* Suppl. 4, 106, 969-974.

[14] De Flora, S., Bagnasco, M., Serra, D. and Zanacchi, P. (1990). Genotoxicity of chromium compounds. *A review: Mut. Res,.* 238, 99-172.

[15] Conett, P.H. and Wetterhahn, K.E. (1983). Metabolism of the carcinogen chromate by cellular constituents. *Struct. Bonding,* 54, 93-124.

[16] Cohen, M. D., Kargacin, B., Klein, C. B. and Costa, M. (1983). Mechanism of chromium carcinogenicity and toxicity. *Crit. Rev. Toxicol.,* 23, 255-281.

[17] Stearns, DM; Wetterhahn, KE. The mechanism of metal carcinogenicity. Chromium (VI) induced genotoxicity: direct and indirect pathways. In: Hadjiliadis N. Cytotoxic, mutagenic and carcinogenic potential of heavy metals including metals related to human environment. *NATO ASI* (Ser.2). Dordrecht: Kluwer Academic; 1997; 107-121.

[18] Zhitkovich, A. (2005). Importance of chromium-DNA adducts in mutagenecity and toxicity of chromium (VI). *Chem. Res. Toxicol.,* 18, 3-11.

[19] Bose, R. N., Moghaddas, S. and Gelerinter, E. (1992). Long-lived chromium (IV) chromium(V) metabolites on the chromium (VI)-glutathione reaction: NMR, ESR, HPLC, and kinetic characterization. *Inorg. Chem.,* 31,1987-1994.

[20] Stearns, D. M. and Wetterhahn, K. E. (1994). Reaction of Cr(VI) with ascorbate produces chromium(V), chromium (IV) and carbon based radicals. *Chem. Res. Toxicol.*, 7, 219-230.

[21] Lay, P. A. and Levina, A. (1998). Activation of molecular oxygen during the reactions of chromium (VI/V/IV) with biological reductants: Implications for chromium-induced genotoxicities. *J. Am. Chem. Soc,.* 120, 6704-6714.

[22] Kasprzak, K. S. (2002). Oxidative DNA and protein damage in metal-induced toxicity and carcinogenesis. *Free Radic. Biol. Med.,* 32, 958-967.

[23] Valko, M., Rhodes, C. J., Moncol, J., Izakovic, M. and Masur, M. (2006). Free radicals, metals and antioxidants in oxidative stress-induced cancer. *Chem. Biol. Interact.*, 160, 1-40.

[24] Halliwell, B. and Gutteridge, J. M. C. Free Radicals in Biology and Medicine. 3rd Edition. Oxford: Oxford Science Publication, Oxford University Press; (1999).

[25] Curtin, J. F., Donovan, M. and Cotter, T. G. (2002). Regulation and measurement of oxidative stress in apoptosis. *J. Immunol. Methods*, 265, 49-72.

[26] Dalton, T. P., Shertzer, H. G. and Puga, A. (1999). Regulation of gene expression by reactive oxygen. *Ann. Rev. Pharmacol. Toxicol.*, 39, 76-101.

[27] Finkel, T. (2000). Redox-dependent signal transduction. *FEBS Lett.*, 476, 52-54.

[28] Finkel, T. (2003). Oxidant signals and oxidative stress. *Curr. Opin. Cell Biol.*, 15, 247-54.

[29] Thunnickal, V. J. and Fanburg, B.L. (2000). Reactive oxygen species in cell signaling. *Am. J. Physiol. Lung Cell Mol. Physiol.*, 279, L1005-1028.

[30] Codd, R., Irwin, J. A. and Lay, P. A. (2003). Sialoglicoprotein and carbohydrate complexes in chromium toxicity. *Curr. Opin. Chem. Biol.*, 7, 213-219.

[31] Shi, X. I., and Dalal, N.S. (1989) ChromiumV and hydroxyl radical formation during glutathione reductase catalazed reduction of chromium (VI). *Biochem. Biophys. Res. Commun.*, 163, 627-634.

[32] Shi, X., Mao, Y., Knapton A. D., Ding, M., Rojanasakul, Y., Gannett, P. M., Dalal, N. S. and Liu, K. (1994). Reaction of Cr(VI) with ascorbate and hydrogen peroxide generates hydroxyl radicals and causes DNA damage: Role of Cr(IV)-mediated Fenton-like reaction. *Carcinogenesis*, 15, 2475-2478.

[33] Zhitkovich, A., Shrager, S. and Messer, J. (2000). Reduction metabolism of Cr(VI) by cysteine leads to the formation of binary and ternary Cr-DNA adducts in the absence of oxidative DNA damage. *Chem. Res. Toxicol.*, 13, 1114-1124.

[34] Zhitkovich, A., Voitkin, V. and Costa, M. (1995). Glutathione and free aminoacids form stable adducts with DNA following exposure of intact mammalian cells to chromate. *Carcinogenesis*, 16, 907-913.

[35] Codd, R., Dillon, C. T., Levina, A. and Lay, P. A. (2001). Studies of the genotoxicity of chromium: from the test tube to the cell. *Coord. Chem. Rev.*, 216/217, 537-582.

[36] Mates, J. M., Perez-Gomez, C. and De Castro, I. N. (1999). Antioxidant enzymes and human diseases. *Clin. Biochem.*, 32, 595-603.

[37] Liu, K., Husler, J., Ye, J., Leonard, S. S., Cutler, D., Chen, F., Wang, S., Zhang, Zh., Ding, M., Wang, L. and Shi, X. (2001). On the mechanism of Cr(VI)-induced carcinogenesis: dose dependence of uptake and cellular responses. *Mol. Cell. Biochem.*, 222, 221-229.

[38] Ding, M. and Shi, X. (2002). Molecular mechanisms of Cr(VI) induced carcinogenesis. *Mol. Cell. Biochem.*, 234/235, 293-300.

[39] Dillon, C. T., Lay, P. A., Bonin, A. M., Choleva, M. and Legge, G. J. F. (2000). Permeability, cytotoxicity and genotoxicity of Cr(III) complexes and some Cr(V) analogues in V79 Chinese hamster lung cells. *Chem. Res. Toxicol.*, 13, 742-748.

[40] Speetjens, J. K., Collins, R. A., Vincent, J. B. and Woski, S. A. (1999). The nutritional supplement chromium(III)-tris(picolinate) cleaves DNA. *Chem. Res. Toxicol.*, 12, 483-487.

[41] Zhitkovich, A., Voitkin, V. and Costa, M. (1996). Formation of the amino acid-DNA complexes by hexavalent and trivalent chromium, in vitro: importance of trivalent chromium and the phosphate groups. *Biochemistry*, 35, 7275-7282.

[42] Zhitkovich, A. and Costa, M. (1991). A simple, sensitive assay to detect DNA-protein crosslinks in intact cells and in vivo. *Carcinogenesis*, 12, 1485-1489.

[43] Tsapakos, M. J. and Wetterhahn, K. E. (1983). The interaction of chromium with nucleic acids. *Chem. Biol. Interact.*, 46, 265-277.

[44] Fridovich, I. (1995). Superoxide radical and superoxide dismutase. *Annu. Rev. Biochem.*, 64, 97-112.

[45] Landis, G. N. and Tower, J. (2005). Superoxide dismutase evolution and life span regulation. *Mech. Ageing Dev.*, 126, 365-379.

[46] Majima, H., Oberlay, T. D., Furukawa, K., Mattson, M. P., Yen, H. C., Szweda, L. I. and St. Clair, D. K. (1998). Prevention of mitochondrial injury by MnSOD reveals a primary mechanism for alkaline-induced cell death. *J. Biol. Chem.*, 273, 8217-8224.

[47] Hsieh, Y., Guan, Y., Tu, C., Bratt, P. J., Angerhofer, A., Lepock, J. R., Hickey, M. J., Tainer, J. A., Nick, H. S. and Silverman, D. N. (1998). Probing the active site of human Mn-SOD: the role of glutamine 143. *Biochemistry*, 37, 4731-4739.

[48] Mates, J. M. (2000). Effects of antioxidant enzymes in the molecular control of reactive oxygen species toxicology. *Toxicology*, 153, 83-104.

[49] Fridovich, I. (1989). Superoxide dismutase. *J. Biol. Chem.*, 264, 7761-7764.

[50] Enghild, J. J., Thogersen, I. B., Oury, T. D., Valnickova, Z., Hojrup, P. and Crapo, J. D. (1999). The heparin-binding domain of Ec-SOD is proteolytically processed intracellularly during biosynthesis. *J. Biol. Chem.*, 274, 14818-14822.

[51] Kroemer, G. and Reed, J. C. (2000). Mitochondrial control of cell death. *Nature Med.*, 6, 513-519.

[52] Meister, A. (1994). Glutathione, ascorbate, and cellular protection. *Cancer Res.*, (Suppl. 7), 54, 1969S-1975S.

[53] Kirlin, W. G., Cai, J., Thompson, S. A., Diaz, D., Kavanagh, T. J. and Jones, D. P. (1999). Glutathione redox potential in response to differentiation and enzyme inducers. *Free Rad. Biol. Med.*, 27, 1208-1218.

[54] Schafer, F. Q. and Buettner, G. R. (2001). Redox environment of the cells as viewed through the redox state of the glutathione disulfide/glutathione couple. *Free Rad. Biol. Med.*, 30, 1191-1212.

[55] Jones, D. P., Go, Y-M., Anderson, C. L., Siegler, T. R., Cincade, Jr, J. M. and Kirlin, W. G. (2004). Cysteine/cystine couple is a newly recognized node in the circuitry for biologic redox signaling and control. *FASEB J.*, 18, 1246-1248.

[56] Lu, S. C. (2000). Regulation of glutathione synthesis. *Curr. Top. Cell Regul.*, 36, 95-116.

[57] Arthur, J. R. (2000). The glutathione peroxidases. *Cell Mol. Life Sci.*, 57, 1825-1835.

[58] Brigelius-Flohe, R. (1999). Tissue specific functions of individual glutathione peroxidases. *Free Rad. Biol. Med.*, 27, 951-965.

[59] Prinz, W. A., Aslund, F., Holmgren, A. and Beckwith, J. (1997). The role of the thioredoxin and glutaredoxin pathways in reducing protein disulfide bonds in the Escherichia coli cytoplasm. *J. Biol. Chem.*, 272, 15661-15667.

[60] Barret, W. C., DeGnore, J. P., Konig, S., Fales, H. M., Keng, Y-F., Zhang, Zh-Y., Yim, M. B. and Chock, P. B. (1999). Regulation of PTP1B via glutathionylation of the active site cysteine 215. *Biochemistry,* 38, 6699-6705.

[61] Holmgren, A. (1979). Glutathione-dependent synthesis of deoxyribonucleotides. Characterization of the enzymatic mechanism of Escherichia coli glutaredoxin. *J. Biol. Chem.*, 254, 3672-3678.

[62] May, J. M., Qu, Z-C. and Morrow, J. D. (1996). Interaction of ascorbate and α–tocopherol in resealed human erythrocyte ghosts. Transmembrane electron transfer and protection from lipid peroxidation. *J. Biol. Chem.*, 271, 10577-10582.

[63] Anderson, M. E. (1997). Glutathione and glutathione delivery compounds. *Adv. Pharmacol.*, 38, 65-78.

[64] Paolicchi, A., Dominici, S., Pieri, L., Maellaro, E. and Pompella, A. (2002). Glutathione catabolism as a signaling mechanism. *Biochem. Pharm.*, 64, 1027-1035.

[65] O'Brien, T. J., Ceryak, S. and Patierno, S. R. (2003). Complexities of chromium carcinogenesis: role of cellular response, repair and recovery mechanisms. *Mut. Res.*, 533, 3-30.

[66] Wu, D. and Cederbaum, A. (2004). Glutathione deplition in CYP2E1-expressing liver cells induced toxicity due to the activation of p38 mitogen-activated protein kinase and reduction of Nuclear Factor-kB DNA binding activity. *Mol. Pharmacol.*, 66, 749-760.

[67] Conour, J. E., Graham, W. E. and Gaskins, H. R. (2004). A combined in vitro/bioinformatic investigation of redox regulatory mechanisms governing cell cycle progression. *Physiol. Genomics,* 18, 196-205.

[68] Fernandes, A. P. and Holmgren. A. (2004). Glutaredoxins: glutathione-dependent redox enzymes with functions far beyond a simple thioredoxin backup system. *Antioxid. Redox Signal.*, 6, 63-74.

[69] Izawa, S., Maedo, K., Miki, T., Mano, J., Inoue, Y. and Kimura, A. (1998). Importance of glucose-6-phosphate dehydrogenase in the adaptive response to hydrogen peroxide in Saccharomyces Serevisiae. *Biochem. J.*, 330, 811-817.

[70] Li, S., Yan, T., Yang, J-Q., Oberley, T. D. and Oberley, L. W. (2000). The role of cellular glutathione reductase redox regulation in the suppression of tumor cell growth by manganese superoxide dismutase. *Cancer Res.*, 60, 3927-3939.

[71] Hassan, H. M. and Fridovich, I. (1978). Regulation of the synthesis of catalase and peroxidase in Escherichia coli. *J. Biol. Chem.*, 253, 6445-6450.

[72] McCormic, M. L., Buettner, G. R. and Britigan, B.E. (1998). Endogenous superoxide dismutase levels regulate iron dependent hydroxyl radical formation in Escherichia coli exposed to hydrogen peroxide. *J. Bacter.*, 180, 622-625.

[73] Li, T., Huang, X., Zhou, R., Liu, Y., Lin, B., Nomura, C. and Zhao, J. (2002). Differential expression and localization of Mn and Fe superoxide dismutases in the Heterocystous Cyanobacterium Anabaena sp strain PCC 7120. *J. Bacter.*, 184, 5096-5103.

[74] Farr, S.B., D'Ari, R. and Touati, D. (1986). Oxygen-dependent mutagenesis in Escherichia coli lacking superoxide dismutase. *Proc. Natl. Acad. Sci.* USA, 83, 8268-8272.

[75] Dubrovskaya, V. A. and Wetterhahn, K. E. (1998). Effects of Cr(VI) on the expression of the oxidative stress genes in human lung cells. *Carcinogenesis*, 19, 1401-1407.

[76] Ye, J. and Shi, X. (2001). Gene expression profile in response to chromium-induced cell stress in A549 cells. *Mol. Cell. Biochem.*, 222, 189-197.

[77] Kahl, R., Kampkotter, A., Watjen, W. and Chovolou, Y. (2004). Antioxidant enzymes and apoptosis. *Drug Metabolism Rev.*, 36, 747-762.

[78] Andersen, M. E., Yang, R. S. H., French, C. T., Chubb, L. S. and Dennison, J. E. (2002). Molecular circuits, biological switches, and non-linear dose-response relationships. *Environm. Health Persp.*, 110, 971-978.

[79] Leqewe, S., Bluthgen, N., Dchafer, R. and Hezzel, H. (2005). Ultrasensitization: switch-like regulation of cellular signaling by transcription induction. *PLoS Comput.Biol.*, 1, 405-414.

[80] Papin, J. A., Hunter, T., Palsson, B. O. and Subramaniam, S. (2005). Reconstruction of cellular signaling networks and analysis of their properties. *Nat. Rev. Mol. Cell Biol.*, 6, 99-111.

[81] White, M. A. and Anderson, G. W. (2005). Signalling networks in living cells. Annu. *Rev. Pharm. Toxicol.*, 45, 587-603.

[82] Keise, S. M. (2000). Protein phosphatases and the regulation of mitogen-activated protein kinase signaling. *Curr. Opin. Cell Biol.*, 12, 186-192.

[83] Ballif, B. A. and Blenis, J. (2001). Molecular mechanisms mediating mammalian mitogen-activating protein kinase (MAPK) kinase (MEK)-MAPK survival signals. *Cell Growth Differ.*, 12, 397-408.

[84] Chang, L. and Karin, M. (2001). Mammalian MAP kinase signaling cascades. *Nature*, 410, 37-40.

[85] Roux, P. D. and Blenis, J. (2004). ERK and p38 MAPK activated protein kinases: a family of protein kinases with diverse biological fuctions. *Microbiol. Mol. Biol. Rev.*, 68, 320-344.

[86] Rubinfeld, H. and Seger, R. (2005). The ERK cascade: a prototype of MAPK signaling. *Mol. Biotechnol.*, 31, 151-174.

[87] Guo, Y. L., Baysal, K., Kang, B., Yang, L. G. and Williamson, J. R. (1998). Correlation between sustained c-Jun N-terminal protein kinase activation and apoptosis induced by tumour necrosis factor a in rat mesenglial cells. *J. Biol. Chem.*, 273, 4027-4034.

[88] Roulston, A., Reinhardt, C., Amiri, P. and Williams, L. T. (1998). Early activation of c–Jun N-terminal kinase and p38 kinase regulate cell survival in response to tumor necrosis factor a. *J. Biol. Chem.*, 273, 10223-10239.

[89] Yen, A., Robertson, M. S., Varvayanis, S. and Lee, A. T. (1998). Retinoic acid induced mitogen-activated protein/extracellular signal-regulated kinase (ERK) kinase-dependent MAP kinase activation needed to elicit HL-60 cell differentiation and growth arrest. *Cancer Res.*, 58, 3163-3172.

[90] Samet, J. M., Graves, L. M., Quay, J., Dailey, L. A., Delvin, R. B., Ghio, A. J., Wu, W., Bromberg, P. A. and Reed, W. (1998). Activation of MAPKs in human bronchial epithelial cells exposed to metals. *Am. J. Physiol.*, 275, L551- L558.

[91] Du, L., Lyle, C. S., Obey, T. B., Gaarde, W. A., Muir, J. A., Bennett, B. L. and Chambers, T. C. (2004). Inhibition of cell proliferation and cell cycle progression by specific inhibition of basal JNK activity. *J. Biol. Chem.*, 279, 11957-11966.

[92] Harris, G. K. and Shi, X. (2003). Signaling by carcinogenic metals and metal-induced reactive oxygen species. *Mut. Res.*, 533, 183-200.

[93] Owuor, E. D. and Kong, A.-N. T. (2002). Antioxidant and oxidant regulated signal transduction pathways. *Biochem. Pharm.*, 64, 765-770.

[94] Noda, T., Iwakiri, R., Fujimoto, K. and Aw, T. Y. (2001). Induction of mild intracellular redox imbalance inhibits proliferation of CaCo-2 cells. *FASEB J.*, 15, 2131-2139.

[95] Klatt, P. and Lamas, S. (2000). Regulation of protein function by S-glutathiolation in response to oxidative and nitrosative stress. *Eur. J. Biochem.*, 267, 4928-4944.

[96] Arnold, R. S., Shi, J., Murad, E., Whalen, A. M., Sun, C. Q., Polavarapu, R., Parthasarathy, R., Petros, J. A. and Lambeth, J. D. (2001). Hydrogen peroxide mediates the cell growth and transformation caused by the mitogenic oxidase *Nox1. Proc. Natl. Acad. Sci.* USA., 98, 5550-5555.

[97] Rhee, S. G., Chang, T. S., Bae, Y. S., Lee, S. R. and Kong, S. W. (2003). Cellular regulation by hydrogen peroxide. *J. Am. Soc. Nephrol.*, 14, S211- S215.

[98] Nindl, G. (2004). Hydrogen peroxide: from oxidative stress to redox regulator. *Cell Science Rev.*, 1, 1-13.

[99] Whisler, R. L. Goyette, M. A., Grants, I. S. and Newhouse, Y. J. (1995). Sublethal levels of oxidant stresss stimulate multiple serine/threonine kinases and suppress protein phosphatases in Jurkat T cells. *Arch. Biochem. Biophys.*, 319, 23-35.

[100] Lee, K. and Esselman, W. J. (2002). Inhibition of PTPs by H_2O_2 regulates the activation of distinct MAPK pathways. *Free Rad. Biol. Med.*, 33,1121-1132.

[101] Salmeen, A. and Barford, D. (2005). Functions and mechanisms of redox regulation of cystein-based phosphatases. *Antioxid. Redox Signal.*, 7, 560-577.

[102] Chen, K., Vita, J. A., Berk, B. C. and Keaney Jr. ,J. K. (2001). cJun N-terminal kinase activation by hydrogen peroxide in endothelial cells involves Src-dependent epidermal growth factor transactivation. *J. Biol. Chem.*, 276, 16045-16050.

[103] Kurata, S. (2000). Selective activation of p38 MAPK cascade and mitotic arrest caused by low level oxidative stress. *J. Biol. Chem.*, 275, 23413-23416.

[104] Chuang, S-M., Liou, G-Y. and Yang, J-L. (2000). Activation of JNK, p38, and ERK mitogen-activated protein kinases by chromium (VI) is mediated through oxidative stress but does not affect cytotoxicity. *Carcinogeneses*, 21, 1491-1500.

[105] Yoshizumi, M., Abe, J., Haendler, J., Huang, Q. and Berk, B. C. (2000). Src and Cas mediate JNK activation but not ERK1/2 and p38 kinases by reactive oxygen species. *J. Biol. Chem.*, 275, 11706-11712.

[106] O'Hara, K., Klei, L .R. and Barchowsky, A. (2003). Selective activation of Src family kinases and JNK by low level of chromium (VI). *Tox. Appl. Pharmacol.*, 190, 214-223.

[107] Balamurugan, K., Rajaram, M., Ramasami, S. and Narayanan, S. (2002). Chromium (III) induced apoptosis of lymphocytes: death decision by ROS and Src-family tyrosine kinases. *Free Radic. Biol. Med.,* 33, 1622-1640.

[108] Filomeni, G., Rotilio, G. and Ciriolo, M. R. (2005). Disulfide relays and phosphorilative cascades: partners in redox-mediated signaling pathways. *Cell Death Differ.,* 12, 1555-1563.

[109] Van Holde, K; Zlatanova, J; Arents, G; Moudrianakis, E. Histones, Nucleosomes, and Fibres. In: Elgin SCR. *Elements of Chromatin Structure.* Oxford, UK: IRL Press; 1995; 1–26.

[110] Annunziato, A. T. (2005). Split decision: what happens to nucleosomes during DNA replication. *Biochemistry,* 280, 12065-12068.

[111] Yuann, J-M. P., Liu, K. J., Hamilton, J. W. and Wetterhahn. K. E. (1999). In vivo effects of ascorbate and glutathione on the uptake of chromium, formation of chromium(V), chromium–DNA binding and 8-hydroxy-2'-deoxyguanosine in liver and kidney of osteogenic disorder Shionogi rats following treatment with chromium(VI). *Carcinogenesis,* 20, 1267-1275.

[112] Wedrychowski, A., Schmidt, W. N. and Hnilica, L. S. (1986). The in vivo cross-linking of proteins and DNA by heavy metals. *J. Biol. Chem.,* 261, 3370-3376.

[113] Sugden, K. D., Campo, C. K. and Martin, B. D. (2001). Direct oxidation of guanine and 7,8-dihydro-8-oxoguanine in DNA by a highvalent chromium complex. *Chem. Res. Toxicol.,* 14, 1315–1322.

[114] Hodges, N. J. and Chipman, J. K. (2002). Down-regulation of the DNA-repair endonuclease 8-oxo-guanine NA glycosylase 1 (hOGG1) by sodium dichromate in cultured human A549 lung carcinoma cells. *Carcinogenesis,* 23, 55– 60.

[115] Sugiyama, M., Patierno, S. R., Cantoni, O. and Costa. M. (1986). Characterization of DNA lesions induced by $CaCrO_4$ in synchronous and asynchronous cultured mammalian cells. *Mol. Pharm.,* 29, 606-613.

[116] Ueno, S., Kashimoto, T., Susa, N., Furukawa, Y., Ishii, M., Yokoi, K., Yasuno, M., Sasaki, Y. F., Ueda, J., Nishimura, Y. and Sugiyama, M. (2001). Detection of dichromate (VI)-induced DNA strand breaks and formation of paramagnetic chromium in multiple mouse organs. *Tox. Appl. Pharm.,* 170, 56-62.

[117] Gambelunghe, A., Piccinin, R., Ambrogi, M., Villarini, M. Moretti, M., Marchetti, C., Abbritti, G. and Muzi, G. (2003). Primary DNA damage in chrome-plating workers. *Toxicology,* 188, 187-195.

[118] Aiyar, J., Borges, K. M., Floyd, R. A. and Wetterhanh, K. E. (1989). Role of Cr(V) in glutathione thiyl radical and hydroxyl radical intermediates in chromium (VI)-induced DNA damage. *Toxicol. Environ. Chem.*, 22, 135-148.

[119] Singh, J., Bridgewater L. C. and Patierno, S. R. (1998). Differential sensitivity of chromium-mediated DNA interstrand crosslinks and DNA-Protein crosslinks to disruption by alkali and EDTA. *Toxicol. Science,* 45, 72-76.

[120] Arakawa, H., Ahmad, R., Naoui, M. and Tajmir-Riahi, H-A. (2000). A comparative study of calf thymus DNA binding to Cr(III) and Cr(VI) ions. *J. Biol. Chem.,* 275, 10150-10153.

[121] Wang, J. F., Bashir, M., Engelsberg, B. N., Witmer, C., Rozmiarek, H. and Billings, P. C. (1997). High mobility group proteins 1 and 2 recognize chromium-damaged DNA. *Carcinogenesis*, 18, 371-375.

[122] Bianchi, M. E., Beltrame, M. and Paonessa, G. (1989). Specific recognition of cruciform DNA by nuclear protein HMG1. *Science*, 243, 1056-1059.

[123] Chow, C. S., Barnes, C. and Lippard, S. J. (1995). A single HMG domain in high-mobility group 1 protein binds to DNAs as small as 20 base pairs containing the major cisplatin adduct. *Biochemistry*, 34, 2956-2964.

[124] Read, C. M., Cary, P. D., Preston, N. S, Lnenicek-Allen, M. and Crane-Robinson, C. (1994). The DNA sequence specificity of HMG boxes lies in the minor wing of the structure. *EMBO J.,* 13, 5639-5646.

[125] Jayaraman, L., Moorthy, N. C., Murthy, K. G., Manley, J. L., Bustin, M. and Prives, C. (1998). High mobility group protein-1 (HMG-1) is a unique activator of p53. *Gen. Dev.,* 12, 462– 472.

[126] Schuler, M. and Green, D. R. (2001). Mechanisms of p53-dependent apoptosis. *Biochem. Soc. Trans.,* 29, 684-688.

[127] Sancar, A., Linsey-Boltz, L. A., Unsal-Kacmaz, K. and Linn, S. (2004). Molecular mechanisms of mammalian DNA repair and the DNA damage checkpoints. *Annu. Rev. Biochem.,* 73, 39-85.

[128] Peterson, C. L. and Cote, J. (2006). Cellular machineries for chromosomal DNA repair. *Gen. Dev.,* 18, 602-616.

[129] Tchou, J., Kasai, H., Shibutani, S., Chung, M., Laval, J., Grollman, A. P. and Nishimura, S. (1991). 8-Oxoguanine (8-Hydroxyguanine) DNA glycosylase and its substrate specificity. *Proc. Natl. Acad. Sci.* USA, 88, 4690–4694.

[130] Tani, M., Shinmura, K., Kohno, T., Shiroishi, T., Wakana, S., Kim, S. R., Nohmi, T., Kasai, H., Takenoshita, S., Nagamachi, Y. and Yokota, J. (1998). Genomic structure and chromosomal localization of the mouse Ogg1 gene that is involved in the repair of 8-hydroxyguanine in DNA damage. *Mammalian Genome,* 9, 32–37.

[131] Maeng, S-H., Chung, H-W., Yu, I-Yu., Kim, H-Y., Lim, Ch-H., Kim, K-J., Kim, S-J., Ootsuyama, Y. and Kasai, H. (2003). Changes of 8-OH-dG levels in DNA and its base excision repair activity in rat lungs after inhalation exposure to hexavalent chromium. *Mut. Res.,* 539, 109-116.

[132] Friedberg, E., Walker, G. C. and Siede, W. (1995). *DNA repair and mutagenesis.* Washington, DC., USA:ASM Press.

[133] O'Donovan, A. and Wood, R. D. (1993). Identical defects in DNA repair in xeroderma pigmentosum group G and rodent ERCC group 5. *Nature*, 363, 185–188.

[134] O'Donovan, A., Scherly, D., Clarkson, S. G. and Wood, R. D. (1994). Isolation of active recombinant XPG protein, a human DNA repair endonuclease. *J. Biol. Chem.,* 269, 15965–15968.

[135] O'Brain, T. J., Brooks, B. R. and Patierno, S. R. (2005). Nucleotide excision repair functions in the removal of chromium-induced DNA damage in mammalian cells. *Mol. Cell. Biochem.,* 279, 85-95.

[136] Reynolds, M., Peterson, E., Quievryn, G. and Zhitkovich, A. (2004). Human nucleotide excision repair efficiently removes chromium-DNA phosphate adducts and protects cells against chromate toxicity. *J. Biol. Chem.*, 279, 30419-30424.

[137] Bryant, H. E., Ying, S. and Helleday, T. (2006). Homologous recombination is involved in repair of chromium-induced DNA damage in mammalian cells. *Mut. Res.*, 599, 116-123.

[138] Osgood, E. E. and Brooke, J. H. (1955). Continuous tissue culture of leucocytes from human leukemic bloods by application of "gradient" principles. *Blood*, 10, 1010-1022.

[139] Solov'ev, V. D. and Gulevich, N. E. (1960). A study of virus resistance using cell culture models: isolation of continuous cell line resistant to poliomyelitis virus. (Russ.) *Acta Virologica*, 4, 220-226.

[140] Mosmann, T. (1983). Rapid coloreimetric assay for cellular growth and survival application. *J. Immun. Methods,* 65, 55-63.

[141] Carmichael, J., DeGraff, W. G., Adi, F., Minna, J. D. and Mitchell, J. B. (1987). Evaluation of a tetrazolium-based semiautomated colorimetric assay: assessment of chemosensitivity testing. *Cancer Res.*, 47, 936-942.

[142] Mueller, H., Kassack, M. and Wiese, M. (2004). Comparison of the usefulness of the MTT, ATP and calcein assays to predict the potency of cytotoxic agents in various human cancer cell lines. *J. Biomol. Screening,* 9, 506-515.

[143] Ormerod, M.G. (2002). Investigating the relationship between the cell cycle and apoptosis using flow cytometry. *J. Immun. Methods,* 265, 73-80.

[144] Wilson, GD. Analysis of DNA-measurement of cell kinetics by the bromodeoxyuridine/antibromodeoxyuridine method. In: Ormerod MG, editor. Flow Cytometry: A Practical Approach. Oxford: IRL Press at Oxford University Press; 2000; 159-177.

[145] Heil, J. and Reifferscheid, G. (1992). Detection of mammalian carcinogens with an immunological DNA synthesis-inhibition test. *Carcinogenesis*, 13, 2389-2394.

[146] Martin, S. J., Reutelingsperger, C. P., McGahon, A. J., Rader, J. A., van Schie, R. C., LaFace, D. M. and Green, D. R. (1995). Early redistribution of plasma membrane phosphatidylserine is a general feature of apoptosis regardless of the initiating stimulus: inhibition by overexpression of Bcl2 and Abl. *J. Exp. Med.*, 182, 1545-1556.

[147] Gurtu, V. and Zhang, G. (1997). Detection of early-stage apoptosis in three adherent mammalian cell lines using the ApoAlert ™Annexin V apoptosis kit. *Clonotechniques*, 11, 24-26.

[148] Diaz, G., Setzu, M. D., Zucca, A., Isola, R., Diana, A., Murru, R., Sogos, V. and Gremo, F. (1999). Subcellular heterogeneity of mitochondrial membrane potential: relationship with organelle distribution and intercellular contacts in normal, hypoxic and apoptotic cells. *J. Cell Sci.*, 112, 1077-1084.

[149] Kinigham, K. K., Oberley, T. D., Lin, S.-M., Mattingly, C. A. and Clair, D. K. S. (1999). Overexpression of manganese superoxide dismutase protects against mitochondrial-initiated poly(ADP-ribose) polymerase-mediated cell death. *FASEB J.*, 13, 1601-1610.

[150] Rottenberg, H. and Wu, S. (1998). Quantitative assay by flow cytometry of the mitochondrial membrane potential in intact cells. *Biochem. Biophys. Acta*, 1404, 393-404.

[151] Castedo, M., Ferri, K., Roumier, T., Metivier, D., Zamzami, N. and Kroemer, G. (2002). Quantitation of mitochondrial alterations associated with apoptosis. *J. Immun. Methods,* 265, 39-47.

[152] Steinman, H. M. (1985). Bacteriocuprein superoxide dismutases in pseudomonads. *J. Bacteriol.*, 162, 1255-1260.

[153] Geller, B. L. and Winge, D. R. (1982). Rat liver Cu,Zn-superoxide dismutase. Subcellular location in lysosomes. *J. Biol. Chem.,* 257, 8945-8952.

[154] Beers, R. F. and Sizer, I. W. (1952). A spectrophotometric method for measuring the breakdown of hydrogen peroxide by catalase. *J. Biol. Chem.,* 195, 133-140.

[155] Laemmli, U. K. (1970). Cleavage of structural proteins during the assembly of the head of bacteriophage T4. *Nature*, 227, 680-685.

[156] Ly, J. D., Grubb, D. R. and Lawen, A. (2003). The mitochondrial membrane potential ($\Delta\psi m$) in apoptosis; an update. *Apoptosis*, 8, 115-128.

[157] Porter, A. G. and Janicke, R. U. (1999). Emerging roles of caspase-3 in apoptosis. *Cell Death Differ.,* 6, 99–104.

[158] Bagchi, D., Bagchi, M. and Stohs, S. J. (2001). Chromium (VI)-induced oxidative stress, apoptotic cell death and modulation of p53 tumor suppressor gene. *Mol. Cell. Biochem.,* 222, 149-158.

[159] Ye, J., Wang, S., Leonard, S. S., Sun, Yi., Butterworth, L., Antonini, J., Ding, M., Rojanasakul, Yo., Vallyathan, V., Castranova, V. and Shi, X. (1999). Role of reactive oxygen species and p53 in chromium(VI)-induced apoptosis. *J. Biol. Chem.,* 274, 34974-34980.

[160] Pardee, A. (1989). G1 events and the regulation of cell proliferation. *Science*, 246, 603-608.

[161] Hartwell, L. H. and Kastan, M. R. (1994). Cell cycle control and cancer. *Science,* 266, 1821-1828.

[162] Bartec, J. and Lukas, J. (2001). Mammalian G1 and S phase checkpoints in response to DNA damage. *Cur. Opin. Cell Biol.,* 13, 738-747.

[163] Bettencourt-Dias, M., Glet, R., Sinka, R., Mazumdar, A., Lock, W. G., Balloux, F., Zafiropoulos, P. S., Yamagouchi, S., Winter, S., Carthew, R. W., Cooper, M., Jones, D., Frenz, L. and Glover, D. M. (2004). Genome-wide survey of protein kinases required for cell cycle progression. *Nature*, 432, 980-987.

[164] Kastan, M. B. and Bartek, J. (2004). Cell cycle checkpoints and cancer. *Nature*, 432, 316-323.

[165] Pritchard, D. E., Ceryak, S., Ha, I., Fornsaglio, J. L., Hartman, S. K., O'Brien, T. J. and Patierno, S. R. (2001). Mechanisms of apoptosis and determination of cell fate in chromium(VI)-exposed populations of telomerase-immortalized human fibroblasts. *Cell Growth Differ.,* 12, 487-496.

[166] Zhang, Z., Leonard, S. S., Wang, S., Vallyathan, V. V., Castranova, V. and Shi, X. (2001). Cr(VI) induces cell growth arrest through hydrogen peroxide-mediated reactions. *Mol. Cell. Biochem.,* 222, 77-83.

[167] Chen, F. and Shi, X. (2002). Intracellular signal transduction of cells in response to carcinogenic metals. *Crit. Rev. Oncology/Hematology,* 42, 105-121.

[168] Du, J., Wang, X., Miereles, Gh., Bailey, J. L., Debigare, R., Zheng, B., Price, S. R. and Mitch, W. (2004). Activation of caspase-3 is an initial step triggering accelerated muscle proteolysis in catabolic conditions. *J. Clin. Invest.,* 113, 115-123.

[169] Keen, N. and Taylor, S. (2004). Aurora-kinase inhibitors as anticancer agents. *Nature Reviews/Cancer,* 4, 927-936.

[170] Castedo, M., Perfettini, J-L., Roumier, T., Valent, A., Feunteun, J., Lenoir, G., Medema, R., Vainchenker, W. and Kroemer, G. (2004). Mitotic catastrope constitutes a special case of apoptosis whose suppression entails aneuploidy. *Oncogene*, 23, 4362-4370.

[171] Roumier, T., Valent, A., Perfettini, J.-L, Mativier, D., Castedo, M. and Kroemer, G. (2005). A cellular machine generating apoptosis-prone cells. *Cell Death Differ.,* 12, 91-93.

[172] Rajodopalan, H. and Lengauer, C. (2004). Aneuploidy and Cancer. *Nature*, 432, 338-341.

[173] Liu, K. and Shi, X. (2001). In vivo reduction of chromium (VI) and its related free radical generation. *Mol. Cell. Biochem.,* 222, 41-47.

[174] Ye, J., Gao, F., Shi, G., Li, Z., Wang, X., Shi, X. and Tang, X. (2002). The inherent cellular level of reactive oxygen species: one of the mechanisms determining apoptotic susceptibility of leukemic cells to arsenite trioxide. *Apoptosis*, 7, 209-215.

[175] Kim, Y-D., An, S.-C., Oyama, T., Kawamoto, T. and Kim, H. (2003). Oxidative stress, hoggl expression and NF-kB activity in cells exposed to low level chromium. *J. Occup.Health,* 45, 271-277.

[176] Frank, L. (1985). Effects of O_2 on the newborn. *Fed. Proc.,* 44, 2328-2335.

[177] Van den Hock, T. L., Becker, L. B., Shao, Z., Li, C. and Schumacker, P. T. (1998). Reactive oxygen species released from mitochondria during brief hypoxia induce preconditioning in cardiomyocytes. *J. Biol. Chem.,* 273, 18092-18098.

[178] Ning, J. and Grant, M. H. (2000). The role of reduced glutathione and glutathione reductase in the cytotoxicity of chromium (VI) in osteoblasts. *Toxicol. in Vitro.,* 14, 329-335.

[179] Lizard, G., Gueldry, S., Sordet, O., Monier, S., Athias, A., Miguet, C., Bessede, G., Lemaire, S., Solary, E. and Gambert, P. (1998). Glutathione is implied in the control of 7-ketocholesterol-induced apoptosis, which is associated with radical oxygen species production. *FASEB. J.,* 12, 1651-1663.

[180] Beatrice, M. C., Stiers, D.L. and Pfeiffer, D. R. (1984). The role of glutathione in the retention of Ca^{+2} by liver mitochondria. *J. Biol. Chem.,* 259, 1279-1287.

[181] Hanigan, M. H. and Pitot, H.C. (1985). Gamma-glutamyl transpeptidase-its role in hepatocarcinogenesis. *Carcinogenesis*, 6, 165-172.

[182] Fukino, H., Hirai, M., Hsueh, Y. M., Moriyasu, S. and Yamane, Y. (1986). Mechanism of protection by zinc against mercuric chloride toxicity in rats: effect of zinc and mercury on glutathione metabolism. *J. Toxicol. Environ. Health,* 19, 75-89.

[183] Domenicotti, C., Paola, D., Vitali, A., Nitti, M., D'Abramo, C., Cottalasso, D., Maloberti, G., Biasi, F., Poli, G., Chiarpotto, E., Marinari, U. M. and Pronzato. M. A.

(2000). Glutathione depletion induces apoptosis of rat hepatocytes through activation of protein kinase C novel isoforms and dependent increase in AP-1 nuclear binding. *Free Rad. Biol. Med.,* 29, 1280-1290.

[184] van den Dobbelsteen, D. J., Nobel, C. S. I., Schlegel, J., Cotgreave, I. A., Orrenius, S. and Slater, A. F. G. (1996). Rapid and specific efflux of reduced glutathione during apoptosis induced by anti-Fas/APO-1 antibody. *J. Biol. Chem.,* 271, 15420-15427.

[185] Gunaratnam, M., Pohlscheidt, M. and Grant, M. H. (2002). Pretreatment of rats with the inducing agents phenobarbitone and 3-methylcholanthrene ameliorates the toxicity of chromium (VI) in hepatocytes. *Toxicol. in Vitro,* 16, 509-516.

[186] Susa, N., Ueno, S., Furukawa, Y., Ueda, J. and M. Sugiyama. (1997). Potent protective effect of melatonin on chromium (VI)-induced DNA single-strand breaks, cytotoxicity, and lipid peroxidation in primary cultures of rat hepatocytes. *Tox. Appl. Pharm.,* 144, 377-384.

[187] Shi, X. and Dalal, N.S. (1988). On the mechanism of the chromate reduction by glutathione: ESR evidence for the glutathionyl radical and isolable Cr(V) intermediate. *Biochem. Biophys. Res. Commun.,* 156, 137-142.

[188] Pourahmad, J; O'Brien, TJ. Biological reactive intermediates that mediate Chromium (VI) toxicity. In: Dansette PM editor. Biological Reactive Intermediates VI: Chemical and Biological Mechanisms in Susceptibility to and Prevention of Environmental Diseases: Advances in Experimental Medicine and Biology. Berlin, New York: Springer; 2001; 203-208.

[189] Trzeciak, A., Kowalik, J., Małecka-Panas, E., Drzewoski, J., Wojewódzka, M., Iwaneńko, T. and Błasiak, J. (2000). Genotoxicity of chromium in human gastric mucosa cells and peripheral blood lymphocytes evaluated by the single cell gel electrophoresis (comet assay). *Med. Sci. Monitor.,* 6, 24-29.

[190] Shi, X., Ding, M., Ye, J., Wang, S., Leonard, S.S., Zang, L., Castranova, V., Vallyathan, V., Chiu, A., Dalal, N. and Liu, K. (1999). Cr(IV) causes activation of nuclear transcription factor-κB, DNA strand breaks and dG hydroxylation via free radical reactions. *J. Inorg. Biochem.,* 75, 37-44.

[191] Hartwig, A. (1998). Carcinogenicity of metal compounds: possible role of DNA reapir inhibition. *Toxicol. Lett.,* 102-103, 235-239.

[192] Monaselidze, J., Abuladze, M., Asatiani, N., Kiziria, E., Barbakadze, Sh., Majagaladze, G., Iobadze, M., Tabatadze, L. Holman, H-Y. and Sapojnikova, N. (2006). Characterization of chromium-induced apoptosis in cultured mammalian cells: A differential scanning calorimetry study. *Themochem. Acta,* 441, 8-15.

[193] Asatiani, N., Sapojnikova, N., Abuladze, M., Kartvelishvili, T., Kulikova, N., Kiziria, E., Namchevadze, E. and Holman H-Y. (2004). Effects of Cr(VI) long-term and low-dose action on mammalian antioxidant enzymes (an in vitro study). *J. Inorg. Biochem.,* 98, 490-496.

[194] Kartvelishvili, T., Abuladze, M., Asatiani, N., Akhvlediani, J., Kiziria, E., Asanishvili, L., Lejava, L., Holman, H-Y. and Sapojnikova, N. (2004). Estimation of the cellular antioxidant response to chromium action using ESR method. *The Scientific World Journal,* 4, 785-794.

In: New Research on Genomic Instability
Editor: Eleanor J. Glascow, pp. 261-283

ISBN: 1-60021-320-0

Chapter IX

Genotoxic Stress, S Phase Checkpoint and Apoptosis

Hideshi Ishii
Center for Molecular Medicine, Jichi Medical School, Tochigi, Japan.

ABSTRACT

Eukaryote cells maintain genomic integrity by surveillance of DNA for damage or incomplete replication. In the event of aberrant structures being detected, checkpoint mechanisms are activated that result in slowing down or arresting cell cycle progression, which allows the DNA damage to be repaired and the replication to be completed. Potentially deleterious mutations can elicit pathways leading to apoptosis. Defects in those pathways, however, seem to induce the susceptibility to DNA damage, genomic defects and resistance to apoptosis, which characterize cancer cells. In response to genotoxic, replicative stress, Chk1 kinase, activated by Atr (ataxia-telangiectasia mutated and Rad3 related), delays cell cycle progression through S to G2 phases by inhibiting Cdc25 phosphatases, which control the timing of early versus late replication origin firing during normal S phase. This pathway maintains DNA replication through structurally unfavorable stretches of DNA, such as chromosome fragile sites in mammalian cells. Common fragile sites are chromosome regions that exhibit gaps or breaks, visible in metaphase chromosomes, when cells are exposed to replicative stress conditions. FRA3B and FRA16D, at chromosome regions 3p14.2 and 16q23.3, are the two most active or most fragile sites in human genome; those are altered frequently in various tumors. In this article, recent findings of the molecular mechanism of DNA damage and replication checkpoint are discussed, linking replication stress surveillance with apoptosis induction, which play a central role in determining if DNA damage is compatible with cell survival or requires cell elimination by apoptosis.

Introduction

DNA can be damaged in a variety of ways. DNA damage is a relatively common event in the life of a cell, which may lead to mutation, cancer and cellular or organismic death. Damage to DNA induces several cellular responses, which enable cells either to cope with the damage or to activate a programmed cell death process, presumably to eliminate cells with potentially catastrophic mutations [1]. Those cellular responses to DNA damage include: activation of a DNA damage checkpoint; removal of DNA damage and restoration of the structure of the DNA duplex, known as DNA repair mechanisms including direct repair, base excision repair, nucleotide excision repair, double-strand break repair, and cross-link repair; transcriptional response; and apoptosis, which eliminates cells with heavily damaged or seriously deregulated genome [1]. In this review, the recent knowledge of the molecular mechanism of checkpoint response to DNA damage linked with apoptosis, especially at S phase of cell cycle, are focused and analyzed.

Checkpoint Responses

In response to damages or cellular stresses, checkpoint at G1/S of cell cycle causes arrest to prevent replication of damaged DNA, and checkpoint at G2/M causes arrest to prevent segregation of damaged chromosomes during mitosis, whereas S-phase checkpoints activated by genotoxic insults delay or arrest cell cycle progression of S phase [1-3]. Atm (ataxia-telangiectasia mutated) and Atr (Atm- and Rad3-related) play an important role in an initiation of the signaling [1].

G1/S Checkpoint

The checkpoint response to DNA damage traversing through G1 depends on the Atm(Atr)/Chk2(Chk1)/p53/Mdm2/p21Cip1/Waf1 pathway, which is capable of inducing sustained, and sometimes even permanent G1 arrest [1, 3, 4]. Atm/Atr directly phosphorylate the p53 protein within its amino-terminal transactivation domain, particularly on serine 15; threonin 18, serine 20 and probably some additional p53 sequence(s) could be targeted by Chk1/Chk2 [3]. The ubiquitin ligase Mdm2 is involved in rapid p53 turnover, which is targeted after DNA damage by Atm/Atr, as well as by Chk2/Chk1 [3]. The modifications of p53 and Mdm2 play a role in regulation of the stabilization and accumulation of the p53 protein, which can lead to an increase of its transcription activity for target genes, including p21Cip1/Waf1, an inhibitor of cyclin-dependent kinases [5]; the p21Cip1/Waf1 is involved in silencing the G1/S-promoting cyclin/Cdk complexes (cyclin E/Cdk2 kinase) and G1 arrest. The signal inhibits the Rb/E2f pathway and prevents G1/S promotion. At G1/S checkpoint, thus Atm(Atr)/Chk2(Chk1) plays a critical role in regulation of p53/Mdm2/p21Cip1/Waf1-dependent checkpoint response. The G1 checkpoint response targets two critical tumor suppressor pathways, p53 and pRB, which are governed commonly deregulated in human cancer [3, 5, 6].

During unperturbed proliferation, the Atr/Chk1, rather than Atm/Chk2, maintains an appropriate abundance of Cdc25A through its ubiquitin-dependent, proteasome-mediated turnover through moderate constitutive phosphorylation of Cdc25A on its several serine residues [3, 4, 7]. The basal turnover of Cdc25A is estimated as T1/2 ~20-30 min [7]. In response to genotoxic stress, increased activity of Chk1 and Chk2, leads to downregulation of Cdc25A and consequently to the inhibition of cyclin E(A)/Cdk2 complexes [3, 4, 7]. The accelerated turnover of Cdc25A is T1/2 ~10 min [7]. Notably, the Chk1/Chk2/Cdc25A mechanism is implemented rapidly, independent of p53, which allows immediate G1/S arrest or relay, without the sustained p53-dependent G1 arrest [3].

S-phase Checkpoints

In response to genotoxic stress in S phase, which can be caused by environmental or endogenous genotoxic insults such as ionizing or ultraviolet radiartion, various chemicals and drugs, and reactive cellular metabolites, S-phase checkpoints can be activated, which delays or arrests cell cycle progression so as to allow for repair and prevention of the transmission of damaged or incompletely replicated chromosomes [1]. In the S phase, genotoxic stress can occur not only from DNA-damaging insults, but also from difficulties with the replication process it self. The creation of the duplicated sister chromatids offers more opportunities for error-free repair by homologous recombination, whereas the double strand breaks can be treated largely via non-homologous end-joining repair [7]. The data that have been obtained from yeast models now clearly indicated that the S phase checkpoints are considerably more significant for preventing genomic instability than the G1 or G2 checkpoint or the mitotic-spindle checkpoint [7].

The processes of S phase checkpoints are classified into three types: (1) the replication-dependent S-phase checkpoint (commonly referred as to the replication checkpoint); (2) the replication-independent intra-S-phase checkpoint (generally referred as to the intra-S-phase checkpoint), which can be induced by double strand breaks-induced; (3) the S-M checkpoint, which, like replication checkpoint, depends on DNA replication [7]. They are mutually closely coordinated and share some components. The molecular effectors remain to be more fully defined. Importantly, none of these three S-phase checkpoints requires p53, which is the key target of the G1/S checkpoint [7].

Replication Checkpoint

The replication checkpoint can be activated when the progression of replication forks is stalled or collapsed in response to stress such as the depletion of deoxyribonucleotide (dNTP) pools, chemical inhibition of DNA polymerases, or as a consequence of the collision of replication forks with aberrant DNA structure [7]. For examples, when cells are exposed to replicative stress conditions, gaps or breaks, visible in metaphase chromosomes can be induced, which are known as common fragile sites [8-11]. The replication checkpoint plays a role in the regulation of common fragile sites, although the other types of S-phase checkpoint seem to be involved, as discussed in below.

The replication checkpoint is shown to has two functions: (1) to inhibit the initiation of DNA replication through targeting the cyclin-Cdk and Cdc7-Dbf4 kinases, which regulate the assembly and firing replication origin, leading to inhibition of S phase; (2) to recover cell-cycle progression and dNTP pool after DNA repair, contributing to maintenance or restoration of the integrity of replication forks [7].

The key components of this machinery include replication protein A (Rpa), Atr, Atr-interacting protein (AtrIP), a mediator protein Claspin, Rad17 and the 9-1-1 complex [7]. Studies in yeast and even human cells have shown that Rad9 interacts with Hus1 and Rad1 in a heterotrimeric complex, the Rad9-Rad1-Hus1 (9-1-1) complex, which acts as a Pcna (Proliferating cell nuclear antigen)-like sliding clamp [12, 13]. In response to DNA damage, the 9-1-1 complex is loaded around DNA lesions by Rad17, which binds to chromatin prior to damage [14], and facilitates Atr-mediated phosphorylation and activation of Chk1 kinase to arrest cell cycle. Rad9 is involved in control of the DNA damage-induced checkpoint [15-18], but Rad9 can also participate in signaling apoptosis, through interaction with anti-apoptotic Bcl-2 family proteins, Bcl-2 and Bcl-X_L, but not with pro-apoptotic Bax and Bad [19], which is also discussed in below.

Intra-S-phase Checkpoint

The DNA double strand breaks, which have occurred in the genome loci outside the active replicons, can activate the intra-S-phase checkpoint [7]. The intra-S-phase checkpoint does not require active replication fork for its initiation. The extension of actively progressing replication forks is largely unaltered by signaling that is mediated by the double strand break-induced checkpoint [7, 20]. Thus this checkpoint is distinguished from replication-dependent checkpoint [7].

Nevertheless the intra-S-phase checkpoint might be overlapped with replication-dependent pathway in a certain physiological process, such as maintenance of chromosome fragility. Previous studies have shown involvement of repair systems in the regulation of replication fork at the common chromosome fragile sites, including Atr, but not Atm protein [21], Brca1 [22], Smc1 [23], and the Fanconi anemia pathway [24], but a recent study also showed that both homologous recombination and non-homologous end-joining repair pathways regulate common fragile site stability [25], indicating that double-strand breaks are formed at common fragile sites as a result of replication perturbation and that the double strand break-dependent process might associate at least partially with replication-dependent process in this case.

S-M Checkpoint

This checkpoint ensures that cells do not attempt to divide before the completion of faithful replication of the entire genome [7]. Like to the replication checkpoint, the S-M checkpoint depends on DNA replication [7]. Failure of the S-M checkpoint results in mitotic catastrophe of cells that incompletely duplicated DNA [7]. The key molecules of this machinery include cyclin B-Cdk1 complex. This process is partially overlapped with the replication checkpoint but also linked with G2/M checkpoint.

G2/M Checkpoint

The G2/M checkpoint prevents cells from initiating mitosis when some DNA damages are unrepaired [3]. The G2/M arrest is largely dependent on Chk1-mediated signaling pathway leading to inhibition of the Cyclin B1/Cdc2 activity. Chk1 is activated by phosphorylation on at least two residues (Ser345 and Ser317) located in a Ser/Thr-Gln-rich domain by the DNA damage-activated Atr and Atm kinases [1-4, 7]. When the G2/M checkpoint is activated, Atm/Atr, Chk1/Chk2 and/or p38-kinase lead to subcellular sequestration, degradation or inhibition of the Cdc25 family of phosphatases, which normally activate Cdc2 at the G2/M boundary [2, 3].

Checkpoint and Therapeutic Strategy

Cell cycle checkpoint and DNA repair pathways allow cells to deal with both endogenous and exogenous sources of DNA damage. These cellular responses are also important for determining toxicities and responses to current cancer therapies, most of which target the DNA [3]. The highly conserved mechanism plays an essential role in the process of mammalian DNA damage checkpoints, embryonic development and tumor suppression. As mentioned above, the G1 checkpoint response targets, p53 and pRB, are deregulated commonly in human cancer [3, 5, 6]; since the G1/S checkpoint is frequently inactivated in cancer due to mutation of p53, it is suggested the G2/M checkpoint may be significant in the response of cancerous cells to DNA-damaging chemotherapeutic agents [1, 2, 4].

Key Components in DNA Damage-dependent Response

PIKKs and the Signal

In general, initiation of the activities of the PI3K (phosphatidyl-inositol-3-OH kinase)-like kinases (PIKKs), Atm and Atr are the first steps characterized to date in the activation of signal transduction pathways that inhibit or arrest cell cycle progression after DNA damage [3]. As damage sensor proteins, not only Atm and Atr but also Rad17-Rfc complex and the 9-1-1 complex are involved in the activation of DNA damage checkpoints, to detect DNA damage and to initiate signal transduction cascades that use Chk1 and Chk2 Ser/Thr kinases and Cdc25 phosphatases [1]. The signal transducers activate p53 and inactivate cyclin-dependent kinases to inhibit cell cycle progression from G1 to S (the G1/S checkpoint), DNA replication (the S-phase checkpoints), or G2 to mitosis (the G2/M checkpoint) [1].

Chk2

Chk2, the mammalian homologue of *Saccharomyces cerevisiae* Rad53 and *Schizosaccharomyces pombe* Cds1, is a kinase whose activation by DNA damage prevents entry into mitosis and into S phase [26, 27]. Chk2 is activated by phosphorylation on Thr68 in an Atm and Nbs1-dependent manner after exposure to ionizing radiation. After DNA damage, Chk2 phosphorylates p53 on Ser20, attenuating the binding of p53 to Mdm2, a protein that targets p53 for degradation, and allowing accumulation of p53 and subsequent up-regulation of p21 [28]. The human Chk2 plays a central role in regulating the cellular response to DNA damage, resulting in cell cycle arrest, DNA repair, or apoptosis, depending on the severity of DNA damage and the cellular context [1].

Chk1

Chk1 in Checkpoints

The G2/M arrest is dependent largely on Chk1-mediated signaling pathways [1]. In response to DNA damage, Atr directly phosphorylates Chk1 at serine 317 and 345 *in vitro*, and phosphorylation of these sites in human cells in response to UV light and hydroxyurea is Atr-dependent [29, 30]. Chk1 also undergoes auto-phosphorylation associated with its full activation [31]. Although Chk1 is believed to play an important role in checkpoint responses to aberrant DNA structures, genetic analysis of Chk1 functions is complicated by lethality of Chk1-deficient embryonic cells. The gene targeting technology using DT40 B lymphoma cells uncovered that Chk1-deficient cells fail to maintain viable replication forks or suppress futile origin firing when DNA polymerase is inhibited, leading to incomplete genome duplication and diminished cell survival after release from replication arrest [32], and that Chk1-dependent processes promote tumor cell survival after perturbations of DNA structure or metabolism [32], suggesting that Chk1-deficient tumor cells are viable but exhibit multiple checkpoint and survival defects [32]. Chk1 heterozygosity results in three distinct haploinsufficient phenotypes that can contribute to tumorigenesis: inappropriate S phase entry, accumulation of DNA damage during replication, and failure to restrain mitotic entry, suggesting that Chk1 is haploinsufficient for multiple functions critical to tumor suppression [33]. Atr disruption leads to chromosomal fragmentation and early embryonic lethality [34, 35]. Thus, defects in the Atr/Chk1 pathway have been implicated in the loss of genomic instability and development of cancer [3]. Genetic ablation of Atr [34, 35] or Chk1 [32, 33] or their inhibition by kinase inhibitors [36-38] sensitizes cells to DNA damage and induces apoptosis, showing that the checkpoint play a critical role in determining cell survival following genotoxic stress.

Chk1 Phyosphorylation

Phosphorylation and activation of Chk1 by Atr require Claspin, a protein first identified by co-purification with Chk1 from *Xenopus* egg extracts [39] and subsequently shown to be involved in Chk1 activation in human cells [40, 41]. The recent study identified human Claspin as a substrate for caspase-7, supporting a distinct role in the control of early events during genotoxic responses [42]. The study shows that Claspin was cleaved by caspase-7 at a single aspartate residue into a large N-terminal fragment and a smaller C-terminal fragment that contain different functional domains [42]. The large N-terminal fragment was heavily phosphorylated in a human cell-free system in response to double-stranded DNA oligonucleotides, and this fragment retained Chk1 binding activity [42]. In contrast, the smaller C-terminal fragment did not bind Chk1, but did associate with DNA and inhibited the DNA-dependent phosphorylation of Chk1 associated with its activation [42]. Thus, the cleavage of Claspin by caspase-7 inactivates the Chk1 signaling pathway, and the inactivation may play a pivotal role in the switch from cell cycle arrest and restraint of apoptosis to inactivation of the checkpoint and induction of apoptosis in response to genotoxic damage [42]. Although the exact mechanism of the induction of apoptosis after blocking the recovery from Chk1-mediated S phase arrest in response to DNA-damage remains to be more fully defined, this mechanism of the cleavage of Claspin by caspase-7 may regulate the balance between cell cycle arrest and induction of apoptosis during the response to genotoxic stress [42].

Chk1 Degradation

Recent study shows that genotoxic stress targets human Chk1 for degradation by the ubiquitin-proteasome pathway [43], which implicates Ser-345 as a nodal site for the regulation of Chk1 activation and inactivation in human cells, and indicates that Atr-mediated phosphorylation of Chk1 at Ser-345 not only leads to Chk1 activation, but also marks this protein for eventual destruction by the ubiquitin-proteasome pathway [43]. The exposure to anti-cancer reagent, camptothecin (CPT), indicate that replication stress triggers Chk1 phosphorylation at Ser-345, leading to dynamic changes in Chk1 subcellular localization and to the ubiquitination of the phosphorylated protein by at least Cul1- and Cul4A-containing E3 ligases [43]. It is suggested that replication stress-induced Chk1 degradation normally functions to limit the duration of Chk1 signaling during low-intensity replication stress [43], and that this response plays an important role in anti-cancer surveillance.

Replication and Related Machineries

Replication Factor C (Rfc)

The Pcna functions at multiple levels in directing DNA metabolic pathways [44]. Unbound to DNA, Pcna can promote the localization of replication factors or components with a consensus Pcna-binding domain to replication machineries [44]. When bound to DNA, Pcna is involved in replication, repair, modification of DNA and chromatin modeling, in concert with various proteins [44]. Loading of Pcna onto DNA at template–primer junctions is performed by Rfc, a heteropentameric AAA+ protein complex consisting of the Rfc1, Rfc2, Rfc3, Rfc4, and Rfc5 subunits [44]. The process is ATP-dependent [44].

Recently a dynamic regulatory function of RFC (p140) was reported [45-48]. Rfc (p140) has been described in a large protein complex containing both Brca1 and Atm, termed the Brca1-associated genome surveillance complex (BASC), postulated to act as a sensor for DNA damage [45]. Rfc (p140) has been found to interact with the Rb tumor suppressor, which was shown to have a prosurvival function following ultraviolet light stimulation [46]. Rfc (p140) was described as a transcriptional coactivator for the bZIP transcription factor C/EBP [47]. Rfc (p140) associates with histone deacetylase activity and interacts with histone deacetylase 1 (Hdac1) via two distinct transcriptional repression domains, showing Rfc (p140) is a deacetylase-dependent repressor of transcription [48].

Alternative Rfc Complexes

Recently, alternative Rfc complexes consisting of the small Rfc subunits together with an alternative Rfc1-like subunit have been identified [44]. In yeast, a complex required for the DNA damage checkpoint contains the Rad24 subunit, a complex required for sister chromatid cohesion contains the Ctf18 subunit, and a complex that aids in genome stability contains the Elg1 subunit. Recently, the study of human and mouse indicates that the Frag1 (Ctf18/Rad24/Elg1-related gene 1) gene, a conserved homolog of Rfc large subunit that is homologous to the alternative Rfc subunits Elg1, Ctf18/Chl12, and Rad24 of budding yeast, plays a role in regulation of genomic integrity [49]. In response to stress, Atr participates in the down-regulation of Frag1, leading to the induction of apoptosis through the release of Rad9 from damaged chromatin during the S phase of the cell cycle, allowing Rad9-Bcl2 association and induction of proapoptotic Bax protein [49]. It is suggested that the Frag1 signal pathway links replication stress surveillance with apoptosis induction, which plays a role in determining whether DNA damage is compatible with cell survival or whether it requires cell elimination by apoptosis [49].

9-1-1 Complex

Studies in yeast and human have shown that Rad9, Hus1, and Rad1 interact in a heterotrimeric complex, namely the 9-1-1 complex, which acts as a Pcna-like sliding clamp [12, 13]. In response to DNA damage, the 9-1-1 complex is loaded around DNA lesions by Rad17, which then facilitates Atr-mediated phosphorylation and activation of Chk1 kinase. It was shown that human Rad17 binds to chromatin prior to damage, and recruits the Rad9 complex onto chromatin after damage [14]. Rad9 and Hus1 were also shown to interact with Hdac1 [50], and Rad9 was shown to interact with TopBP1, a protein containing multiple BRCT domains [51]. Taken together with the recent reports showing that the 9-1-1 complex is also involved in DNA repair [12], it suggested that the 9-1-1 clamp coordinates multiple functions of chromatin regulation, checkpoint activation, DNA repair, and apoptosis.

H2ax

The phosphorylated histone H2ax foci show DNA double strand breaks and co-localize with double strand break repair factors such as Atm, 53BP1, and Mdc1 [1, 3]. Complete or partial deficiency of H2ax results in a dramatically increased incidence of both lymphoma and solid tumors in mice [52, 53]. When H2ax levels are reduced to 50% of wild-type levels, the resultant decrease in H2ax foci formation after ionizing radiation is insufficient to maintain genomic stability and leads to increased levels of chromosomal aberrations, reduced growth rates, and radiation sensitivity, indicating that the H2ax functions in a dosage-dependent or haploinsufficient manner [52, 53].

Apoptosis Pathway

Whereas DNA damage activates checkpoint mechanisms, which slow down or arrest cell cycle progression, allowing the cell to be repaired, it can also initiate pathways leading to apoptosis and the removal of a damaged cell from a tissue [54]. The balance between cell cycle arrest and damage repair on one hand and the initiation of apoptosis on the other hand determines if cellular or DNA damage is compatible with cell survival or requires cell elimination by apoptosis, which is important for maintenance of genomic stability. Defects of those processes may lead to hypersensitivity to cellular or DNA damage, which induces the susceptibility to DNA damage, genomic defects and resistance to apoptosis, which characterize cancer cells. The exact mechanism how apoptosis is induced in cells after their cell cycle was arrested remains to be more fully defined.

Genotoxic stress or potentially deleterious mutations can elicit checkpoint-signaling pathways that block cell cycle progression, regulate DNA repair, and trigger apoptosis [13], through both membrane death receptors and the endogenous mitochondrial death pathways [55]. The mitochondrial pathway is supposed to play a critical role in response to genotoxic stress [1]. In this article, recent findings are discussed.

Rad9

It is reported that Rad9 exerts multiple functions. Rad9 is seems to be involved in activation of the checkpoint, which can contribute to the regulation of cell survival. Although biological effects of the protein function may vary and might depend on type of stresses and cells, and experimental conditions, the differential roles of Rad9 or the mechanism for switching the biological effects remain to be more fully defined.

Pro-apoptosis Aspect of Rad9

Recent study of yeast two-hybrid and co-immunoprecipitation techniques showed Rad9, a component of a heterotrimeric 9-1-1 complex, which is involved in the control of damage-induced checkpoint, interacting with anti-apoptotic Bcl-2 family proteins Bcl-2 and Bcl-X_L, but not with the pro-apoptotic Bax and Bad [19]; in this model, Rad9 protein is involved in control of the DNA damage-induced checkpoint [15-18], but Rad9 can also participate in signaling apoptosis, through interaction with anti-apoptotic Bcl-2 family proteins, Bcl-2 and Bcl-X_L, but not with pro-apoptotic Bax and Bad [19]. Rad9 translocates to mitochondria, where it triggers apoptosis by binding to and inhibiting Bcl-2 and Bcl-xL [19]. The interaction of Bcl2 with Bax prevents Bax from inducing cytochrome c release and cell death, and the Bax:Bcl2 ratio is crucial for regulation of apoptosis [56, 57]. Since the 9-1-1 clamp is also involved in DNA repair [12], the Rad9 complex is thought to play a key role in coordinating multiple functions of checkpoint activation, DNA repair and apoptosis.

Anti-apoptosis Function of Rad9

On the other hand, recent study showed another possible role of Rad9 in regulation of response to replication stress or DNA damage; a Rad9-containing clamp is loaded onto damaged DNA, where it participates in Chk1 activation and subsequent events that contribute to cell survival [58]. The study of Rad9-/- murine embryonic stem (ES) cells indicates that the predominant role of Rad9 in ES cells is to promote survival after replicative stress and topoisomerase-mediated DNA damage [58]. In this study, ES cells were treated with camptothecin (CPT), etoposide, or cytarabine, all prototypic examples of three classes of widely used anticancer agents [58]. All three agents induced Rad9 chromatin binding. Each of these agents also triggered S-phase checkpoint activation in parental ES cells, whereas the ability of cytarabine, but not camptothecin and etoposide, to activate the S-phase checkpoint was severely compromised in Rad9-/- cells [58]. Nonetheless, Rad9 deletion sensitized ES cells to the cytotoxic effects of all three agents [58]. In the report, the possibilities are discussed about rad9 effect including that the amount of mitochondrial Rad9 was below the limit of detection with the reagents utilized or that mitochondrial trafficking of Rad9 plays a more important role in other cell lines (but not ES calls) [58].

The balance between pro-apoptotic function via mitochondrial pathway on one hand and the activation of checkpoint leading to cell survival on the other hand determines if cellular is compatible with cell survival or requires cell elimination by apoptosis. Recent studies indicate that human Rad9 can activate transcription of p21 [59], that Rad9 is required for activation of Chk1 pathway [60-62], and that human Rad9 is required for the activation of S-phase checkpoints and the maintenance of chromosomal stability, via possible involvement of the phosphorylation of Rad17, Chk1 and Smc1 [18]. Analyses of the Rad9 gene from yeast, mouse and human have uncovered roles in multiple, fundamental biological processes, which are primarily but not exclusively important for regulating genomic integrity [63]. It is proposed that the analyses of the functions of Rad9, and in particular its role in regulating and coordinating numerous fundamental biological activities, should not only provide information about the molecular mechanisms of individual cellular processes, but might also lend insight into the more global coordination of what at least superficially present as independent pathways [63].

Cooperation of Chk1 and p21

Using thymidine treatment in culture, it was shown that Chk1 plays a primary role in the protection of cells from death induced by replication fork stress, whereas p21 mediates its role in regulating entry into S phase [64]. The model of a role of Chk1 and p21 in prevention of cell death after replication fork stress proposed that: in normal cells, DNA synthesis inhibitors activate Chk1 that protects replication forks and prevents S-phase apoptosis; in the absence of Chk1, the lack of Chk1-induced protection favors the induction of apoptosis in the S phase, whereas p21 activation triggers an alternative protective mechanism by preventing cells from entering S phase; the depletion of both Chk1 and p21 eliminates the protection of G1 phase cells and results in apoptosis of virtually the entire cell population [64]. It is suggested that Chk1 and p21 cooperate to prevent apoptosis during DNA replication fork stress, which is does not require p53-dependent mechanism [64].

Bcl-2 Family

BCL-2 gene was originally identified through the study of human follicular lymphoma cells with the t(14:18) chromosome translocation [65, 66]; numerous pro-apoptotic or anti-apoptotic signal transducing molecules are studied, which participate in pathological stimuli from the Bcl-2 family, including mitochondrial lipid-proteins that regulate bioenergetic metabolite flux, are putative components of the permeability transition pore, and converge on mitochondria in the local regulation and execution of mitochondrial outer membrane permeabilization (MOMP) [67]. It is known that the blc-2 family of proteins consists not only of promoters but also of suppressors of programmed cells death [67]. Among evolutionally conserved Bcl-2 homology (BH) domains, 1-4, the BH3 domain is critical for pro-apoptotic Bcl-2 family member, which appear to function essentially as trans-dominant inhibitors by binding to anti-apoptotic Bcl-2 family and neutralizing their cell-survival activity [68, 69]. Collectively, it is suggested that DNA damage induces apoptosis though a signaling pathway involving proteins with the BH3 domain, which can play a role in regulation of the Bcl-2 function [19].

GENOMIC INSTABILITY AND CANCER

In response to genotoxic stress or DNA damage, which can be caused by environmental or endogenous insults, cell cycle checkpoints can be activated. Defects in those pathways seem to induce the genomic instability, which characterize cancer cells. The DNA damage response underlies several important aspects in cancer research, including carcinogenesis, the therapeutic modalities, and the side effect of therapy, which DNA damage is responsible for [3].

Coupling DNA Replication with S Phase Checkpoint

DNA replication is an error-prone process that requires continuous surveillance of DNA for damage or incomplete replication before the cell enters mitosis [1]. In response to aberrant structures being detected, checkpoint mechanisms are activated that result in slowing down or arresting cell cycle progression, which allow the DNA damage to be repaired and the replication to be completed. As mentioned above, replicative stress activates a checkpoint pathway that includes Atr and its downstream substrates, Chk1 kinase [70]. This pathway is responsible for maintenance of DNA replication through structurally unfavorable stretches of DNA, termed chromosome fragile sites in mammalian cells. The common chromosome fragile sites exhibit gaps or breaks, visible in metaphase chromosomes, when cells are exposed to replicative stress conditions.

Different forms of genotoxic stress can lead to activation of the Atr-dependent DNA damage checkpoint [71]. It has been uncertain whether each distinct lesion is directly recognized by checkpoint proteins or whether these lesions are processed into a common intermediate. Recently it was shown that functional uncoupling of MCM helicase and DNA polymerase activities occurs in response to multiple forms of DNA damage [72] and that this event is necessary to generate a common DNA intermediate required for activation of Chk1 during S phase [72]. As the model, it was shown that functional uncoupling of MCM helicase and DNA polymerase activities serves to increase the amount of checkpoint-activating signal generated at each replication fork [72].

Alterations of DNA Damage-dependent Checkpoint Response in Cancer

In early stages of cancer development, precancerous cells experience selective pressure to escape from blocked cell cycle progression, induced by checkpoint responses to DNA damage. Frequent mutations of DNA damage checkpoint genes contribute to overcoming the cell cycle block during cancer development. Two recent studies of lung and skin hyperplasias and dysplasias [73, 74] showed that an Atr/Atm-regulated DNA damage response network is activated from early dysplastic stages. Mutations compromising this checkpoint, including defects in the Atm-Chk2-p53 and Atr-Chk1 pathways, would allow cell proliferation, survival and tumor progression, leading to further DNA replication stress, DNA double-strand breaks, increased genomic instability and selective pressure for accumulating mutations [73, 74]. Chromosomal instability is a common feature of many human cancers, and most of the instability syndromes characterized by sensitivity to different types of DNA damage also show increased cancer susceptibility [3, 75]. Better understanding of these syndromes and their links with familial cancer would provide new insights into the association between defects in DNA damage response, cell cycle control, DNA repair and cancer [3, 75]. The elucidation of the damage response repair networks will have important implications for the development of cancer management and treatment [3, 75].

Chromosome Fragile Sites

Under conditions that inhibit DNA replication, culture of cells leads to formation of non-random gaps, breaks and rearrangements in metaphase chromosomes, which is known as chromosome fragile sites [76], as mentioned above. Studies of the fragile X syndrome have been extended to the classification of fragile sites into two large sub-groups, 28 rare [77] and 89 common fragile sites [78], based on the biological basis of induction and the relative frequency of observed specific fragile sites [76]. Whereas rare fragile sites, such as FRAXA within Fragile X Mental Retardation (FMR)- 1 gene, arise from mutation at di- or tri-nucleotide repeats and present in less than 5% of the population [79-81], the common fragile sites are found in all individuals and represent a normal component of chromosome structure [8, 9, 76, 78].

Rare Fragile Sites

A positional cloning approach allowed the identification of FRAXA fragile sites, responsible for fragile X syndrome, which is one of the leading causes of familial mental retardation [82]. The variation of a (CGG)n repeat was found within the 5' untranslated region in the first exon of the FMR-1 gene, which is increased to over 200 repeats and associated with chromosomal fragility, abnormal methylation of adjoining CpG isoland, and the absence of expression of the FMR-1 gene [83]. FMR-1 is implicated in controlling translation of microtubule-associated protein 1B and microtubule stability in brain neuron development [84]. Also, other four folate-sensitive fragile sites, FRAXE, FRAXF, FRA16A and FRA11B have been cloned, which contain an increased repeat of (CGG)n with abnormal methylation in individuals demonstrating fragility at those sites [85-88]. Despite molecular characterization of the rare fragile sites, the relationship between the chemistry of fragile sites induction and their DNA sequence position remain to be more fully defined.

Common Fragile Sites

Common fragile sites are chromosome regions that exhibit gaps or breaks, visible in metaphase chromosomes, when cells are exposed to replicative stress conditions, and FRA3B and FRA16D, at chromosome regions 3p14.2 and 16q23.3, are the two most active or most fragile sites in human genome; those are altered frequently in various tumors [8-11].

The common fragile sites extend over large regions (> several hundreds bps), exhibit late or delayed replication, and presumably correspond to, or are as a consequence of stalled or collapsed replication forks; they contain regions of high flexibility but are otherwise unremarkable in sequence [78]. While the rare common fragile sites are more or less sensitive to low levels of folic acids in the culture media, the common ones are not apparently sensitive to folate deficiency, but most of them are induced by aphidicolin [8, 89]. Aphidicolin is a specific inhibitor of the replicative DNA polymerase alpha and has a high specificity for breakage at the constitutive fragile site, FRA3B, at human chromosomal band 3p14.2 [90], which stands out as the most active fragile site in the human genome, as well as the other aphidicolin-inducible common fragile sites, 16q23.2 (*FRA16D*), 6q26 (*FRA6E*), 7q31.2 (*FRA7G*), and Xp22.31 (*FRAXB*) [10, 21, 76, 91]. As such, common fragile sites provide an appropriate cytological assay for studying the pathway affecting stalled replication.

Molecular Mechanism of Regulation of Common Fragile Sites

Previous studies have shown involvement of repair systems in the regulation of the common chromosome fragile sites, including Atr, but not Atm protein [21], Brca1 [22], Smc1 [23], and the Fanconi anemia pathway [24]. A recent study showed that both homologous recombination and non-homologous end-joining repair pathways regulate FRA3B and FRA16D fragile site stability, indicating that double-strand breaks are formed at common fragile sites as a result of replication perturbation [25].

Although fragile sites are genomic regions that are significantly more sensitive to replication perturbation or inhibition than other regions in the genome, which are likely "difficult" regions for replication, the reason for their sensitivity is not fully understood. The previous studies have shown that the fragile regions are enriched in AT-dinucleotide runs [92], which may lead to perturbed or inhibit DNA replication process due to their high DNA flexibility and their potential to form secondary structures [11].

A model for the molecular events leading from replication stress to the induction of common fragile sites is proposed [25]. Under conditions of replication stress, which only slow the replication of the entire genome, the replication forks tend to stall and collapse at the fragile regions, leading to double strand break formation [25]. The stalled replication forks are recognized by the DNA damage response mechanism, which activates cell cycle checkpoints through the Atr cascade and other components, including histone H2ax, Brca1, Smc1, Mdc1, and Fancd2 [25]. In parallel to the checkpoint activation, the DNA repair pathways are recruited to the damage sites [25]. The stalled forks may be restarted or repaired by the homologous recombination pathway; or, in the case of fork collapse and double strand break formation, repaired by the homologous recombination and/or non-homologous end-joining pathways [25]. Most of these replication-induced lesions are repaired and the chromosome structure at metaphase is normal [25]. However, in cases where the lesion fails to be repaired and persists into mitosis, or if the repair occurs late in S phase or in G2, the chromatin condensation of a large region will be disrupted [25]; in this case, gaps and constrictions will appear in metaphase at the fragile regions [25].

Fragile sites are relatively unstable in normal cells under the condition with replicative stress, such as exposure to aphidicolin, an inhibitor of DNA polymerases. It remains to be more fully defined how alterations of DNA structure or DNA damage, which can be visible as gaps and breaks in fragile sites, result in the activation of checkpoint to allow further time for those repairs before mitosis; and whether during or after mitosis, they elicit the induction of apoptosis or mitotic catastrophe of cells that incompletely duplicated DNA.

Common Fragile Sites and Cancer

Unrepaired double strand breaks that accumulate at the regions such as the fragile sites following replication stress *in vivo* predispose the genome to chromosomal rearrangements that can promote cancer or lead to inherited diseases [25]. Since the common fragile sites are found in all individuals and represent a normal component of chromosome structure [8, 9, 76, 78], the significance in carcinogenesis has been investigated. The common chromosomal fragile sites apparently coincide with chromosomal bands that are frequently altered in cancer [89], and some fragile sites lie within tumor suppressor genes, such as FHIT at FRA3B [93, 94] and WWOX at FRA16D [95-97], which led to hypothesize that the pathology of them may be associated with cancer development [10].

The FHIT gene locates at the FRA3B region. The 1.7 MB FHIT gene spans the most active common fragile site FRA3B in the human genome, encodes a 1.2 kb mRNA and 16.8 kDA protein, and is frequently involved in biallelic loss and other cytogenetic abnormalities in tumors [93, 98]. FHIT deletions, abnormal transcripts, promoter hypermethylation and associated loss of Fhit expression are common in human malignancies, including those of the head and neck, esophagus, stomach, breast, lung and bladder [10, 93, 98-102]. Indeed, loss of Fhit expression is observed in premalignant lesions as well as advanced cancers, strongly suggesting that loss of Fhit expression, as a result of the exquisite susceptibility of FHIT/FRA3B to damage caused by environmental carcinogens, plays a role in initial stages of multi-step carcinogenesis [103].

Fhit overexpression experiments following plasmid, adenoviral or adenoassociated virus gene transfer into cancer cell lines of lung, head and neck, kidney, esophagus, cervix and pancreas [104-108], have shown that Fhit effectively suppresses cell growth and induces caspase-dependent apoptosis in some cancers and cancer-derived cell lines [106]. Introduction of Fhit into esophageal and renal cancer cells showed that susceptibility to apoptosis or cell growth inhibition is not restricted to cancer cells with complete loss of endogenous Fhit expression [106, 109]; in contrast, it was observed that Fhit exerts little effect on noncancerous cells [105, 106, 110] and some cancer cells stably expressing Fhit [111, 112]. How Fhit induces susceptibility to apoptosis or cell growth inhibition in cancer but not normal cells is poorly understood. Fhit-/- and Fhit+/- mice exhibit increased susceptibility to spontaneous and carcinogen-induced tumors [113-115] and viral mediated Fhit expression effectively inhibited or prevented the development of induced tumors [115, 116]. The animal experiments strongly suggest a selective Fhit suppressive effect on abnormally proliferating precancerous and cancerous lesions with altered cellular functions.

Conclusion

In all eukaryotic organisms, checkpoint responses to genotoxic stress or DNA damage plays a central role in determining if DNA damage is compatible with cell survival or requires cell elimination by apoptosis. This system is important at least for preventing carcinogenesis. In cancer, dysfunctions of checkpoint responses contribute to accumulated alterations of specific chromosome loci, leading to gains of oncogens or losses of tumor suppressor genes. Those alterations associated with evading apoptosis by down- or up-regulation of pro-apoptotic or anti-apoptotic proteins. Ongoing genomic instability leads to cumulative genetic errors and thereby the generation of multiple clones of transformed cells, which are generally refractory to conventional multidisciplinary interventions. Understanding the damage response and repair networks will have important implications for the development of management and treatment of human diseases.

References

[1] Sancar, A., Lindsey-Boltz, L.A., Unsal-Kacmaz, K., and Linn, S. (2004). Molecular mechanisms of mammalian DNA repair and the DNA damage checkpoints. *Annu. Rev. Biochem., 73*:39-85.

[2] Nyberg, K.A., Michelson, R.J., Putnam, C.W., and Weinert, T.A. (2002). Toward maintaining the genome: DNA damage and replication checkpoints. *Annu. Rev. Genet., 36*:617-56.

[3] Kastan, M.B., and Bartek, J. (2004). Cell-cycle checkpoints and cancer. *Nature, 432*:316-23.

[4] Bartek, J., and Lukas, J. (2003). Chk1 and Chk2 kinases in checkpoint control and cancer. *Cancer Cell, 3*:421-9.

[5] Wahl, G.M., and Carr, A.M. (2001). The evolution of diverse biological responses to DNA damage: insights from yeast and p53. *Nat. Cell Biol., 3*:E277-86.

[6] Bartek, J., Bartkova, J., and Lukas, J. (1997). The retinoblastoma protein pathway in cell cycle control and cancer. *Exp. Cell Res., 237*:1-6.

[7] Bartek, J., Lukas, C., and Lukas, J. (2004) Checking on DNA damage in S phase. *Nat. Rev. Mol. Cell. Biol., 5*:792-804.

[8] Glover, T.W., Berger, C., Coyle, J., and Echo, B. (1984). DNA polymerase alpha inhibition by aphidicolin induces gaps and breaks at common fragile sites in human chromosomes. *Human Genetics, 67*:136-42.

[9] Glover, T.W., and Stein, C.K. (1988). Chromosome breakage and recombination at fragile sites. *Am. J. Hum. Genet., 43*:265-73.

[10] Huebner, K., and Croce, C.M. (2001). FRA3B and other common fragile sites: the weakest link. *Nat. Rev. Cancer, 1*:214-221.

[11] Zlotorynski, E., Rahat, A., Skaug, J., Ben-Porat, N., Ozeri, E., Hershberg, R., Levi, A., Scherer, S.W., Margalit, H., and Kerem, B. (2003). Molecular basis for expression of common and rare fragile sites. *Mol. Cell. Biol., 23*:7143-51.

[12] Hang, H., and Lieberman, H.B. (2000). Physical interactions among human checkpoint control proteins HUS1p, RAD1p, and RAD9p, and implications for the regulation of cell cycle progression. *Genomics, 65*:24-33.

[13] Parrilla-Castellar, E.R., Arlander, S.J., and Karnitz, L. (2004). Dial 9-1-1 for DNA damage: the Rad9-Hus1-Rad1 (9-1-1) clamp complex. *DNA Repair (Amst), 3*:1009-14.

[14] Zou, L., Cortez, D., and Elledge, S.J. (2002). Regulation of ATR substrate selection by Rad17-dependent loading of Rad9 complexes onto chromatin. *Genes Dev., 16*:198-208.

[15] Lindsey-Boltz, L.A., Bermudez, V.P., Hurwitz, J., and Sancar, A. (2001). Purification and characterization of human DNA damage checkpoint Rad complexes. *Proc. Natl. Acad. Sci.* USA., *98*:11236-41.

[16] St Onge, R.P., Besley, B.D., Pelley, J.L., and Davey, S. (2003). A role for the phosphorylation of hRad9 in checkpoint signaling. *J. Biol. Chem., 278*:26620-8.

[17] Roos-Mattjus, P., Hopkins, K.M., Oestreich, A.J., Vroman, B.T., Johnson, K.L., Naylor, S., Lieberman, H.B., and Karnitz, L.M. (2003). Phosphorylation of human Rad9 is required for genotoxin-activated checkpoint signaling. *J. Biol. Chem., 278*:24428-37.

[18] Dang, T., Bao, S., and Wang, X.F. (2005). Human Rad9 is required for the activation of S-phase checkpoint and the maintenance of chromosomal stability. *Genes Cell., 10*:287-95.

[19] Komatsu, K., Miyashita, T., Hang, H., Hopkins, K.M., Zheng, W., Cuddeback, S., Yamada, M., Lieberman, H.B., and Wang, H.G. (2000). Human homologue of S. pombe Rad9 interacts with BCL-2/BCL-xL and promotes apoptosis. *Nat. Cell Biol., 2*:1-6.

[20] Merrick, C.J., Jackson, D., and Diffley, J.F. (2004). Visualization of altered replication dynamics after DNA damage in human cells. *J. Biol. Chem., 279*:20067-75.

[21] Casper, A.M., Nghiem, P., Arlt, M.F., and Glover, T.W. (2002). ATR regulates fragile site stability. *Cell, 111*:779-789.

[22] Arlt, M.F., Xu, B., Durkin, S.G., Casper, A.M., Kastan, M.B., and Glover. T.W. (2004). BRCA1 is required for common-fragile-site stability via its G2/M checkpoint function. *Mol. Cell. Biol., 24*:6701-9.

[23] Musio, A., Montagna, C., Mariani, T., Tilenni, M., Focarelli, M.L., Brait, L., Indino, E., Benedetti, P.A., Chessa, L., Albertini, A., Ried, T., and Vezzoni, P. (2005). SMC1 involvement in fragile site expression. *Hum. Mol. Genet., 14*:525-33.

[24] Howlett, N.G., Taniguchi, T., Durkin, S.G., D'Andrea, A.D., and Glover, T.W. (2005). The Fanconi anemia pathway is required for the DNA replication stress response and for the regulation of common fragile site stability. *Hum. Mol. Genet., 14*:693-701.

[25] Schwartz, M., Zlotorynski, E., Goldberg, M., Ozeri, E., Rahat, A., le Sage, C., Chen, B.P., Chen, D.J., Agami, R., and Kerem, B. (2005). Homologous recombination and nonhomologous end-joining repair pathways regulate fragile site stability. *Genes Dev., 19*:2715-26.

[26] Matsuoka, S., Huang, M., and Elledge, S.J. (1998). Linkage of ATM to cell cycle regulation by the Chk2 protein kinase. *Science, 282*:1893-7.

[27] Matsuoka, S., Rotman, G., Ogawa, A., Shiloh, Y., Tamai, K., and Elledge, S.J. (2000). Ataxia telangiectasia-mutated phosphorylates Chk2 in vivo and in vitro. *Proc. Natl. Acad. Sci.* USA., *97*:10389-94.

[28] Hirao, A., Cheung, A., Duncan, G., Girard, P.M., Elia, A.J., Wakeham, A., Okada, H., Sarkissian, T., Wong, J.A., Sakai, T., De Stanchina, E., Bristow, R.G., Suda, T., Lowe, S.W., Jeggo, P.A., Elledge, S.J., and Mak, T.W. (2002). Chk2 is a tumor suppressor that regulates apoptosis in both an ataxia telangiectasia mutated (ATM)-dependent and an ATM-independent manner. *Mol. Cell. Biol., 22*: 6521-32.

[29] Liu, Q., Guntuku, S., Cui, X.S., Matsuoka, S., Cortez, D., Tamai, K., Luo, G., Carattini-Rivera, S., DeMayo, F., Bradley, A., Donehower, L.A., and Elledge, S.J. (2000). Chk1 is an essential kinase that is regulated by Atr and required for the G(2)/M DNA damage checkpoint. *Genes Dev., 14*: 1448-59.

[30] Zhao, H., and Piwnica-Worms, H. (2001). ATR-mediated checkpoint pathways regulate phosphorylation and activation of human Chk1. *Mol. Cell. Biol., 21*:4129-39.

[31] Walworth, N.C., and Bernards, R. (1996). rad-dependent response of the chk1-encoded protein kinase at the DNA damage checkpoint. *Science, 271*:353-6.

[32] Zachos, G., Rainey, M.D., and Gillespie, D.A. (2003). Chk1-deficient tumour cells are viable but exhibit multiple checkpoint and survival defects. *EMBO J., 22*:713-23.

[33] Lam, M.H., Liu, Q., Elledge, S.J., and Rosen, J.M. (2004). Chk1 is haploinsufficient for multiple functions critical to tumor suppression. *Cancer Cell, 6*:45-59.

[34] Brown, E.J., and Baltimore, D. (2000). ATR disruption leads to chromosomal fragmentation and early embryonic lethality. *Genes Dev., 14*:397-402.

[35] de Klein, A., Muijtjens, M., van Os, R., Verhoeven, Y., Smit, B., Carr, A.M., Lehmann, A.R., and Hoeijmakers, J.H. (2000). Targeted disruption of the cell-cycle checkpoint gene ATR leads to early embryonic lethality in mice. *Curr. Biol., 10*:479-82.

[36] Sarkaria, J.N., Tibbetts, R.S., Busby, E.C., Kennedy, A.P., Hill, D.E., and Abraham, R.T. (1998). Inhibition of phosphoinositide 3-kinase related kinases by the radiosensitizing agent wortmannin. *Cancer Res., 58*:4375-82.

[37] Graves, P.R., Yu, L., Schwarz, J.K., Gales, J., Sausville, E.A., O'Connor, P.M., and Piwnica-Worms, H. (2000). The Chk1 protein kinase and the Cdc25C regulatory pathways are targets of the anticancer agent UCN-01. *J. Biol. Chem., 275*:5600-5.

[38] Busby, E.C., Leistritz, D.F., Abraham, R.T., Karnitz, L.M., and Sarkaria, J.N. (2000). The radiosensitizing agent 7-hydroxystaurosporine (UCN-01) inhibits the DNA damage checkpoint kinase hChk1. *Cancer Res., 60*:2108-12.

[39] Kumagai, A., and Dunphy, W.G. (2000). Claspin, a novel protein required for the activation of Chk1 during a DNA replication checkpoint response in Xenopus egg extracts. *Mol. Cell., 6*: 839-49.

[40] Chini, C.C., and Chen, J. (2003). Human Claspin Is Required for Replication Checkpoint Control. *J. Biol. Chem., 278*:30057-62.

[41] Lin, S.Y., Li, K., Stewart, G. S., and Elledge, S. J. (2004). Human Claspin works with BRCA1 to both positively and negatively regulate cell proliferation. *Proc. Natl. Acad. Sci.* USA., *101*: 6484-89.

[42] Clarke, C.A., Bennett, L.N., and Clarke, P.R. (2005). Cleavage of claspin by caspase-7 during apoptosis inhibits the Chk1 pathway. *J. Biol. Chem., 280*:35337-45.

[43] Zhang, Y.W., Otterness, D.M., Chiang, G.G., Xie, W., Liu, Y.C., Mercurio, F., and Abraham, R.T. (2005). Genotoxic stress targets human Chk1 for degradation by the ubiquitin-proteasome pathway. *Mol. Cell., 19*:607-18.

[44] Majka, J., and Burgers, P.M. (2004). The PCNA-RFC families of DNA clamps and clamp loaders. *Prog. Nucl. Acid Res. Mol. Biol., 78*:227-60.

[45] Wang, Y., Cortez, D., Yazdi, P., Neff, N., Elledge, S.J., and Qin, J. (2000). BASC, a super complex of BRCA1-associated proteins involved in the recognition and repair of aberrant DNA structures. *Genes Dev., 14*:927-39.

[46] Pennaneach, V., Salles-Passador, I., Munshi, A., Brickner, H., Regazzoni, K., Dick, F., Dyson, N., Chen, T.T., Wang, J.Y., Fotedar, R., and Fotedar, A. (2001). The large subunit of replication factor C promotes cell survival after DNA damage in an LxCxE motif- and Rb-dependent manner. *Mol. Cell, 7*:715-27.

[47] Hong, S., Park, S.J., Kong, H.J., Shuman, J.D., and Cheong, J. (2001). Functional interaction of bZIP proteins and the large subunit of replication factor C in liver and adipose cells. *J. Biol. Chem., 276*:28098-105.

[48] Anderson, L.A., and Perkins, N.D. (2002). The large subunit of replication factor C interacts with the histone deacetylase, HDAC1. *J. Biol. Chem., 277*:29550-4.

[49] Ishii, H., Inageta, T., Mimori, K., Saito, T., Sasaki, H., Isobe, M., Mori, M., Croce, C.M., Huebner, K., Ozawa, K., and Furukawa, Y. (2005). Frag1, a homolog of alternative replication factor C subunits, links replication stress surveillance with apoptosis. *Proc. Natl. Acad. Sci.* USA., *102*: 9655-60.

[50] Cai, R.L., Yan-Neale, Y., Cueto, M.A., Xu, H., and Cohen, D. (2000). HDAC1, a histone deacetylase, forms a complex with Hus1 and Rad9, two G2/M checkpoint Rad proteins. *J. Biol. Chem. 275*: 27909-16.

[51] Makiniemi, M., Hillukkala, T., Tuusa, J., Reini, K., Vaara, M., Huang, D., Pospiech, H., Majuri, I., Westerling, T., Makela, T.P., and Syvaoja, J.E. (2001). BRCT domain-containing protein TopBP1 functions in DNA replication and damage response. *J. Biol. Chem., 276*:30399-406.

[52] Celeste, A., Difilippantonio, S., Difilippantonio, M.J., Fernandez-Capetillo, O., Pilch, D.R., Sedelnikova, O.A., Eckhaus, M., Ried, T., Bonner, W.M., and Nussenzweig, A. (2003). H2AX haploinsufficiency modifies genomic stability and tumor susceptibility. *Cell, 114*:371-83.

[53] Bassing, C.H., Suh, H., Ferguson, D.O., Chua, K.F., Manis, J., Eckersdorff, M., Gleason, M., Bronson, R., Lee, C., and Alt, F.W. (2003). Histone H2AX: a dosage-dependent suppressor of oncogenic translocations and tumors. *Cell, 114*:359-70.

[54] Norbury, C.J., and Zhivotovsky, B. (2004). DNA damage-induced apoptosis. *Oncogene, 23*: 2797-808.

[55] Kaina, B. (2003). DNA damage-triggered apoptosis: critical role of DNA repair, double-strand breaks, cell proliferation and signaling. *Biochem. Pharmacol. 66*:1547-54.

[56] Kirkin, V., Joos, S., and Zornig, M. (2004). The role of Bcl-2 family members in tumorigenesis. *Biochim. Biophys. Acta, 1644*:229-49.

[57] Sharpe, J.C., Arnoult, D., and Youle, R.J. (2004). Control of mitochondrial permeability by Bcl-2 family members. *Biochim. Biophys Acta, 1644*:107-13.

[58] Loegering, D., Arlander, S.J., Hackbarth, J., Vroman, B.T., Roos-Mattjus, P., Hopkins, K.M., Lieberman, H.B., Karnitz, L.M., and Kaufmann, S.H. (2004). Rad9 protects cells from topoisomerase poison-induced cell death. *J. Biol. Chem., 279*:18641-7.

[59] Yin, Y., Zhu, A., Jin, Y.J., Liu , Y.X., Zhang, X., Hopkins, K.M., and Lieberman, H.B. (2004). Human RAD9 checkpoint control/proapoptotic protein can activate transcription of p21. *Proc. Natl. Acad. Sci.* USA., *101*:8864-9.

[60] Blankley, R.T., and Lydall, D. (2004). A domain of Rad9 specifically required for activation of Chk1 in budding yeast. *J. Cell. Sci., 117*(Pt 4):601-8.

[61] Furuya, K., Poitelea, M., Guo, L., Caspari, T., and Carr, A.M. (2004). Chk1 activation requires Rad9 S/TQ-site phosphorylation to promote association with C-terminal BRCT domains of Rad4TOPBP1. *Genes Dev., 18*:1154-64.

[62] Sorensen, C.S., Syljuasen, R.G., Lukas, J., and Bartek, J. (2004). ATR, Claspin and the Rad9-Rad1-Hus1 complex regulate Chk1 and Cdc25A in the absence of DNA damage. *Cell Cycle, 3*: 941-5.

[63] Lieberman, H.B. (2006). Rad9, an evolutionarily conserved gene with multiple functions for preserving genomic Integrity. *J. Cell. Biochem.,97*: 690-7.

[64] Rodriguez, R., and Meuth, M. (2006). Chk1 and p21 cooperate to prevent apoptosis during DNA replication fork stress. *Mol. Biol. Cell., 17*:402-12.

[65] Tsujimoto, Y., Finger, L.R., Yunis, J., Nowell, P.C., and Croce, C.M. (1984). Cloning of the chromosome breakpoint of neoplastic B cells with the t(14;18) chromosome translocation. *Science, 226*:1097-9.

[66] Tsujimoto, Y., Cossman, J., Jaffe, E., and Croce, C.M. (1985). Involvement of the bcl-2 gene in human follicular lymphoma. *Science, 228*:1440-3.

[67] Green, D.R., and Kroemer, G. (2004). The pathophysiology of mitochondrial cell death. *Science, 305*:626-9.

[68] Kelekar, A., and Thompson, C.B. (1998). Bcl-2-family proteins: the role of the BH3 domain in apoptosis. *Trends Cell Biol., 8*:324-30.

[69] Adams, J.M., and Cory, S. (1998). The Bcl-2 protein family: arbiters of cell survival. *Science, 281*: 1322-6.

[70] Abraham, R.T. (2001). Cell cycle checkpoint signaling through the ATM and ATR kinases. *Genes Dev., 15*:2177-96.

[71] Melo, J., and Toczyski, D. (2002). A unified view of the DNA-damage checkpoint. *Curr. Opin. Cell Biol., 14*:237-45.

[72] Byun, T.S., Pacek, M., Yee, M.C., Walter, J.C., and Cimprich, K.A. (2005). Functional uncoupling of MCM helicase and DNA polymerase activities activates the ATR-dependent checkpoint. *Genes Dev., 19*:1040-52.

[73] Bartkova, J., Horejsi, Z., Koed, K., Kramer, A., Tort, F., Zieger, K., Guldberg, P., Sehested, M., Nesland, J.M., Lukas, C., Orntoft, T., Lukas, J., and Bartek, J. (2005). DNA damage response as a candidate anti-cancer barrier in early human tumorigenesis. *Nature, 434*:864-70.

[74] Gorgoulis, V.G., Vassiliou, L.V., Karakaidos, P., Zacharatos, P., Kotsinas, A., Liloglou, T., Venere, M., Ditullio, R.A. Jr., Kastrinakis, N.G., Levy, B., Kletsas, D., Yoneta, A., Herlyn, M., Kittas, C., and Halazonetis, T.D. (2005). Activation of the DNA damage checkpoint and genomic instability in human precancerous lesions. *Nature, 434*:907-13.

[75] Eyfjord, J.E., and Bodvarsdottir, S.K. (2005). Genomic instability and cancer: Networks involved in response to DNA damage. *Mutat. Res, 592*:18-28.

[76] Smith, D.I., Huang, H., and Wang, L. (1998). Common fragile sites and cancer (review). *Int. J. Oncol., 12*:187-196.

[77] Sutherland, G.R. (2003). Rare fragile sites. *Cytogenet. Genome Res., 100*:77-84.

[78] Arlt, M.F., Casper, A.M., and Glover, T.W. (2003). Common fragile sites. *Cytogenet. Genome Res., 100*:92-100.

[79] Sutherland, G.R. (1977). Fragile sites on human chromosomes: demonstration of their dependence on the type of tissue culture medium. *Science, 197*:265-6.

[80] Sutherland, G.R., Parslow, M.I., and Baker, E. (1985). New classes of common fragile sites induced by 5-azacytidine and bromodeoxyuridine. *Hum. Genet., 69*:233-7.

[81] Tassone, F., and Hagerman, P.J. (2003). Expression of the FMR1 gene. *Cytogenet. Genome Res., 100*:124-8.

[82] Verkerk, A.J., Pieretti, M., Sutcliffe, J.S., Fu, Y.H., Kuhl, D.P., Pizzuti, A., Reiner, O., Richards, S., Victoria, M.F., and Zhang, F.P. (1991). Identification of a gene (FMR-1) containing a CGG repeat coincident with a breakpoint cluster region exhibiting length variation in fragile X syndrome. *Cell, 65*:905-14.

[83] Pieretti, M., Zhang, F.P., Fu, Y.H., Warren, S.T., Oostra, B.A., Caskey, C.T., and Nelson, D.L. (1991). Absence of expression of the FMR-1 gene in fragile X syndrome. *Cell, 66*:817-22.

[84] Lu, R., Wang, H., Liang, Z., Ku, L., O'donnell, W.T., Li, W., Warren, S.T., and Feng, Y. (2004). The fragile X protein controls microtubule-associated protein 1B translation and microtubule stability in brain neuron development. *Proc. Natl. Acad. Sci.* USA., *101*:15201-6.

[85] Knight, S.J., Flannery, A.V., Hirst, M.C., Campbell, L., Christodoulou, Z., Phelps, S.R., Pointon, J., Middleton-Price, H.R., Barnicoat, A., and Pembrey, M.E. (1993). Trinucleotide repeat amplification and hypermethylation of a CpG island in FRAXE mental retardation. *Cell, 74*: 127-34.

[86] Ritchie, R.J., Knight, S.J., Hirst, M.C., Grewal, P.K., Bobrow, M., Cross, G.S., and Davies, K.E. (1994). The cloning of FRAXF: trinucleotide repeat expansion and methylation at a third fragile site in distal Xqter. *Hum. Mol. Genet.*, *3*:2115-21.

[87] Jones, C., Slijepcevic, P., Marsh, S., Baker, E., Langdon, W.Y., Richards, R.I., and Tunnacliffe, A. (1994). Physical linkage of the fragile site FRA11B and a Jacobsen syndrome chromosome deletion breakpoint in 11q23.3. *Hum. Mol. Genet.*, *3*:2123-30.

[88] Nancarrow, J.K., Kremer, E., Holman, K., Eyre, H., Doggett, N.A., Le Paslier, D., Callen, D.F., Sutherland, G.R., and Richards, R.I. (1994). Implications of FRA16A structure for the mechanism of chromosomal fragile site genesis. *Science*, *264*:1938-41.

[89] Yunis, J.J., and Soreng, A.L. (1984). Constitutive fragile sites and cancer. *Science*, *226*:1199-204.

[90] Smeets, D.F., Scheres, J.M., and Hustinx, T.W. (1986). The most common fragile site in man is 3p14. *Hum. Genet.*, *72*:215-20.

[91] Arlt, M.F., Miller, D.E., Beer, D.G., and Glover, T.W. (2002). Molecular characterization of FRAXB and comparative common fragile site instability in cancer cells. *Genes Chromosom. Cancer.*, *33*:82-92.

[92] Shiraishi, T., Druck, T., Mimori, K., Flomenberg, J., Berk, L., Alder, H., Miller, W., Huebner, K., and Croce, C.M. (2001). Sequence conservation at human and mouse orthologous common fragile regions, FRA3B/FHIT and Fra14A2/Fhit. *Proc. Natl. Acad .Sci.* USA., *98*:5722-7.

[93] Ohta, M., Inoue, H., Cotticelli, M.G., Kastury, K., Baffa, R., Palazzo, J., Siprashvili, Z., Mori, M., McCue, P., Druck, T., Croce, C.M., and Huebner, K. (1996). The FHIT gene, spanning the chromosome 3p14.2 fragile site and renal carcinoma-associated t(3;8) breakpoint, is abnormal in digestive tract cancers. *Cell*, *84*:587-597.

[94] Pekarsky, Y., Zanesi, N., Palamarchuk, A., Huebner, K., and Croce, C.M. (2002). FHIT: from gene discovery to cancer treatment and prevention. *Lancet Oncol.*, *3*:748-54.

[95] Bednarek, A.K., Laflin, K.J., Daniel, R.L., Liao, Q., Hawkins, K.A., and Aldaz, C.M. (2000). WWOX, a novel WW domain-containing protein mapping to human chromosome 16q23.3-24.1, a region frequently affected in breast cancer. *Cancer Res.*, *60*:2140-5.

[96] Ried, K., Finnis, M., Hobson, L., Mangelsdorf, M., Dayan, S., Nancarrow, J.K., Woollatt, E., Kremmidiotis, G., Gardner, A., Venter, D., Baker, E., and Richards, R.I. (2000). Common chromosomal fragile site FRA16D sequence: identification of the FOR gene spanning FRA16D and homozygous deletions and translocation breakpoints in cancer cells. *Hum. Mol. Genet.*, *9*: 1651-63.

[97] Ludes-Meyers, J.H., Bednarek, A.K., Popescu, N.C., Bedford, M., and Aldaz, C.M. (2003). WWOX, the common chromosomal fragile site, FRA16D, cancer gene. *Cytogenet. Genome Res.*, *100*: 101-110.

[98] Sozzi, G., Huebner, K., and Croce, C.M. (1998). FHIT in human cancer. *Adv. Cancer Res.*, *74*: 141-66.

[99] Sozzi, G., Veronese, ML., Negrini, M., Baffa, R., Cotticelli, MG., Inoue, H., Tornielli, S., Pilotti, S., De Gregorio, L., Pastorino, U., Pierotti, MA., Ohta, M., Huebner, K.,

and Croce, CM. (1996). The FHIT gene 3p14.2 is abnormal in lung cancer. *Cell, 85*:17-26.

[100] Baffa, R., Veronese, M.L., Santoro, R., Mandes, B., Palazzo, J.P., Rugge, M., Santoro, E., Croce, C.M., and Huebner, K. (1998). Loss of FHIT expression in gastric carcinoma. *Cancer Res., 58*: 4708-14.

[101] Baffa, R., Gomella, L.G., Vecchione, A., Bassi, P., Mimori, K., Sedor, J., Calviello, C.M., Gardiman, M., Minimo, C., Strup, S.E., McCue, P.A., Kovatich, A.J., Pagano, F., Huebner, K., and Croce, C.M. (2000). Loss of FHIT expression in trasitional cell carcinoma of the urinary bladder. *Am. J. Pathol., 156*:419-24.

[102] Ishii, H., Dumon, K.R., Vecchione, A., Fong, L.Y., Baffa, R., Huebner, K., and Croce, C.M. (2001). Potential cancer therapy with the fragile histidine triad gene: review of the preclinical studies. *JAMA, 286*:2441-2449.

[103] Huebner, K., and Croce, C.M. (2003). Cancer and the FRA3B/FHIT fragile locus: it's a HIT. *Br. J. Cancer, 88*:1501-6.

[104] Siprashvili, Z., Sozzi, G., Barnes, L.D., McCue, P., Robinson, A.K., Eryomin, V., Sard, L., Tagliabue, E., Greco, A., Fusetti, L., Schwartz, G., Pierotti, M.A., Croce, C.M., and Huebner, K. (1997). Replacement of Fhit in cancer cells suppresses tumorigenicity. *Proc. Natl. Acad. Sci.* USA., *94*:13771-6.

[105] Ji, L., Fang, B., Yen, N., Fong, K., Minna, J.D., and Roth, J.A. (1999). Induction of apoptosis and inhibition of tumorigenicity and tumor growth by adenovirus vector-mediated fragile histidine triad (FHIT) gene overexpression. *Cancer Res., 59*:3333-9.

[106] Ishii, H., Dumon, K., Vecchione, A., Trapasso, F., Mimori, K., Alder, H., Mori, M., Sozzi, G., Baffa, R., Huebner, K., and Croce, C.M. (2001). Effect of adenoviral transduction of FHIT into esophageal cancer cells. *Cancer Res., 61*:1578-84.

[107] Roz, L., Gramegna, M., Ishii, H., Croce, C.M., and Sozzi, G. (2002). Restoration of fragile histidine triad (FHIT) expression induces apoptosis and suppresses tumorigenicity in lung and cervical cancer cell lines. *Proc. Natl. Acad. Sci.* USA., *99*:3615-20.

[108] Dumon, K.R., Ishii, H., Vecchione, A., Trapasso, F., Baldassarre, G., Chakrani, F., Druck, T., Rosato, E.F., Williams, N.N., Baffa, R., During, M.J., Huebner, K., and Croce, C.M. (2001). Fragile histidine triad expression delays tumor development and induces apoptosis in human pancreatic cancer. *Cancer Res., 61*:4827-36.

[109] Werner, N.S., Siprashvili, Z., Fong, L.Y.Y., Marquitan, G., Schroder, J.K., Bardenheuer, W., Seeber, S., Huebner, K., Schutte, J., and Opalka, B. (2000). Differential susceptibility of renal carcinoma cell lines to tumor suppression by exogeneous Fhit expression. *Cancer Res., 60*:2780-5.

[110] Guo, Z., and Vishwanatha, J.K. (2000). Effect of regulated expression of the fragile histidine triad gene on cell cycle and proliferation. *Mol. Cell. Biochem., 204*:83-88.

[111] Wu, R., Connolly, D.C., Dunn, R.L., and Cho, K.R. (2000). Restored expression of fragile histidine triad protein and tumorigenicity of cervical carcinoma cells. *J. Natl. Cancer Inst., 92*:338-44.

[112] Otterson, G.A., Xiao, G.-H., Geradts, J., Jin, F., Chen, W.-d., Niklinska, W., Kaye, F.J., and Yeung, R.S. (1998). Protein expression and functional analysis of the FHIT gene in human tumor cells. *J. Natl. Cancer Inst., 90*:426-32.

[113] Fong, L.Y.Y., Fidanza, V., Zanesi, N., Lock, L.F., Siracusa, L.D., Mancini, R., Siprashvili, Z., Ottey, M., Martin, S.E., Druck, T., McCue, P.A., and Croce, C.M. (2000). Muir-Torre-like syndrome in Fhit-deficient mice. *Proc. Natl. Acad. Sci.* USA., *97*:4742-7.

[114] Zanesi, N., Fidanza, V., Fong, L.Y., Mancini, R., Druck, T., Valtieri, M., Rudiger, T., McCue, P.A., Croce, C.M., and Huebner, K. (2001). The tumor spectrum in FHIT-deficient mice. *Proc. Natl, Acad. Sci.* USA., *98*:10250-5.

[115] Dumon, K.R., Ishii, H., Fong, L.Y.Y., Zanesi, N., Fidanza, V., Mancini, R., Vecchione, A., Baffa, R., Trapasso, F., During, M.J., Huebner, K., and Croce, C.M. (2001). FHIT gene therapy prevents tumor development in Fhit-deficient mice. *Proc. Natl. Acad. Sci.* USA., *98*:3346-51.

[116] Ishii, H., Zapesi, N., Vecchione, A., Trapasso, F., Yendamuri, S., Sardi, M., Baffa, R., During, M.J., Huebner, K., Fong, Y.Y.L., and Croce, C.M. (2003). Prevention and regression of upper gastric cancer in mice by FHIT gene therapy. *FASEB J., 17*:1768-70.

Index

A

H

I

T

U

V

W

X

Y

Z